Programmed Cell Death in Animals and Plants

EXPERIMENTAL BIOLOGY REVIEWS

Environmental Stress and Gene Regulation
Sex Determination in Plants
Plant Carbohydrate Biochemistry
Programmed Cell Death in Animals and Plants

Forthcoming titles include:

Biomechanics in Animal Behaviour
Cell and Molecular Biology of Wood Formation

Programmed Cell Death in Animals and Plants

J.A. BRYANT
School of Biological Sciences, University of Exeter, Exeter EX4 4QG, UK

S.G. HUGHES
School of Biological Sciences, University of Exeter, Exeter EX4 4QG, UK

J.M. GARLAND
Institute of Clinical Science, University of Exeter, Exeter EX5 2SR, UK

© BIOS Scientific Publishers Limited, 2000

First published in 2000

A CIP catalogue record for this book is available from the British Library.

ISBN 1 85996 167 3

BIOS Scientific Publishers Ltd
9 Newtec Place, Magdalen Road, Oxford OX4 1RE, UK.
Tel. +44 (0) 1865 726286. Fax. +44 (0) 1865 246823
World Wide Web home page: http://www.bios.co.uk/

Production Editor: Fran Kingston.
Typeset by Saxon Graphics Ltd, Derby, UK.
Printed by Biddles Ltd, Guildford, UK.

Contents

Contributors

Akbar, A.N., Department of Clinical Immunology, The Royal Free Hospital, London, NW3 2PF, UK

Amin, F., School of Biosciences, Cardiff University, P.O. Box 915, Cardiff, CF1 3TL, UK

Bentley, P., Department of Haematology, Llandough Hospital, Penlan Road, Penarth, South Glamorgan, CF64 2XX, UK

Bowen, I.D., School of Biosciences, Cardiff University, P.O. Box 915, Cardiff, CF1 3TL, UK

Brown, R.J., Cell Biology Unit, GlaxoWellcome Medicines Research Centre, Gunnels Wood Road, Stevenage, SG1 2NY, UK

Bryant, J.A., School of Biological Sciences, University of Exeter, Exeter, EX4 4QG, UK

Buchanan-Wollaston, V., Department of Plant Genetics & Biotechnology, Horticulture Research International, Wellesbourne, Warwick, CV35 9EF, UK

Buckley, C.D., MRC Centre for Immune Regulation, Division of Immunity and Infection, Rheumatology Research Group, Birmingham, B15 2TT, UK

Clarke, A.R., Edinburgh University Medical School, Edinburgh, EH8 9AG, UK

Coughlin, P., Monash University, Department of Medicine, Box Hill Hospital, Nelson Road, Box Hill, Australia 3128

Degenhardt, K., Center for Biotechnology and Medicine, Rutgers University, 679 Hoes Lane, Piscataway, NJ08854, USA

Donnison, I., Cell Biology Department, Institute of Grassland and Environmental Research, Plas Goggerddan, Aberystwyth, Ceredigion, SY23 3EB, UK

Doran, E., Department of Biochemistry, School of Medical Studies, University of Bristol, BS8 1TD, UK

Drew, M.C., Department of Horticultural Sciences, Texas A&M University, College Station, Texas 77843, USA

Garland, J.M., Institute of Clinical Sciences, University of Exeter, Exeter, EX5 2SR, UK

Gillespie, J.P., CONTECH Laboratories UK Ltd., Unit 2 Intec, 2 Wade Road, Basingstoke, Hampshire, RG24 8NE, UK

Griffin, M., Department of Life Sciences, Nottingham Trent University, Clifton Lane, Nottingham, NG11 8NS, UK

Halestrap, A.P., Department of Biochemistry, School of Medical Studies, University of Bristol, BS8 1TD, UK

Hampson, L., Paterson Institute for Cancer Research, Wilmslow Road, Withlington, Manchester, M20 4BX, UK

He, C-J., Department of Horticultural Sciences, Texas A&M University, College Station, Texas 77843, USA

Hengartner, M.O., Cold Spring Harbor Laboratory, 1 Bungtown Road, Cold Spring Harbor, New York, NY11724, USA

Hooley, R., IACR-Long Ashton Research Station, Department of Agricultural Sciences, University of Bristol, Long Ashton, Bristol, BS41 9AF, UK

Hughes, S.G., School of Biological Sciences, University of Exeter, Exeter, EX4 4QG, UK

Jones, C.J., Department of Pathology, University of Wales College of Medicine, Heath Park, Cardiff, CF14 4XN, UK

Kingston-Smith, A.H., Department of Animal Science and Microbiology, Institute of Grassland and Environmental Research, Plas Goggerddan, Aberystwyth, Ceredigion, SY23 3EB, UK

Lord, J.M., MRC Centre for Immune Regulation, Division of Immunity and Infection, Department of Immunology, Birmingham, B15 2TT, UK

McCann, M.C., Department of Cell Biology, John Innes Centre, Norwich Research Park, Colney, Norwich, NR4 7UH, UK

Morgan, P.W., Department of Soil and Crop Sciences, Texas A&M University, College Station, Texas 77843, USA

Morris, E., Department of Haematology, University of Cambridge, Cambridge Institute for Medical Research, Wellcome Trust/MRC Building, Addenbrooke's Hospital, Hills Road, Cambridge, CB2 2XY, UK

Morris, K., Department of Plant Genetics & Biotechnology, Horticulture Research International, Wellesbourne, Warwick, CV35 9EF, UK

Norbury, C., Imperial Cancer Research Fund, Molecular Oncology Laboratory, University of Oxford, Institute of Molecular Medicine, John Radcliffe Hospital, Oxford, OX3 9DS, UK

O'Toole, A., Department of Biochemistry, School of Medical Studies, University of Bristol, BS8 1TD, UK

Orteu, C.H., Department of Clinical Immunology, The Royal Free Hospital, London, NW3 2PF, UK

Pepper, C., Department of Haematology, Llandough Hospital, Penlan Road, Penarth, South Glamorgan, CF64 2XX, UK

Perez, D., Center for Biotechnology and Medicine, Rutgers University, Piscataway, NJ08854, USA

Pilling, D., MRC Centre for Immune Regulation, Division of Immunity and Infection, Rheumatology Research Group, Birmingham, B15 2TT, UK

Pun, K.T., Cell Biology Unit, GlaxoWellcome Medicines Research Centre, Gunnels Wood Road, Stevenage, SG1 2NY, UK

Roberts, J.A., School of Biological Sciences, University of Nottingham, Sutton Bonington Campus, Loughborough, Leicestershire, LE12 5RD, UK

Roberts, K., Department of Cell Biology, John Innes Centre, Norwich Research Park, Colney, Norwich, NR4 7UH, UK

Salmon, M., MRC Centre for Immune Regulation, Division of Immunity and Infection, Rheumatology Research Group, Birmingham, B15 2TT, UK

Scheel-Toellner, D., MRC Centre for Immune Regulation, Division of Immunity and Infection, Rheumatology Research Group, Birmingham, B15 2TT, UK

Schneider, P., Institute of Biochemistry, University of Lausanne, Ch. des Boveresses 155, CH-1066 Epalinges, Switzerland

Stacey, N.J., Department of Cell Biology, John Innes Centre, Norwich Research Park, Colney, Norwich, NR4 7UH, UK

Stennicke, H.R., The Burnham Institute, Program for Apoptosis and Cell Death Research, 10901 North Torrey Pines Road, La Jolla, CA92037, USA

Theodorou, M.K., Department of Animal Science and Microbiology, Institute of Grassland and Environmental Research, Plas Goggerddan, Aberystwyth, Ceredigion, SY23 3EB, UK

Thomas, H., Cell Biology Department, Institute of Grassland and Environmental Research, Plas Gogerddan, Aberystwyth, Ceredigion, SY23 3EB, UK

Tschopp, J., Institute of Biochemistry, University of Lausanne, Ch. des Boveresses 155, CH-1066 Epalinges, Switzerland

Verderio, E., Department of Life Sciences, Nottingham Trent University, Clifton Lane, Nottingham, NG11 8NS, UK

Wang, K., MRC Centre for Immune Regulation, Division of Immunity and Infection, Department of Immunology, Birmingham, B15 2TT, UK

Webb, P., MRC Centre for Immune Regulation, Division of Immunity and Infection, Department of Immunology, Birmingham, B15 2TT, UK

White, E., Howard Hughes Medical Institute, Center for Biotechnology, and Medicine, Department of Molecular Biology and Biochemistry, and Cancer Institute of New Jersey, Rutgers University, 679 Hoes Lane, Piscataway, NJ08854, USA

Abbreviations

[Ca^{2+}]	calcium ion concentration
[Na$^+$]	sodium ion concentration
AER	apical ectodermal ridge
AK	adenylate kinase
ANT	adenine nucleotide translocator
ANZ	anterior necrotic zone
Apc gene	*adenomatous polyposis coli* gene
AT	ataxia telangiectasia
B-CLL	B-cell chronic lymphocytic leukaemia
BenzdR	benzodiazepine receptor
BH domains	Bcl-2 homology domains
CARD	caspase recruitment domain
ced	cell death abnormal
ces	cell death specification
CsA	cyclosporin A
CSFs	colony stimulating factors
CTL	cytotoxic lymphocytes
CyP	cyclophilin
Dad	defender against apoptic death
DAPI	4′,6′-diamidino-2-phenylindole
DED	death effector domain
DEF	death effector filaments
DISC	death-inducing signalling complex
DNP	dinitrophenol
DOG	deoxyglucose
DTT	dithiothreitol
EGF-R	epidermal growth factor receptor
egl	egg-laying defective
ER	endoplasmic reticulum
ERSE	ER-stress response element
FAP	familial adenomatous polyposis
FDC-P	factor dependent cells – Paterson
GPCR	G protein coupled receptor
GPI anchor	glucosyl phosphatidylinositol anchor
GST	glutathione-S-transferase
GUS	β-glucuronidase
HR	hypersensitive response
IAP	inhibitor of apoptosis
ICE	interleukin 1β converting enzyme
IDW	interdigital webs
IL	interleukin
ipt	iso-pentenyl transferase
LAK	lymphokine-activated natural killer

LLL	carbobenzoyl-leucinyl-leucinyl-leucinal
LTBMC	long-term bone marrow cultures
LVDP	left ventricular developed pressure
MAPKs	mitogen-activated protein kinases
MARKs	microtubule affinity-regulating kinases
MCM2	minichromosome maintenance protein 2
mdm	mitochondrial distribution and morphology
MFI	mean fluorescent intensity
mgm	genome maintenance
MLC	myosin light chain
mmm	morphology and movement
MMR	mismatch repair
MPTP	mitochondrial permeability transition pore
NHL	non-Hodgkin's lymphoma
NK cells	natural killer cells
NMR	nuclear magnetic resonance
NO	nitric oxide
NOS	nitric oxide synthase
nuc	nuclease
ORF	open reading frame
PAI	plasminogen activator inhibitor
PaO	phaeophorbide *a* oxygenase
PARP	poly (ADP-ribose) polymerase
PCD	programmed cell death
pdl	population doublings
PG	polygalacturonase
PI-PLC	phosphatidylinositol-specific phospholipase C
PLC	phospholipase C
PPIase	peptidyl-prolyl *cis-trans* isomerase
PR proteins	pathogenesis-related proteins
PT pore	permeability transition pore
Q-FISH	quantitative fluorescence *in situ* hybridization
RNase	ribonuclease
RUBISCO	ribulose bisphosphate carboxylase-oxygenase
SAH hydrolase	*S*-adenosyl-L-homocysteine hydrolase
SAPKs	stress-activated kinases
SCA	spinal cerebellar ataxia
SCID	severe combined immunodeficiency disease
SDS-PAGE	sodium dodecyl sulphate polyacrylamide gel electrophoresis
TEs	tracheary elements
TIL	tumour-infiltrating lymphocytes
TMRH	tetramethylrhodamine
TNF	tumour necrosis factor
TNF-R	tumour necrosis factor receptor
TRF	telomeric repeat binding factor
tTgase	tissue transglutaminase
UPR	unfolded protein response
VDAC	voltage-dependent anion channel

VEGF-R	vascular endothelial growth factor receptor
XET	xyloglucan *endo*-transglycosylase (XET)
ZFA-fmk	benzyloxycarboxyl–Phe–Ala(*O*-methyl)-fluoromethylketone

Preface

The seminal paper by Kerr, Wyllie and Currie (1972) in which, for the first time, a particular form of cell death, apoptosis[1], was clearly defined, did not initially attract much attention. However, a realization slowly developed that several phenomena, including the turnover of blood cells, the loss of specific cells during animal development, and the removal of certain types of damaged cells, shared the features that would allow them to be classified as apoptosis, a type of cell death quite different from necrosis. Further, it was clear that this was *programmed* cell death in that cell demise appeared to take place within a clear programme of defined steps. This realization and its implications for human disease, particularly cancer, slowly brought the topic of apoptosis to a wider research community and to readier sources of research funds. This has led to a vast increase in the rate of publication on apoptosis such that the annual number of papers now runs into thousands with little sign of any decline in interest.

The obvious medical applications of apoptosis have tended to overshadow its developmental role and yet it is clear from many systems, ranging from the growth of the nematode *Caenorhabditis elegans* to the development of neuronal connections in the vertebrate brain, that apoptosis has a clear role in animal development. It is equally clear that the key elements of apoptosis are very ancient in evolutionary terms: programmed cell death was apparently a part of life from the times of the earliest multicellular animals. However, we must not overlook plants. They, like animals, need in the course of development to remove specific cells and to respond to cellular damage and disease. Furthermore, many plants undergo the senescence and loss of whole organs, a process that is controlled tightly in space and time and which again involves programmed cell death. The question then arises: to what extent are the processes of programmed cell death in plants and animals comparable?

The widespread occurrence of programmed cell death as part of life itself and its role in responses to damage and disease make it an ideal topic for a detailed comparative approach between the plant and animal kingdoms. Thus, when the cell biology section committee of the Society for Experimental Biology (SEB) suggested that cell death might be an appropriate subject for its annual symposium meeting, we readily agreed to organize such a meeting, and a very stimulating meeting it turned out to be. Under the title 'Cell Death in Health and Disease' we spent three days in December 1998 at the University of Exeter in stimulating discussion of the basic processes and their regulation and pathology in plants and animals. It was a time for genuine learning, with delegates from the worlds of animal and plant biology and medicine contributing to the proceedings and opening each others' minds to the different systems under investigation.

This book comes from that symposium but it is not directly 'the book of the meeting'. Rather, it is our attempt to produce a book that will introduce the reader to this topic in an accessible, logical and exciting way. The overall plan is to deal with the occurrence of programmed cell death in a variety of systems, both 'normal' and pathological, and to discuss the various regulatory mechanisms that operate in those

---

[1] From a Greek word meaning 'falling away', as applied to leaves.

systems. We are very grateful to those speakers who shared this vision and who produced chapters for us. We are also grateful to the three authors that we recruited who did not actually speak at the meeting (Ifor Bowen, Martin Griffin and Henning Stennicke) whose chapters contribute significantly to the value of this book and to Howard Thomas for providing the photograph of the dying leaf used in the cover montage. Thus, we have been able to assemble a coherent and stimulating text, a text with which we are pleased and that does indeed deal with some of the basic questions we posed when we set up the meeting.

In running the meeting we had a great deal of help from many people. The other members of the symposium steering committee, Dennis Francis and Hilary Rogers, contributed many excellent and helpful ideas. The officers of the SEB, especially Jerry Roberts (Publications Officer) and Tony Stead (Treasurer) were always ready with help and advice before, during and after the meeting. Michelle Laurence in the SEB office deserves our special thanks, being thrown in at the deep end on conference organization. Neither must we forget Karen Moore and Sue Downs who temporarily deserted their research laboratories to assist at the conference desk. We wish to record our gratitude to all those who participated in the meeting, not just the speakers, but all the delegates, for the lively and enthusiastic atmosphere of learning, enquiry and discussion that was so apparent throughout the meeting. Finally, in the context of the meeting, we must mention the excellent work of the staff at the University of Exeter – the staff of the Conference Office, the porters in the Peter Chalk Centre, the catering and ancilliary staff – without whom such a meeting would not be possible.

In preparing the book, we have been very fortunate in working with BIOS Scientific Publications. On the editorial side, Rachel Offord and her successor, Will Sansom, and on the production side, Fran Kingston have all been both patient and encouraging as we have suffered the inevitable delays in receipt of manuscripts. Even with these delays (the last manuscript was not delivered until late September 1999) they have produced the book in very good time. We are also pleased, as always with BIOS, with the standards of production and it is our hope that the text itself is equally good. The decision on that though, lies not with us but with you, the readers.

John Bryant
Steve Hughes
John Garland
Exeter

Apoptotic cell death: from worms to wombats … but what about the weeds?

Michael O. Hengartner and John A. Bryant

1. *Caenorhabditis elegans*: the worm that turned … into a model multicellular eukaryote

Caenorhabditis elegans is a nematode (roundworm) whose potential for use as a model for genetic regulation of development was first recognized by Sydney Brenner as long ago as the 1960s (Brenner, 1974; Wood, 1988). The hermaphrodite worms can be grown easily in culture on agar in petri dishes, feeding on *E. coli* (model eats model!). The adult hermaphrodite is about 1 mm long (*Figure 1*) and has exactly 959 somatic cells (so it is certainly a very simple metazoan); the origins, lineages and developmental pathways of each of these cells have been precisely described (Sulston and Horvitz,

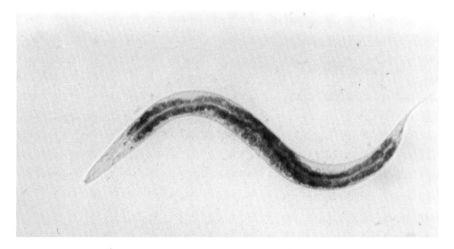

Figure 1. The nematode worm, Caenorhabditis elegans.

Programmed Cell Death in Animals and Plants, edited by J.A. Bryant, S.G. Hughes and J.M. Garland.
© 2000 BIOS Scientific Publishers Ltd, Oxford.

1977; Kimble and Hirsch, 1979; Sulston *et al.*, 1983). The relevance of those developmental pathways for this book is that 1090 somatic cells are actually generated during development; the final number of 959 is achieved by the elimination *via* programmed cell death of 131 specific cells. Thus, programmed cell death is a normal part of development and because it always occurs in the same place and at the same stage of development, the process is readily studied.

As research on development of *C. elegans* progressed, mutants were discovered that had very specific effects on cell fate and function (as discussed in respect of cell death in subsequent sections of this chapter). The obvious usefulness of the genetic approach to studying development, taken with the relatively small genome size of *C. elegans*, led to its inclusion as one of the specific targets for genome sequencing. The first eukaryotic genome to be sequenced was that of budding yeast, *Saccharomyces cerevisiae*, completed in 1996 (Goffeau *et al.*, 1996). The full genome sequence of its very distant cousin, the fission yeast, *Schizosaccharomyces pombe*, is likely to be known as this book is published in early 2000. However, between those two events, and very much to the credit of the *C. elegans* sequencing consortium, the essentially complete sequence of the genome of what is affectionately known as 'the worm', was known at the end of 1998 (The *C. elegans* Sequencing Consortium, 1998; Hodgkin *et al.*, 1998). This was a very significant event, being the first complete elucidation of the genome sequence of a multicellular organism. The genome consists of 97×10^6 base pairs (about eight times the size of the *S. cerevisiae* genome and just under one thirtieth the size of the human genome). These 97 million base pairs make up about 19 000 predicted protein-coding genes (humans have nearly five times as many), over 40% of which have significant matches with genes in other organisms. The genes that regulate programmed cell death in *C. elegans* are within this 40%. The value of this organism in understanding the essential features of cell death is thus apparent.

2. The molecular mechanisms of programmed cell death in *Caenorhabditis elegans*

The detailed knowledge of the location and timing of programmed cell death during development of the adult hermaphrodite *C. elegans* has led in turn to screening for mutations that affect particular steps in the regulation and actuation of the cell death pathway. Analysis of mutant phenotypes suggests that the genes identified in the mutant screens fall into four general groups (*Figure 2*; Hengartner and Horvitz, 1994b; Hengartner, 1999) which, in order of action, are as follows:

1. There are several genes involved in the selection of the particular cell types that are to die, including *ces-1* and *-2* (ces = cell death specification).
2. Five genes control the actual killing of the selected cells. These are *ced-3, -4, -8, -9* and *egl-1* (ced = cell death abnormal and egl = egg-laying defective).
3. Six genes code for proteins required for the recognition, engulfment and removal of the dying cells. Functionally they fall into two groups, *ced-1, -6* and *-7* and *ced-2, -5,* and *-10*.
4. The product of the *nuc-1* (nuc = nuclease) gene completes degradation of the genomic DNA of dead cells.

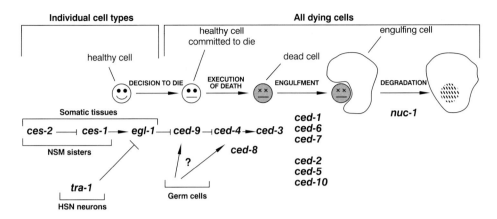

Figure 2. Genes involved in programmed cell death in C. elegans.

2.1 *The 'killer genes' and their regulation*

Two genes, *ced-3* and *ced-4* are both absolutely required for essentially all programmed cell deaths in *C. elegans*. Mutational inactivation of either of these genes leads to the survival of all the 131 cells that normally die during development of the hermaphrodite adult (Ellis and Horwitz, 1986). A key question here is whether the products of these genes act from outside or within the cells that are killed, i.e., are the dying cells 'murdered' or do they 'commit suicide'? The answer comes from mosaic analysis and from recent experiments using ectopic expression that clearly show that both these genes act within the dying cells themselves and thus the cells bring about their own death (Yuan and Horvitz, 1990; Shaham and Horvitz, 1996).

If *ced-3* and *ced-4* may be called 'killer' genes then *ced-9* is a 'moderator' or 'regulator' gene (Hengartner *et al.*, 1992). In *ced-9* mutants exhibiting loss-of-function, many cells that should survive instead become apoptotic and die. In contrast to this, transgenic worms that over-express *ced-9* or worms with *ced-9* mutations causing gain-of-function fail to activate the apoptosis programme and so cells that should die do not. These data suggest that the normal role of *ced-9* is to prevent the inappropriate action of *ced-3* and *ced-4* *via* an interaction of the CED-9 protein with the CED-3 and/or CED-4 proteins. In fact, interaction with the latter is strongly indicated by several lines of evidence: genetic analysis suggests that the *ced-4* gene acts upstream of the *ced-3* gene (Shaham and Horvitz, 1996). Interaction between the CED-9 and the CED-4 proteins has been demonstrated directly *in vitro* and in the yeast two-hybrid system (Chinnaiyan *et al.*, 1997b; Irmler *et al.*, 1997; James *et al.*, 1997; Ottilie *et al.*, 1997; Wu *et al.*, 1997), whilst mutation analysis has identified the domains in CED-9 that facilitate the interactions with CED-4 (Spector *et al.*, 1997). CED-4 also interacts with the pro-form of the CED-3 caspase (a caspase is an apoptotic cysteine protease); this interaction leads indirectly to the processing of pro-CED-3 and hence the production of active CED-3 (Chinnaiyan *et al.*, 1997a; Seshagiri and Miller, 1997; Wu *et al.*, 1997; see also Chapter 2 for more on caspases). Further, CED-4 that is complexed with CED-9 is unable to activate the processing of pro-CED-3, again emphasizing that CED-9 acts a moderator protein 'upstream' of CED-4 (Chaudhary *et al.*, 1998; Chinnaiyan *et al.*, 1997a). However, although CED-4 that is bound to CED-9 cannot initiate the processing of pro-CED-3, it can still bind to it because different domains in CED-4 are involved in the two different protein–protein interactions. The significance of this is discussed in Section 2.3.

Gain-of-function mutations in the *egl-1* gene lead to the death of cells in the nervous system that normally survive whereas loss-of-function mutations cause the survival of all cells that normally die during hermaphrodite development (Trent *et al.*, 1983; Desai *et al.*, 1988; Conradt and Horvitz, 1998). However, *egl-1* loss-of-function mutants are not like *ced-3* and *ced-4* loss-of-function mutants, despite the apparent similarity of the phenotypes. For example, worms carrying loss-of-function mutations in both *ced-9* and in *ced-3* or *-4* exhibit inappropriate survival; worms carrying loss-of-function mutations in both *ced-9* and *egl-1* exhibit massive cell death: loss-of-function in *egl-1* cannot suppress loss-of-function in *ced-9* (Conradt and Horvitz, 1998). Thus, it is likely that *egl-1* acts upstream of *ced-9* (*Figure 2*). Finally it should be noted that the place of the *ced-8* gene and the specific role of its protein product are not yet clear; the placing of *ced-8* in the pathway in *Figure 2* is an indication that it may be involved in some way at approximately that point.

2.2 *Bodies of death: apoptosomes*

In the previous section we have demonstrated that the proteins involved in bringing about developmentally regulated cell death in the somatic cells of *C. elegans* interact with each other in a very ordered manner. The question now is 'how do these interacting proteins actually bring about apoptosis?' One of us has recently developed a possible model that starts to answer this question (Hengartner, 1997, 1998a, 1998b, 1999). It is suggested, based on gene expression and protein distribution patterns, that CED-3, CED-4 and CED-9 are present in all cells whether or not they are destined to die. It is a key feature of the model that in most cells these proteins are associated together in a pre-assembled but inactive multiprotein 'body of death' or apoptosome. The apoptosome has been likened to a cocked gun which is safe unless it is disturbed in some way (Hengartner, 1999); it is safe because CED-9 is preventing CED-4 from initiating the activation of CED-3. So, what would make the complex 'unsafe'? It is likely that CED-4 must be released from its binding by CED-9. The EGL-1 protein is a good candidate for being such a release factor (Conrad and Horvitz, 1998) as overexpression of EGL-1 results in the dissociation of CED-9/CED-4 complexes, possibly by competing with CED-4 for the binding sites on CED-9 (Wu *et al.*, 1997). This would then leave CED-4 free to participate in the activation of CED-3. This model is illustrated in *Figure 3* which also shows that data from mammalian apoptosis are consistent with the operation of a similar set of interactions.

However elegant the apoptosome model may be, it still does not give any indication of how CED-4 may activate CED-3, particularly since CED-4 is not a protease enzyme (see Section 3). Some recent results provide a clue to this problem (Yang *et al.*, 1998a, 1998b). Release of CED-4 from its binding by CED-9 allows it to oligomerize with other CED-4 molecules; each CED-4 molecule can bind a molecule of pro-CED-3 thus bringing several of the latter into close proximity. As noted already, the caspase activity of CED-3 is not properly exhibited until the extra peptide has been removed. However, in common with mammalian pro-caspases, pro-CED-3 does exhibit a very low level of caspase activity (Muzio *et al.*, 1998) and under conditions of 'induced proximity' as in the formation of an oligomer of pro-CED-3, this low level

(a)

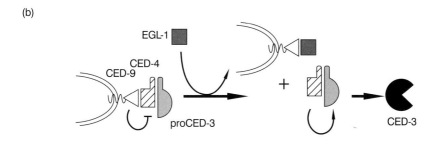

(b)

Figure 3. The pathway for programmed cell death is conserved between C. elegans and mammals. *(a) **Genetic pathways**. Genetic analysis in C. elegans has led to the elucidation of a genetic pathway that controls all developmental cell deaths. Each of the key nematode cell death regulators has functional homologues in mammals: CED-3 is a nematode caspase, CED-4 is the nematode equivalent of the adaptor protein Apaf-1, CED-9 is a member of the Bcl-2 family of cell death regulators and the BH3-domain protein EGL-1 is similar in structure and function to the pro-apoptotic BH3-domain proteins. (b) **Molecular pathways**. In C. elegans, the cell death regulators CED-9, CED-4 and pro-CED-3 are present in living cells as an inactive tri-molecular complex, the apoptosome. In cells destined to die, the EGL-1 protein binds to CED-9 and promotes its dissociation from the CED-4/pro-CED-3 complex. Once free, CED-4 oligomerizes, resulting, via a poorly characterized process, in the activation of CED-3. In mammals, an apoptosome-like complex has been reported which might contain Bcl-Xl, Apaf-1 and pro-caspase 9. Dissociation of this complex might also be mediated by BH3-domain-only proteins such as Bad, Bid and Bik. The situation in mammals is complicated by the fact that Bcl-2 family members also regulate exit of cytochrome c from mitochondria. The biological relevance of the interaction reported between Bcl-2 family members and Apaf-1 has still to be confirmed.*

of activity may be sufficient to enable pro-CED-3 molecules to process and thus activate each other. Indeed, such a mechanism has been demonstrated for mammalian caspase 8 (Muzio *et al.*, 1998) and it has been suggested that 'induced proximity' may be a general feature of caspase activation in animal cells (Hengartner, 1998a).

2.3 *Engulfment of the dead and dying cells*

Removal of the dead and dying cells and the recycling of the degradation products by the surrounding living cells is a key phase in the apoptotic process. This is essentially a process of phagocytosis *(Figure 2)* but in the context of apoptosis is known as engulfment. Although the genes involved in specifying the cells that are destined to die are different from cell to cell, all steps from the action of CED-9 onwards, including engulfment, are regulated in the same way in all dying cells.

As already indicated, there are six genes involved in regulating engulfment, namely *ced-1, -6* and *-7* and *ced-2, -5* and *-10* (Hedgecock *et al.*, 1983; Ellis *et al.*, 1991; Horvitz *et al.*,

1994). Mutations in any one of these genes causes a failure of engulfment in some but not all of the apoptotic cells. The fact that none of the mutations leads to a complete block to engulfment means that none of these genes is absolutely essential; in other words there is a partial redundancy. This makes it difficult to define the precise order in which these genes work. However, the use of double mutants suggests that the engulfment genes fall into the two groups listed immediately above (Ellis et al., 1991). In double mutants in which the mutant genes are one from each group, the number of persisting cell corpses is very much greater than when the two mutant genes are in the same group.

In order to further understand the order of genetic events in engulfment it is necessary to elucidate the specific functions of the proteins encoded by these genes. This analysis reveals that the CED-6 protein is likely to be a key regulator in this process (Liu and Hengartner, 1998). Genetic mosaic analysis indicates that the gene is expressed in the engulfing cells and functional analysis of its sequence indicates that the protein contains domains typically found in certain adaptor proteins that function in some signal transduction pathways. In particular, the CED-6 protein contains at its N-terminus, a phosphotyrosine-binding (PTB) domain while the C-terminal end is rich in proline and serine and has several Pro–X–X–Pro motifs that are implicated in protein–protein interactions in signalling pathways (Ren et al., 1993; Feng et al., 1994; Yu et al., 1994). The presence of the PTB domain is strongly suggestive that the initiation of engulfment involves a signalling pathway that includes tyrosine phosphorylation (Yenush and White, 1997; Liu and Hengartner, 1998). This in turn implies that upstream of CED-6 there should be a tyrosine kinase. Such a kinase has not been identified but there is preliminary evidence that the CED-7 protein, which is predicted to encode an ABC transporter with both membrane-spanning and cytoplasmic domains, possesses an accessible tyrosine in one of its putative cytoplasmic domains (Wu and Horvitz, 1998a). It is tempting to suggest, albeit on the basis of little concrete evidence, that phosphorylation of this tyrosine promotes the binding of CED-6 as part of the signalling pathway that leads to engulfment (Liu and Hengartner, 1998).

The apparent redundancy of function amongst these 'engulfment genes' has already been mentioned, as has the resulting difficulty in ascertaining the order of action. However, the evidence from protein structure, mentioned above, is consistent with the view that the ced-7 gene acts upstream of the ced-6 gene. Further evidence for this comes from study of engulfment in transgenic worms over-expressing the ced-6 gene (Liu and Hengartner, 1998). The rationale for this is that overexpression is equivalent to a weak gain-of-function and as such might override upstream defects but not downstream defects. Over-expression of ced-6 partly suppresses the effects of mutations in ced-1 and ced-7, slightly suppresses the effect of mutations in ced-10 but does not suppress the effects of mutations in ced-2 or ced-5. The most straightforward interpretation of this is that ced-1 and ced-7 act upstream of ced-6 (as already suggested from the protein structure evidence for ced-7) and that ced-2 and ced-5 act downstream of or in parallel with ced-6. Action in parallel is the more likely since, as we have already noted, ced-2, -5 and -10 are in a different functional group from ced-1, -6 and -7. Putting these data together it is suggested that there are two parallel but functionally overlapping signalling pathways that induce engulfment of the dead and dying cells (Liu and Hengartner, 1998; Figure 4). There is still a lot to be done before these pathways are understood, not least of which is the need to identify the signal(s) that come from the dying cells to set off the signal transduction chain.

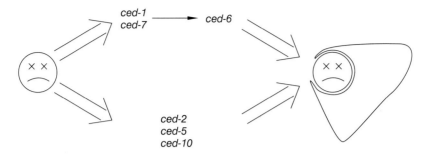

Figure 4. *Genes involved in the engulfment stage of programmed cell death in* C. elegans: *two possible parallel pathways.*

3. Comparison with mammalian apoptosis

Aravind *et al.* (1999), in a recent extensive survey of gene and protein sequences, have made comparisons between protein domains involved in cell death right across the living world. These studies lead to the suggestion that the cell death pathway described in *C. elegans* is the 'core' apoptotic pathway, certainly pre-dating the origin of the coelomates and possibly pre-dating the separation of the plant and animal evolutionary lines (see Section 4). This core pathway is apparent if we focus on the main phase of cell death as described above in Sections 2.2 and 2.3.

In the genetic pathway for *C. elegans* cell death, *ced-9* acts as an inhibitor of or at least a moderator of the action of *ced-3* and *-4*. Sequence analysis of *ced-9* shows that it encodes a protein of the Bcl-2 family of cell death regulators (see Chapter 4). Within that family it is most like the mammalian pro-survival proteins Bcl-2 and Bcl-xL, sharing with them features such as four BH domains and a C-terminal hydrophobic tail (Hengartner and Horvitz, 1994a; Muchmore *et al.*, 1996). Furthermore, the mammalian Bcl-2 can protect from death those cells in *C. elegans* that are mutant in respect of *ced-9* (Vaux *et al.*, 1994; Hengartner and Horvitz, 1994a). It has already been noted that *ced-3* encodes a caspase, a specific type of cysteine protease; indeed, CED-3 was the first caspase for which a role in programmed cell death was demonstrated (Yuan *et al.*, 1993) but we now know that caspases are extensively involved in mammalian cell death (Chapter 2). Immediately preceding *ced-3* in the genetic pathway is *ced-4* and sequence comparisons (Aravind *et al.*, 1999; Yuan and Horvitz, 1992; Zou *et al.*, 1997) show a very close similarity between the CED-4 protein and Apaf-1, an ATPase that is involved in the auto-catalytic activation of caspase-9 in mammalian apoptosis (see Chapter 2). These comparisons between nematode and mammalian programmed cell death have already been illustrated in *Figure 3.*

Although the cell death pathway that operates in *C. elegans* may be regarded as a 'core' pathway of very ancient evolutionary origin, it is also clear that apoptosis in mammals is much more complex. Much of this complexity is discussed in subsequent chapters. Here, we wish to point out just three features. First, programmed cell death in *C. elegans* has so far been reported to occur only as part of developmental programmes; in mammals, developmental cell death certainly occurs but there are other triggers too, such as certain categories of stress and disease. This leads on to the second point, namely that the variety of inducing signals that lead to cell death means a variety of signal transduction pathways and thus different ways of setting in motion the cell death pathway. We have already noted for example, the role of extrusion of cytochrome c from the mitochondria as an inducing signal for the apoptosome (*Figure 3*). Third, as might be expected, mammals exhibit a much greater complexity in the controls operating within the cell death pathway itself. For example, the single Bcl-2-type protein in *C. elegans* has its equivalent in mammals as a family of proteins, some promoters of survival and some of death (Chapter 4). Similarly, while only a single apoptotic caspase has so far been characterized in the worm, a cascade of caspases acts in mammals, passing the activation down in a chain reaction eventually to those caspases that actually participate directly in killing the targeted cell(s) (Chapter 2). All this implies much more sophisticated means of regulation in mammals compared to nematodes (see, for example, Chapter 7).

Finally, in this comparison between nematodes and mammals, we also note that there is at least come degree of evolutionary conservation in the engulfment pathway. Thus, CED-7 is homologous to the macrophage surface protein ABC1 that has been reported to promote the recognition and clearance of apoptotic cells (Luciani and Chimini, 1996); at least partial sequence homologues of the *ced-6* gene have been identified from expressed sequence tags (i.e., partially sequenced cDNA molecules) in human and mouse cells (Liu and Hengartner, 1998) while CED-5 is homologous to the mammalian protein Dock180, implicated in cytoskeleton rearrangement during phagocytosis (Wu and Horvitz, 1998b).

4. Plants, from *Arabidopsis* to *Zinnia*

It is clear from several other chapters in this book that plants undergo programmed cell death in a variety of situations. Some of these are developmentally regulated, such as leaf senescence (Chapters 10 and 11), the ablation of pollen tapetum cells (for example, Papini *et al.*, 1999) and the formation of conducting tissues (Chapter 14), while others result from biotic and abiotic stresses, such as the hypersensitive reaction to pathogens (Chapter 11) and the formation of aerenchyma (Chapter 13). To what extent are any of these cell death programmes equivalent to those described for animals?

An increasing number of researchers working on plants (Ryerson *et al.*, 1996) take the view that a positive TUNEL assay (an indication of chromatin fragmentation) or the presence of DNA 'ladders' on agarose gels (indicative of inter-nucleosome breaks in DNA; see Chapter 7 for a discussion of these methods) are absolute criteria for the identification of plant cell death as being directly similar to animal apoptosis. We believe, rather, that these are indicative merely of a relatively ordered process of chromatin degradation in which the DNA has been broken down by an endonuclease with the nucleosomes still in place. Nevertheless, it is clear from the recent extensive survey of protein and gene sequences that several of the proteins involved in animal cell death have their homologues in plants (Aravind *et al.*, 1999). Sequence homology does not necessarily mean functional homology, however, particularly since there is evidence that, in animal evolution, some of the protein domains involved in cell death seem to have been 'recruited' from other pathways.

So can plant cell death be directly compared with animal cell death? Survey of plant genes and proteins that resemble those involved in the core apoptosis pathway (as seen in *C. elegans*) reveals the presence of Apaf-1-like proteins (CED-4 homologues) (Aravind *et al.* 1999; Chinnaiyan *et al.*, 1997a; Dineshkumar *et al.*, 1995) and of proteases with caspase-like substrate specificities (del Pozo and Lam, 1998). Cysteine protease activity is involved in the localized cell death that occurs in the hypersensitive response to pathogens (del Pozo and Lam, 1998; de Silva *et al.*, 1998) and in the late stages of formation of water-conducting cells (Chapter 14). Indeed, caspase inhibitors actually prevent cell death in the hypersensitive response (del Pozo and Lam, 1998) and also prevent the late stages of the formation of water-conducting cells (Chapter 14) (although there is also evidence for a specific non-caspase protease in the latter process: Groover and Jones, 1999). However, there is no evidence for the involvement of the Apaf-1-like ATPases in the activation of the plant caspase-like proteases. The Apaf-1 homologues in plants certainly participate in inducing the hypersensitive response (i.e., programmed cell death) in disease resistance: these proteins are the products of the N-gene family of resistance genes, first identified in tobacco but now known to be spread throughout the plant kingdom (Dineshkumar *et al.*, 1995; Kanazin *et al.*, 1996; Speulman *et al.*, 1998). Their role in the hypersensitive response is to induce or activate a protein kinase of the MAPK (mitogen-activated protein kinase) type in a signalling pathway (Zhang and Klessig, 1998), a very different role from that of Apaf-1 in animal cell apoptosis.

Coming one further step upstream in the core cell death pathway, there is as yet no evidence at all that plants possess proteins of the Bcl-2 family. Nevertheless this has not deterred Ohashi's group from transferring the mammalian $Bcl-X_L$ gene and the *C. elegans ced-9* gene (both are pro-survival genes) into tobacco where they are found to inhibit cell death in the hypersensitive response (Mitsuhara *et al.*, 1999). The basis for this inhibition is not clear since, as we have already stated, there is no obvious down-stream pathway to link the Bcl-2 family of proteins to cell death in plants.

So, although certain proteins (or domains thereof) associated with animal cell death are present in plants, indicating a very ancient evolutionary origin for these sequences, there is only limited evidence for a direct equivalence of function. Furthermore, there are programmed cell death pathways in plants, for example the death of endosperm cells in cereal seed development, in which the signalling, induction and actual mechanisms do not in any way resemble animal cell apoptosis (Bethke *et al.*, 1999). The point must also be made that the engulfment phase of cell death in plants must differ from that in animals. Plants, like animals, mobilize the contents of dying cells but because plant cells are constrained by their cell walls and by their immobility, phagocytosis of one cell by another is not possible. However, this does not preclude the possibility of signal molecules, breakdown products and even enzymes moving between cells. In many cases, the mobilization of cell contents is controlled both from within and from outside the dying cell, with exquisite coordination of the timing of the various events (see, for example, Groover and Jones, 1999 and Chapters 13 and 14). Cereal seeds by contrast, mobilize the contents of endosperm totally from the outside. In these seeds, the dead endosperm cells remain intact through the seed desiccation period and the contents are mobilized for embryo growth during germination. This is achieved by the secretion of enzymes into the endosperm from the surrounding layer of living cells (the aleurone) and the uptake of the mobilized reserves by transfer cells in the embryo adjacent to the endosperm (see Bryant and Cuming, 1999).

5. General conclusions

The nematode worm *Caenorhabditis elegans* has proved an excellent model for defining and understanding the core pathway of programmed cell death or apoptosis. This pathway, which appears to be of very ancient evolutionary origin, has become more complex, both in its mechanisms and in its regulation, in more 'advanced' animals, particularly mammals. Plants possess some features of the core pathway (as might be predicted from the widespread distribution of some of the core domains involved in apoptosis), but despite this there is only limited similarity between some forms of plant cell death and apoptosis, while other forms of plant cell death do not in any way resemble apoptosis. Thus we conclude that while programmed cell death is a key element of life in both plants and animals, the two kingdoms have evolved mainly different mechanisms to bring this about. These mechanisms and their regulation are the subject of the following chapters in this book.

Acknowledgements

Research on *C. elegans* apoptosis in Michael Hengartner's laboratory is supported by the National Institutes of Health and by the Donaldson Charitable Trust. Michael O. Hengartner is a Rita Allen Foundation Scholar. Research on the plant cell division cycle in John Bryant's laboratory is supported by the Biotechnology and Biological Sciences Research Council and by NATO.

References

Aravind, L., Dixit, V.M. and Koonin, E.V. (1999) The domains of death: evolution of the apoptosis machinery. *Trends Biochem. Sci.* **24**: 47–53.

Bethke, P.C., Lonsdale, J.E., Fath, A. and Jones, R.L. (1999) Hormonally regulated programmed cell death in barley aleurone cells. *Plant Cell* **11**: 1033–1045.

Brenner, S. (1974) The genetics of *Caenorhabditis elegans*. *Genetics* **77**: 71–94.

Bryant, J.A. and Cuming, A.C. (1999) Molecular control of development. In *Plant Biochemistry and Molecular Biology*, 2nd edn (eds P.J. Lea and R.C. Leegood). Wiley, Chichester, UK, pp. 288–333.

Chaudhary, D., O'Rourke, K., Chinnaiyan, A.M. and Dixit, V.M. (1998) The death-inhibitory molecules CED-9 and CED-4L use a common mechanism to inhibit the CED-3 death protease. *J.Biol.Chem.* **273**: 17708–17712.

Chinnaiyan, A.M., Chaudhary, D, O'Rourke, K., Koonin, E.V. and Dixit, V.M. (1997a) Role of CED-4 in the activation of CED-3. *Nature* **388**: 728–729.

Chinnaiyan, A.M., O'Rourke, K., Lane, B.R. and Dixit, V.M. (1997b) Interaction of CED-4 with CED-3 and CED-9: a molecular framework for cell death. *Science* **275**: 1122–1126.

Conradt, B. and Horvitz, H.R. (1998) The *C. elegans* protein EGL-1 is required for programmed cell death and interacts with the Bcl-2-like protein, CED-9. *Cell* **93**: 519–529.

de Silva, I., Poirier, G.G. and Heath, M.C. (1998) Activation of cysteine proteases in cowpea plants during the hypersensitive response – a form of programmed cell death. *Expl. Cell Res.* **245**: 389–399.

del Pozo, O. and Lam, E. (1998) Caspases and programmed cell death in the hypersensitive response of plants to pathogens. *Curr. Biol.* **8**: 1129–1132.

Desai, C., Garriga, G., McIntire, S.L. and Horvitz, H.R. (1988) A genetic pathway for the development of the *Caenorhabditis elegans* HSN motor neurons. *Nature* **336**: 638–646.

Dineshkumar, S.P., Whitham, S., Choi, D., Hehl, R., Corr, C. and Baker, B. (1995) Transposon tagging of tobacco mosaic virus resistance gene-N: its possible role in the TMV-N-mediated signal transduction pathway. *Proc. Natl Acad. Sci. USA* **92**: 4175–4180.

Ellis, H.M. and Horvitz, H.R. (1986) Genetic control of programmed cell death in the nematode *C. elegans*. *Cell* **44**: 817–829.

Ellis, R.E., Jacobsen, D.M. and Horvitz, H.R. (1991) Genes required for the engulfment of cell corpses during programmed cell death in *Caenorhabditis elegans*. *Genetics* **129**: 79–94.

Feng, S., Chen, J.K., Yu, H., Simon, J.A. and Schreiber, S.L. (1994) Two binding orientations for peptides to the Src SH3 domain: development of a general model for SH3-ligand interactions. *Science* **266**: 1241–1247.

Goffeau, A., Barrell, B.G., Bussey, H., Davis, R.W., Dujon, B., Feldman, H., Galibert, F., Hoheisel, J.D., Jacq, C., Johnston, M., Louis, E.J., Mewes, H.W., Murakimi, Y., Philippsen, P., Tettelin, H. and Oliver, S.G. (1996) Life with 6000 genes. *Science* **274**: 546–551.

Groover, A. and Jones, A.M. (1999) Tracheary element differentiation uses a novel mechanism coordinating programmed cell death and secondary wall synthesis. *Plant Physiol.* **119**: 375–384.

Hedgecock, E.M., Sulston, J.E. and Thomson, J.N. (1983) Mutations affecting programmed cell deaths in the nematode *Caenorhabditis elegans*. *Science* **220**: 1277–1279.

Hengartner, M.O. (1997) Apoptosis: CED-4 is a stranger no more. *Nature* **388**: 714–715.

Hengartner, M.O. (1998a) Apoptosis: Death by crowd control. *Science* **281**: 1298–1299.

Hengartner, M.O. (1998b) Apoptosis: Death cycle and Swiss army knives. *Nature* **391**: 441–442.

Hengartner, M.O. (1999) Programmed cell death in the nematode *C. elegans*. *Rec. Prog. Horm. Res.* **54**: 213 – 222.

Hengartner, M.O. and Horvitz, H.R. (1994a) *C. elegans* cell-survival gene *ced-9* encodes a functional homolog of the mammalian proto-oncogene *Bcl-2*. *Cell* **76**: 665–676.

Hengartner, M.O. and Horvitz, H.R. (1994b) The ins and outs of programmed cell death during *C. elegans* development. *Phil. Trans. Roy. Soc. Lond. Ser. B.* **345**: 243–246.

Hengartner, M.O., Ellis, R.E. and Horvitz, H.R. (1992) *Caenorhabditis elegans* gene *ced-9* protects cells from programmed cell death. *Nature* **356**: 494–499.

Hodgkin, J., Horvitz, H.R., Jasny, B.R. and Kimble, J. (1998) *C. elegans*: sequence to biology. *Science* **282**: 2011.

Horvitz, H.R., Shaham, S. and Hengartner, M.O. (1994) The genetics of programmed cell death in the nematode *Caenorhabditis elegans*. *Cold Spring Harb. Quant. Biol.* **59**: 377–385.

Irmler, M., Hofmann, K., Vaux, D. and Tschopp, J. (1997) Direct physical interaction between the *Caenorhabditis elegans* 'death proteins', CED-3 and CED-4. *FEBS Lett.* **406**: 189–190.

James, C., Gschmeissner, S., Fraser, A. and Evan, G.I. (1997) CED-4 induces chromatin condensation in *Schizosaccharomyces pombe* and is inhibited by direct physical association with CED-9. *Curr. Biol.* **7**: 246–252.

Kanazin, V., Marck, L.F. and Shoemaker, R.C. (1996) Resistance gene analogs are conserved and clustered in soybean. *Proc. Natl Acad. Sci. USA* **93**: 11746–11750.

Kimble, J. and Hirsch, D. (1979) The post-embryonic cell lineages of the hermaphrodite and male gonads in *Caenorhabditis elegans*. *Devel. Biol.* **70**: 396–417.

Liu, Q.A. and Hengartner, M.O. (1998) Candidate adaptor protein CED-6 promotes the engulfment of apoptotic cells in *C. elegans*. *Cell* **93**: 961–972.

Luciani, M.F. and Chimini, G. (1996) The ATP-binding cassette transporter ABC1 is required for the engulfment of corpses generated by apoptotic cell death. *EMBO J.* **15**: 226–235.

Mitsuhara, I., Malik, K.A., Miura, M. and Ohashi, Y. (1999) Animal cell-death suppressors *Bcl-x(L)* and *ced-9* inhibit cell death in tobacco plants. *Curr. Biol.* **9**: 775–778.

Muchmore, S.W., Sattler, M., Liang, H., Meadows, R.P., Harlan, J.E., Yoon, H.S., Nettesheim, D., Chang, B.S., Thompson, C.B., Wong, S.L., Ng, S.L. and Fesik, S.W. (1996) X-ray and NMR structure of human Bcl-x(L), an inhibitor of programmed cell death. *Nature* **381**: 335–341.

Muzio, M., Stockwell, B.R., Stennicke, H.R., Salvesen, G.S. and Dixit, V.M. (1998) An induced proximity model for caspase-8 activation. *J. Biol. Chem.* **273**: 2926- 2930.

Ottilie, S., Wang, Y., Banks, S., Chang, J., Vigna, N.J., Weeks, S., Armstrong, R.C., Fritz, L.C. and Oltersdorf, T. (1997) Mutational analysis of the interacting cell death regulators CED-9 and CED-4. *Cell Death Diff.* **4**: 526–533.

Papini, A., Mosti, S. and Brighigna, L. (1999) Programmed cell death events during tapetum development of angiosperms. *Protoplasma* **207**: 213–221.

Ren, R., Mayer, B.J., Cicchetti, P. and Baltimore, D. (1993) Identification of a ten-amino acid proline-rich SH3 binding site. *Science* 259: 1157–1161.

Ryerson, D.E., Heath, M.C., Wang, H., Li, J., Bostock, R.M. and Gilchrist, D.G. (1996) DNA ladders connecting animals and plants. *Trends Plant Sci.* 7: 212.

Seshagiri, S. and Miller, L.K. (1997) *Caenorhabditis elegans* CED-4 stimulates CED-3 processing and CED-3-induced apoptosis. *Curr. Biol.* 7: 455–460.

Shaham, S. and Horvitz, H.R. (1996) Developing *Caenorhabditis elegans* neurons may contain both cell-death protective and killer activities. *Genes Dev.* 10: 578–591.

Spector, M.S., Desnoyers, S., Hoeppner, D.J. and Hengartner, M.O. (1997) Interaction between the *C. elegans* cell-death regulators CED-9 and CED-4. *Nature* 385: 653–656.

Speulman, E., Bouchez, D., Holub, E.B. and Beynon, J.L. (1998) Disease resistance gene homologues correlate with disease resistance loci of *Arabidopsis thaliana*. *Plant J.* 14: 467–479.

Sulston, J.E. and Horvitz, H.R. (1977) Post-embryonic cell lineages of the nematode, *Caenorhabditis elegans*. *Dev. Biol.* 56: 110–156.

Sulston, J.E., Schierenberg, E., White, J.G. and Thomson, J.N. (1983) The embryonic cell lineage of the nematode, *Caenorhabditis elegans*. *Dev. Biol.* 100: 64–119.

The *C. elegans* sequencing consortium (1998) Genome sequence of the nematode *C. elegans*: a platform for investigating biology. *Science* 282: 2012–2018.

Trent, C., Tsung, N. and Horvitz, H.R. (1983) Egg-laying defective mutants of the nematode *Caenorhabditis elegans*. *Genetics* 104: 619–647.

Vaux, D.L., Haecker, G. and Strasser, A. (1994) An evolutionary perspective on apoptosis. *Cell* 76: 777–779.

Wood, W.B. (1988) The nematode *C. elegans*. Cold Spring Harbor Laboratory, Cold Spring Harbor, NY, USA.

Wu, Y.C. and Horvitz, H.R. (1998a) The *C. elegans* cell corpse engulfment gene *ced-7* encodes a protein similar to ABC transporter. *Cell* 93: 951–960.

Wu, Y.C. and Horvitz, H.R. (1998b) *C. elegans* phagocytosis and cell-migration protein CED-5 is similar to human DOCK-180. *Nature* 392: 501–504.

Wu, D.Y., Wallen, H.D., Inohara, N. and Nunez, G. (1997) Interaction and regulation of the *Caenorhabditis elegans* death protease CED-3 by CED-4 and CED-9. *J. Biol. Chem.* 34: 21449–21454.

Yang, X.L., Chang, H.Y. and Baltimore, D. (1998a) Autoproteolytic activation of pro-caspases by oligomerization. *Molec. Cell* 2: 319–325

Yang, X.L., Chang, H.Y. and Baltimore, D. (1998b) Essential role of CED-4 oligomerization in CED-3 activation and apoptosis. *Science* 281: 1355–1357.

Yenush, L. and White, M.F. (1997) The IRS-signaling system during insulin and cytokine action. *BioEssays* 19: 491–500.

Yu, H., Chen, J.K, Feng, S., Dalgarno, D.C., Brauer, A.W. and Schreiber, S.L. (1994) Structural basis for the binding of proline-rich peptides to SH3 domains. *Cell* 76: 933–945.

Yuan, J. and Horvitz, H.R. (1990) The *Caenorhabditis elegans* genes *ced-3* and *ced-4* act cell autonomously to cause programmed cell death. *Dev. Biol.* 138: 33–41.

Yuan, J.Y and Horvitz, H.R. (1992) The *Caenorhabditis elegans* cell-death gene *ced-4* encodes a novel protein and is expressed during the period of extensive programmed cell death. *Development* 116: 309–320.

Yuan, J.Y., Shaham, S., Ledoux, S., Ellis, H.M. and Horvitz, H.R. (1993) The *C. elegans* cell-death gene *ced-3* encodes a protein similar to mammalian interleukin-1-β-converting enzyme. *Cell* 75: 641–652.

Zhang, S.Q. and Klessig, D.F. (1998) Resistance gene-N-mediated *de novo* synthesis and activation of a tobacco mitogen-activated protein kinase by tobacco mosaic virus infection. *Proc. Natl Acad. Sci. USA* 95: 7433–7438.

Zou, H., Henzel, W.J., Liu, X., Lutschg, A. and Wang, X. (1997) Apaf-1, a human protein homologous to *C. elegans* CED-4, participates in cytochrome c-dependent activation of caspase-3. *Cell* 90: 405–413.

2

Caspases – at the cutting edge of cell death

Henning R. Stennicke

1. Introduction

The pathologies associated with defects in the cell death pathway are described in detail elsewhere in this volume; this chapter focuses on the mechanism and regulation of cell death in mammalian cells. Cell death is often described as having two main mechanisms: necrosis and apoptosis, with a prime distinction being the involvement of caspases in apoptosis (Samali *et al.*, 1999). The caspases were originally identified by genetic analysis of *C. elegans* mutants in which 131 of the total 1090 somatic cells failed to die correctly because they did not express the two 'killer' genes, ced-3 and ced-4 (Yuan and Horvitz, 1990; see also Chapter 1). Nucleotide sequence data reveal that CED-3 is a homologue of interleukin 1β converting enzyme (ICE) and thus made the link between proteases and cell death (Yuan *et al.*, 1993). Caspase stands for cysteine-dependent *asp*artate specific prote*ase*, and is a term coined to define proteases related to ICE and CED-3 (Alnemri *et al.*, 1996). Thus their enzymatic properties are governed by a dominant specificity for substrates containing Asp, and by the use of a Cys side-chain for catalysing peptide bond cleavage. While the use of a Cys side-chain as a nucleophile during peptide bond hydrolysis is common to several protease families, the primary specificity for Asp turns out to be very rare among proteases. Currently the only other known mammalian protease with the same primary specificity is serine protease granzyme B, which is a physiological caspase activator. In addition to this unusual primary specificity, caspases are remarkable in that certain of their zymogens have intrinsic proteolytic activity. This latter property is essential to trigger the proteolytic pathways that lead to apoptosis.

2. Caspases and their properties

2.1 *Known human caspases and their primary structure*

Eleven human caspases have been identified so far, but this may not be all those present in the human cell. The human caspases share a large degree of similarity in what is referred to as the catalytic domain, while there is little similarity in the N-peptide.

Programmed Cell Death in Animals and Plants, edited by J.A. Bryant, S.G. Hughes and J.M. Garland.
© 2000 BIOS Scientific Publishers Ltd, Oxford.

The catalytic domain consists of a large ~20 kDa subunit and a small ~10 kDa sub-unit, both of which contain residues that are essential for the function of the activity of the caspases (*Figure 1a*). The caspases are normally activated by a single proteolytic cleavage in the linker region situated between these two subunits although sometimes a short linker segment can be released. The N-peptide of the caspases varies in length

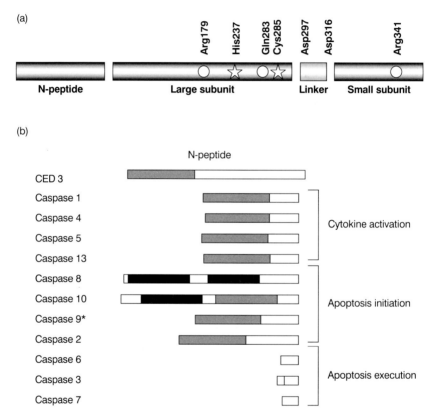

Figure 1. Diagrammatic structure of the human caspases. (a) The zymogens of the caspases all comprise an N-peptide and a catalytic domain. The caspases are activated by proteolytic cleavage at Asp 297 (caspase 1 numbering) inside the catalytic domain to form the characteristic large and small subunits. Both of the subunits contain components that are essential for catalysis as indicated by stars for catalytic residues and circles for residues involved in the determination of the primary specificity. Often an additional cleavage takes place in the inter-domain linker region to release a short linker peptide. (b) The caspases may be divided into three major groups depending on their biological function. These three groups each perform a highly specialized function and there appears to be little cross-over between the groups. The length of the N-peptide varies from 22 amino acids in caspase 6 to over 200 in caspase 8 and as a result there are significant variations between the sequences of the N-peptides. In contrast, the amino acid sequence of the caspases is highly conserved within the catalytic domain although there are significant differences in the inter-domain linker region where the activating cleavage occurs. The N-peptides contain different embedded elements that are important for specific protein interactions required for the activation of the caspases. Thus, caspases 8 and 10 contain tandem DED domains (in black) which are required for binding to adapter proteins during death receptor ligation. Caspases 1, 2, 4, 5, 9 and 13, as well as Ced-3, contain a CARD domain (in grey) which is recognized by sequence similarity and is believed to be important for the assembly of a number of activator complexes (Hofmann et al., 1997). The * indicates that the N-peptide is not removed subsequent to activation of caspase 9.

from 22 amino acid residues in caspase 6 to over 200 in caspases 8 and 10 (*Figure 1b*).

Embedded within the longer N-peptides of the caspases are distinct domains that are required for the activation of the caspases while the function of the short N-peptides remains unknown. There are two different types of elements known as the death effector domain (DED domain) and the caspase recruitment domain (CARD domain). The DED domains are found exclusively in the caspases 8 and 10, which are associated with death receptor initiated cell death and are essential for the recruitment of these caspases to the receptor via specific adapter molecules. The role of the CARD domain is more promiscuous since it is found in both caspases involved in cytokine activation and in the mediation of cell death. Thus, caspases 2 and 9 as well as caspases 1, 4, 5 and 13 all contain CARD domains. For convenience the known caspases are often divided into three major groups based on their presumed biological function: cytokine activators, initiators of apoptosis and executioners of apoptosis – as reviewed by Cohen (1997), Nicholson and Thornberry (1997) and Salvesen and Dixit (1997). Although the classification provides a useful way of thinking about the caspases there still are many open questions associated with the biological function of caspases 2, 4, 5 and 13. Caspase 2 is primarily grouped with the initiator caspases, based on the observation that it can bind to the tumour necrosis factor (TNF) receptor via the proteins known as RIP and RAIDD, and therefore may induce apoptosis in response to TNF. The exact roles of caspases 4, 5 and 13 remain unknown; these are mainly grouped as cytokine activators due to their extended sequence similarity with caspase-1. However, at least caspase 13 is activated by caspase 8 *in vitro* although the physiological significance of this remains unclear (Humke *et al.*, 1998).

2.2 *Biological role of caspases*

One way of determining the biological role of the caspases has been ectopic expression of the individual proteases in various cell lines. Using this approach almost all known caspases are capable of inducing apoptosis to some extent. In contrast, there are significant differences when the caspases are ablated in mice. One example of this is the observation that ectopic expression of caspase 1 induces apoptosis, but an actual role in developmental cell death is unlikely given the normal phenotype of the knockout mice (Kuida *et al.*, 1995). The caspase 1 knockout mice are, however, disabled in their ability to activate cytokines in response to septic shock. The developmental phenotypes of mice ablated of caspases 3, 8 or 9 are in stark contrast to those observed for the caspase 1 knockouts. These particular knockouts all show severe developmental defects with early lethality due to failure of developmental cell death of neuronal cell precursors (Kuida *et al.*, 1998; Kuida *et al.*, 1996), or impaired heart muscle development (Varfolomeev *et al.*, 1998). Unfortunately, the effect of knockouts are not always that clear, as caspase 2 apparently can act as both a positive and negative cell death effector, depending upon cell lineage and stage of development (Bergeron *et al.*, 1998). These conflicting effects are possibly associated with the different functions of the various spliced forms of caspase 2 (Jiang *et al.*, 1998; Wang *et al.*, 1994).

The different biological functions of the caspases are reflected in their substrate specificity and preference for distinct natural protein substrates. The pool of known natural substrates is steadily increasing as more proteins get tested as substrates for the caspases. As it is apparent from even a limited number of substrates there is a wide range of events associated with the activation of the caspases (*Table 1*). The distinctions between the

Table 1. Selected natural caspase substrates, cleavage sites and effects

Caspase	Substrate	Cleavage site	Effect
Caspase 1	proIL-1β	YVHD/A (Howard et al., 1991)	Activates IL-1β
	proIGIF	LESD/N (Ghayur et al., 1997)	Activates IGIF
Caspase 3 and 7	DFF	DETD/S (Liu et al., 1997)	Initiates DNA fragmentation
	PARP	DEVD/G (Lazebnik et al., 1994)	Inactivates DNA repair
	PKCδ	DMQD/N (Ghayur et al., 1996)	Activates kinase
	MEKK	DTVD/G (Cardone et al., 1997)	Activates kinase
	U1–70kDa	DGPD/G (Casciola-Rosen et al., 1996)	Inactivates mRNA splicing
Caspase 6	LaminA	VEID/N (Takahashi et al., 1996)	Nuclear collapse
	Keratin 18	VEVD/A (Caulin et al., 1997)	Cytoskeletal collapse
Caspase 8	Caspase 3	IETD/S (H.R. Stennicke, unpublished)	Activates protease
	Caspase 7	IQAD/S (Zhou and Salvesen, 1997)	Activates protease
	BID	LQTD/A (Luo et al., 1998)	Induces cytochrome c release

known substrates are largely based on the interactions in the S_4 binding site of the caspases as they all depend on the presence of an Asp in the P_1 position of the substrate.

2.3 *Extended substrate specificity and the basis for synthetic inhibitors*

While the natural substrates do provide an indication of the specificity of the caspases it is basically impossible to perform a rigorous analysis of their specificity without implementing synthetic peptide substrates. All known caspases clearly cleave after Asp residues in synthetic substrates, but the degree of discrimination versus the next best side-chain at this position in most caspases remains uncharacterized. Since the initial studies using caspase 1 revealed no gain in catalysis beyond occupancy of S_4 (Thornberry et al., 1992), further experimental studies on substrate specificity of the caspases have concentrated on occupancy of subsites S_2–S_4. Within this range, all the known caspases seem to prefer Glu at P_3 and differences in substrate specificities are largely dominated by individual preferences at P_2 and P_4. Based on subsite mapping using combinatorial libraries exploring the P_2–P_4 positions Thornberry et al. (1997) have divided caspases 1–9 and *C. elegans* CED-3 into three specificity groups according to P_4 preferences. Of these, group II is the most distinct with a high degree of selectivity for Asp at this position. Significantly this group contains the executioner caspases 3 and 7, whose natural substrates cleaved during apoptosis frequently contain Asp at P_4. Groups I and III are less distinct since they both accept branched aliphatic side-chains fairly well, and the key distinction is that group I, including caspases 1, 4 and 5, the presumed cytokine activators, will also accept aromatic side-chains. Division into these three groups is useful in understanding the natural substrates and activation pathway, but a note of caution is added by Talanian et al. (1997) who demonstrated that occupancy of S_5 was necessary for efficient cleavage of substrates by caspase 2.

Based on the specificity of different caspases for synthetic substrates (Nicholson *et al.*, 1995; Thornberry *et al.*, 1992) a number of peptidyl inhibitors have been synthesized, and some have been used by numerous authors in attempts to discriminate the role of individual caspases *in vivo* and in cell-free systems. Because caspases are cysteine proteases, reactive groups successful with cysteine proteases of the papain family have been tagged onto simple peptides in an attempt to convey specificity. These reactive groups include reversible aldehydes, and the less reversible halomethyl ketones, various acyloxymethyl ketones, and diazomethanes (Fernandes-Alnemri *et al.*, 1995; Margolin *et al.*, 1997; Mittl *et al.*, 1997; Prasad *et al.*, 1995; Rano *et al.*, 1997; Talanian *et al.*, 1997; Thornberry *et al.*, 1992; Thornberry *et al.*, 1994). Fluoromethyl ketone and aldehyde derivatives of the tetrapeptide YVAD (based on an early caspase 1 substrate) and DEVD (based on an early caspase 3 substrate) have been used by many researchers attempting to relate specific apoptotic events or cytokine activation to specific caspases. However, it should be noted that though the inhibitors containing the YVAD frame are reasonably selective for caspases 1 and 4, those based on DEVD show little selectivity (Garcia-Calvo *et al.*, 1998). Nevertheless the general trend seems to indicate that tetrapeptide-based inhibitors are unlikely to achieve the specificity required to allow inhibition of individual caspases.

2.4 *Enzymatic characteristics*

In addition to the substrate specificity, one of the major determinants of the activity of a protease, or any enzyme for that matter, is the pH dependence. This is clearly demonstrated by the protease pepsin, with optimum activity around pH 2, making it suitable for working in the gut (Szecsi, 1992). The same is true for various lysosomal/vacuolar proteases which are most often characterized by a pH optimum between 4 and 5, making them ideally suited for their particular environment. The caspases are optimally active at cytosolic pH; thus, the pH optimum of the caspases varies from 6.5 for caspase 2 to 7.4 for caspase 3 (Garcia-Calvo *et al.*, 1999; Stennicke and Salvesen, 1997). The pH dependences of the caspases, however, are notably different from other thiol proteases in that they are represented by a relatively narrow bell shaped curve, essentially limiting their activity to the pH range between 5.5 and 9.0 (Stennicke and Salvesen, 1997).

As with other cysteine proteases the caspases require a reducing environment in order to retain full activity, presumably because the catalytic thiol is susceptible to oxidation. This would tend to limit caspase activity to the reducing environment found inside cells, and argue against an extracellular role. It is clear from work on the recombinant caspases that they require quite high concentrations of reducing agents such as DTT (dithiothrietol) to achieve maximal activity (probably due to oxidation of the catalytic Cys during purification: H.R. Stennicke & G.S. Salvesen, unpublished results). The endogenous caspases in cytosol appear to be fully active in the presence of low concentrations of reducing agents and recombinant protein, once activated, also retains full activity in these conditions, providing that there are metal chelators such as EDTA present (H.R. Stennicke and G.S. Salvesen, unpublished results). This reveals a related aspect to the oxidation, which is the effect of transition metals on the activity. The influence of such metal ions on the activity of other cysteine proteases is well established. It is predominantly due to the direct interaction with the catalytic thiol but also to the ability of these metals to catalyse oxidation. It is therefore not surprising that the caspases are sensitive to Zn^{2+}, being completely inhibited in the mM range (Stennicke and Salvesen, 1997), although there are significant differences in their affinity. Caspase 6 is most readily inhibited by Zn^{2+}, becoming completely inactivated at 0.1 mM, and caspase 3 is the least sensitive requiring more than 1 mM for complete inactivation in the presence of 20 mM 2-mercaptoethanol. The inhibition of caspases by Zn^{2+} may explain the inhibitory action of this metal on apoptosis (Perry *et al.*, 1997; Takahashi *et al.*, 1996), though the interaction is blocked by thiol compounds (Stennicke and Salvesen, 1997), and therefore presumably highly dependent on the redox potential of the cell. Ca^{2+} has little effect on the activity of the caspases at concentrations up to 100 mM, although minor effects on the activity of caspase 7 have been reported (Garcia-Calvo *et al.*, 1999). Thus, the reported role of Ca^{2+} in apoptosis (see for example Bian *et al.*, 1997), is unlikely to be due to any effect on the caspases.

3. Structure of the caspases

3.1 *Folding and catalytic mechanism*

The three-dimensional structures have thus far been determined for caspases 1 (Margolin *et al.*, 1997; Walker *et al.*, 1994; Wilson *et al.*, 1994) and 3 (Mittl *et al.*, 1997; Rotonda *et al.*, 1996). The two structures are similar, both demonstrating a molecule composed of two large and two small subunits interacting with inverse two-fold symmetry at the interface of the small subunits. Each symmetry-related unit of the tetramer contains a large six-stranded parallel β-sheet with a single anti-parallel strand near the C-terminus; this anti-parallel strand is one of the main contributors in the formation of the tetramer, which (based on the crystal structure) is hypothesized to be the active species. The reason that this seems reasonable is the formation of a β-sheet extending all the way through the tetramer (*Figures 2a and b*).

From a mechanistic point of view the caspase contains a catalytic Cys–His pair, equivalent to that found in all other cysteine proteases, with Cys285 acting as the nucleophile and His237 acting as the general base to abstract the proton from the catalytic Cys, thus promoting the nucleophile. In other cysteine proteases, such as papain, it is believed that the Cys–His dyad exists as an ion-pair in which the thiol proton of the catalytic Cys has been transferred to the catalytic His prior to substrate binding (Lewis *et al.*, 1981). This mechanism of action represents one of the major differences between the serine and cysteine proteases. Further comparisons between papain and the caspases reveal distinctive components of catalysis.

The majority of the members of the cysteine proteases contain an Asn side-chain that is believed to orient the histidine in the Cys–His ion-pair, thereby determining the pH-dependence of the enzyme (Vernet *et al.*, 1995). A significant deviation from the papain family is demonstrated in the identity of the third component of the catalytic triad, which in caspases is the backbone carbonyl of residue 177 which in caspase 1 is a proline and in caspase 3 a threonine (Rotonda *et al.*, 1996; Wilson *et al.*, 1994). Unfortunately, the importance of this third member is difficult to establish in the caspases since the interactions are with a backbone moiety, and thus cannot easily be experimentally verified by mutagenesis. Furthermore, alignment of the known mammalian caspases (Accession PF00656; Bateman *et al.*, 1999) does not reveal any significant conservation in the region surrounding the putative third member of the

(a)   (b)

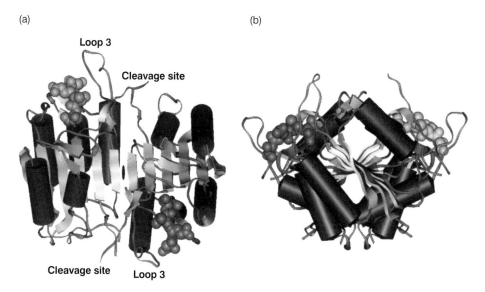

Figure 2. The three-dimensional structure of caspase 3 complexed with DEVD aldehyde
(Rotonda et al., 1996) (in space filling) shows it exists as hetero-tetramer composed of two small
and two large subunits. Each tetramer contains two catalytic sites that appears to function
independently, since there is no evidence of allosteric control of the activity of the caspases. The
dimerization of the caspase is believed to occur by formation of the large central β-sheet that
extends the entire length of the dimer. The central β-sheet is sandwiched between five α-helixes, in
addition to these five helixes caspase 1 contains an additional helix located at the end opposite the
catalytic site. (**a**) shows caspase 3 in a side view along the symmetry axis of the active tetramer. The
large S_4 specificity determining loop 3 is clearly visible in this view as are the inter-domain
cleavage sites. (**b**) shows caspase 3 in an end view, looking down, and highlighting the central
β-sheet extending all the way through the enzyme.

triad; thus, the putative third member may not be important for caspase structure and
mechanism.

3.2 *Structural determinants of substrate specificity*

Another key feature in the function of any enzyme is the ability to discriminate
between substrates. In the caspases this discrimination is primarily based on the
recognition of Asp in the P_1 position. In both the structures solved so far the S_1 subsite contains three residues interacting directly with the P_1 Asp – Arg179, Gln283 and
Arg341. Together these three residues form the most prominent binding site in the
caspases, which is apparent as a hole in the surface of the active site. Of the three
residues involved in P_1 binding, Arg179 appears to be the primary hydrogen bond
donor in the binding of the negatively charged P_1 residue side-chain, whereas Gln283
and Arg341 each donate a single hydrogen bond. Arg341 also contributes to the binding of the P_3 Glu in the caspase 3 structures as well as backbone interactions with the
P_3 residue. The S_2 and S_3 binding sites are rather poorly defined and merely show up as
shallow cavities in the surface of the enzyme.

Comparison of the structures of caspase 1 and 3 provides increased insight into
the determinants of the P_4 specificity. Inspection of these two structures reveals the

presence of an additional large loop (loop3) in caspase 3 as compared to caspase 1 (*Figure 2a*). This additional loop is the key determinant of the differences in the substrate specificity of these two enzymes since the loop in caspase 3 forms a closed cavity with multiple specific interaction to the carboxylate of the P_4 Asp residue of the substrate. At the same time this loop imposes steric hindrance on other larger residues providing additional restriction on the substituents allowed at the P_4 position (Rotonda *et al.*, 1996). By comparison the same region in caspase 1 appears as a narrow crevice suitable for binding the large aromatic residues (Walker *et al.*, 1994; Wilson *et al.*, 1994). Another interesting feature of the substrate binding in the caspases is the formation of a short anti-parallel β-sheet between the enzyme and the substrate. This is distinct from other thiol proteases belonging to the papain family, which do not contain such a structure. Formation of such structures as part of the substrate binding has previously been observed for the chymotrypsin family of serine proteases.

To summarize, the mode of substrate binding observed in the caspases shows numerous features that bear more resemblance to the serine proteases than to other well described thiol proteases. The caspases have a dominant P_1 preference whereas the papain family of thiol proteases has a dominant P_2 preference. Furthermore, the caspases form a secondary structure with the substrate backbone whereas the papain-like proteases form interactions with the substrate backbone by multiple side-chains. Thus, the caspases may be considered to be thiol proteases evolved on a serine protease-like scaffold.

4. Activation and regulation of the caspases

Caspases can be activated by two different mechanisms: homo-activation, which requires recruitment of the zymogens to select factors, and hetero-activation, which occurs when another protease acts upon the caspase zymogen (Stennicke and Salvesen, 1998). Often the latter is the result of the first since the caspases belonging to the activator group are most frequently activated by homo-activation. Apart from regulation by activation there are also a number of well described specific inhibitors of the caspases. The endogenous inhibitors presumably serve to prevent undesired activation of the apoptotic system, whereas the viral inhibitors are most likely there to prevent the cell from committing suicide upon infection.

4.1 *Homo-activation of the caspases*

Caspases 8 and 10 are activated by recruitment to the death initiator complex (*Figure 3*), or the DISC, which is formed upon ligation by FAS_L or TNF and consists of a death receptor of the TNF-R1 family, a modifier molecule and/or an adaptor molecule such as FADD (reviewed by Ware *et al.*, 1998). The activation of the caspases in this complex is believed to occur by homo-activation or by the ability of the caspases to process themselves. While most recombinant caspases process themselves during expression, this is highly dependent upon the expression conditions used and thus, one can obtain either processed active caspase or unprocessed zymogen from the same construct, at least for caspases 3, 7 and 9 (Stennicke *et al.*, 1998, 1999; Zhou and Salvesen, 1997). Significantly, caspase 8 processes itself extremely rapidly upon heterologous expression in *E. coli*, suggesting that the zymogen must possess significant intrinsic proteolytic activity, allowing for auto-processing (H.R. Stennicke and G.S.

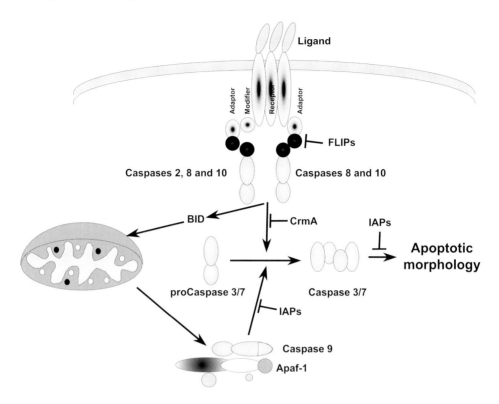

Figure 3. *Currently there exist two recognized points at which apical caspases are activated to initiate apoptosis. Following TNFR-1 or Fas ligation, the initiator caspase 8 is activated by adapter-mediated recruitment to the receptor's cytosolic face (Muzio et al., 1998; Yang et al., 1998). Alternatively, the initiator caspase 9 is activated following release of mitochondrial components to form the Apaf complex (Li et al., 1997; Liu et al., 1996; Zou et al., 1997). Both activated initiators converge on the proteolytic activation of caspase 3. In death receptor triggered apoptosis, the main pathway is direct activation of pro-caspase 3 by caspase 8 (Stennicke et al., 1998). In some cell types an additional pathway may operate by caspase 8-mediated cleavage of BID resulting in release of cytochrome c from mitochondria (Li et al., 1998; Luo et al., 1998). The importance of the mitochondrial pathway in death receptor triggered apoptosis is unknown, but is apparently subordinate to the dominant, direct pathway in most cell types. This model is in agreement with the observation that caspase 9 deficiency does not affect death receptor apoptosis that has been triggered by caspase 8 activation (Kuida et al., 1998).*

Salvesen, unpublished data). Indeed, a non-processable caspase 8 zymogen possesses a very substantial activity compared with most protease zymogens (Muzio *et al.*, 1998). Furthermore, apoptosis may be induced by artificial clustering of caspase 8 zymogens following ectopic expression of chimeric caspase 8 that can be oligomerized by a synthetic cell-permeable 'dimerizer' (Martin *et al.*, 1998; Muzio *et al.*, 1998; Yang *et al.*, 1998a). These observations are the basis for the induced proximity hypothesis for the *in vivo* activation of caspase 8, whose assembly is forced by ligation of death receptors in a process mediated by specific adapter proteins (Ware *et al.*, 1998). This clustering of zymogens possessing intrinsic enzymatic activity forces processing in *trans* and activation of the first protease in the death receptor pathway to apoptosis (Martin *et al.*, 1998; Muzio *et al.*, 1998; Yang *et al.*, 1998a; see also Chapter 1).

This mechanism used for activation of caspases 8 and 10 also opens the possibility for a very special way of regulating their activation. As with most other inhibitors of apoptosis, the proteins known as v-FLIPs were first identified in viruses, where they have been found in several γ-herpesviruses and molluscipox-viruses (Thome et al., 1997). The v-FLIPs contains two DED domains which interact with FADD, thereby preventing recruitment and activation of the caspases 8 and 10. The human analogue of v-FLIP is even more elaborate since it is an inactive homologue of caspases 8 and 10 (Goltsev et al., 1997; Han et al., 1997; Hu et al., 1997; Inohara et al., 1997; Irmler et al., 1997; Shu et al., 1997; Srinivasula et al., 1997). Thus, in addition to the DED domains it contains the entire caspase domain mutated in the catalytic site as well as the S_1 binding site. The role of the human analogue of v-FLIP is not clear since in some cases it protects cells against apoptosis when over-expressed (Hu et al., 1997; Irmler et al., 1997; Srinivasula et al., 1997) while in others it induces apoptosis (Goltsev et al., 1997; Han et al., 1997; Inohara et al., 1997; Shu et al., 1997). However, the FLIPs most likely act as competitive inhibitors, preventing binding of caspase 8 and 10 to the adapter molecule and consequently, the clustering of caspase zymogen at the cytosolic face of the receptor required for activation as discussed above.

Since the activation of caspase 9 can occur independently of added proteases it must be considered another example of homo-activation. The activation of caspase 9 occurs upon release of cytochrome c from the mitochondria (see Chapters 6 and 7) which trigger the formation of the caspase 9 activator complex consisting of Apaf-1, the mammalian analogue of the *C. elegans* CED-4, cytochrome c and caspase 9 (Li et al., 1997; Zou et al., 1997). Like the CED-3 activator complex in *C. elegans*, the caspase 9 activator complex depends on proteins containing ATP binding motifs and most likely also on ATP hydrolysis, since non-cleavable analogues fail to promote activation (Chinnaiyan et al., 1997; Li et al., 1997; Zou et al., 1997). While the mechanism of clustering and activation in the complex remains unclear, it does appear that the caspase 9 activator complex exists as a large oligomer or **apoptosome** (Zou et al., 1999; see also Chapter 1). However, the exact mechanism of caspase 9 may be more complex than that of caspase 8, since proteolytic processing in the inter-domain linker does not appear to be sufficient for activation of caspase 9 (Stennicke et al., 1999). Thus, the activity of both the zymogen and the processed form of caspase 9 is stimulated significantly upon binding to the caspase 9 activator complex. Furthermore, in contrast to most other caspases, caspase 9 does not have its N-peptide removed once activated and thus, may remain associated with the activator complex and activate the executioner caspases as part of the apoptosome.

4.2 *Hetero-activation – activation by caspases and other proteases*

The group of caspases that are referred to as the executioner caspases cannot be activated efficiently at physiological conditions via a homo-activation pathway, since they have no means of being recruited and activated like the previously discussed proteases. This group of caspases relies on the action of the initiator caspases for their activation *in vivo*. The activation by caspases 8 and 10 as well as granzyme B is best characterized for caspase 3 and 7, because the zymogens of these caspases easily can be obtained. All the tested initiator caspases are potent activators of caspase 3 and 7, activating them with rate constants in the 10^6 M^{-1} s^{-1} range (*Table 2*). While the activation of caspase 3 can be described by a direct and fast mechanism both *in vivo*

Table 2. Activation rates of executioner caspase zymogens by selected initiators ($M^{-1}\ s^{-1}$).

	Caspase 8	Caspase 10	Granzyme B	Subtilisin	Cathepsin G
Pro-caspase 3	2.2×10^6	0.7×10^6	4.8×10^6	n.d.	n.d.
Pro-caspase 7	4.4×10^6	1.2×10^6	6.4×10^6	0.7×10^6	0.3×10^6

Purified recombinant zymogens of the executioner caspases 3 and 7 are activated most efficiently by the serine protease granzyme B, followed by the apical caspase 8, and then caspase 10. Data for caspase 3 are taken from Stennicke *et al.* (1998) and for caspase 7 from H.R. Stennicke and G.S. Salvesen (unpublished results). Subtilisin and cathepsin G activation data are from Zhou and Salvesen (1997).

and *in vitro*, the activation pattern of caspase 7 is not as easily interpreted: in the case of caspase 7 there appears to be an additional requirement *in vivo*. This is apparent from the observation that in TNF or granzyme B treated MCF-7 cells not containing caspase 3, caspase 7 is not activated (Yang *et al.*, 1998b) whereas in the same cell line transfected with caspase 3 the processing is normal. The apparent role of caspase 3 in the activation of caspase 7 is to remove the N-peptide; however, the reason for this removal and how it influences the activation remains unresolved.

There have been several suggestions that the signal initiated by activation of caspase 8 or 10 is amplified by an additional step (Scaffidi *et al.*, 1998) such as feedback by caspase 3 onto caspase 9 (Srinivasula *et al.*, 1996) or by the cleavage of BID and initiation of the mitochondrial pathway (Li *et al.*, 1998; Luo *et al.*, 1998). In the majority of cells these two pathways do not appear to have any significant influence. First, the direct feedback of caspase 3 onto caspase 9 seems unlikely since caspase 9 requires more than simple cleavage for activation (Stennicke *et al.*, 1999). Second, the rate of cleavage by caspase 8 of a protein called BID appears to be slightly slower than the rate by which caspase 3 is activated by caspase 8 (Bossy-Wetzel and Green, 1999; H.R. Stennicke and G.S. Salvesen, unpublished results).

More surprising is the demonstration that proteases without specificity for Asp are able to activate caspase zymogens, at least *in vitro*. For example, the serine proteases subtilisin Carlsberg and cathepsin G are both potent activators of pro-caspase 7 (Zhou and Salvesen, 1997). This observation indicates that the linker segment between the large and small subunits is an unusually susceptible inter-domain connector designed to be used for rapid proteolysis. Normally, within the hierarchy of caspase cascades, zymogen activation takes place at conserved Asp residues, but the proteolytic sensitivity of the caspase inter-domain link may allow non-Asp-specific proteases such as those from lysosomes or viruses to engage the apoptotic apparatus under pathological conditions.

4.3 *Regulation of caspase activity by natural inhibitors*

Not surprisingly, inhibition of caspases is a strategy adopted by viruses in their attempt to elude the apoptotic response of the cell to infection (see Chapter 18). Thus, much of our understanding of the pivotal role played by proteases in apoptosis and cytokine activation comes from understanding that the poxvirus protein CrmA (Ray *et al.*, 1992) and the baculovirus protein p35 (Bump *et al.*, 1995) specifically target caspases. Both inhibitors are believed to use a serpin-like mechanism for inhibiting the caspases (Zhou *et al.*, 1998) although there are significant differences in the three dimensional structure (Fisher *et al.*, 1999). However, both CrmA and p35 appear to be cleaved during the inhibition and the binding affinity to the catalytic mutants of caspases 3 and 7 are significantly reduced, indicating that the mechanisms of inhibition are of the suicide substrate type (Zhou *et al.*, 1998; Zhou *et al.*, 1997). Interestingly, no homologues of p35 are known in mammals, and though mammals possess many homologues of the serpin CrmA, none of the known ones seem to be targeted against caspases. Thus, the only endogenous human serpin with a well-characterized target in apotosis is PI-9 which is a potent inhibitor of the caspase activator granzyme B (Sun *et al.*, 1996) but does not inhibit the caspases (G.S. Salvesen, personal communication). So far the only demonstrated caspase inhibitors endogenous to mammals are members of the 'inhibitor of apoptosis' (IAP) family, including the human proteins XIAP, cIAP-1 and cIAP-2. All three of these IAPs appear to specifically target the executioner caspases 3 and 7 while XIAP also targets caspase 9 (Deveraux *et al.*, 1997, 1998; Roy *et al.*, 1997). The selectivity of the IAPs clearly distinguishes them from CrmA which is an efficient inhibitor of caspases 1 and 8, but very weak with caspases 3 and 7 (Zhou *et al.*, 1997). Furthermore, in contrast to observations made with purified recombinant caspases 9, CrmA does not appear to efficiently inhibit caspase 9 *in vivo* since it fails to block cytochrome C/dATP induced activation of caspases 3, 6 and 7 in cell free lysates (Deveraux *et al.*, 1998).

Thus, the IAPs and CrmA appear to have very distinct specificities with only a very limited overlap. Furthermore, the apparent differences in the affinity of CrmA for caspase 9 *in vivo* and *in vitro* suggests that the caspase 9 found inside the cell is quite different from the truncated recombinant form. This is supported by the observation that caspase 9 requires binding to Apaf-1 for full activity (Stennicke *et al.*, 1999). The baculovirus inhibitor p35 constitutes an entirely different type of inhibitor which seems to inhibit most caspases with comparable efficiency (Bertin *et al.*, 1996; Bump *et al.*, 1995; Xue and Horvitz, 1995; Zhou *et al.*, 1998). However, caspase 9 appears to deviate from this general observation since it is poorly inhibited by p35 *in vivo* (H.R. Stennicke and G.S. Salvesen, unpublished results).

5. Evolution of the caspases

From an evolutionary point of view, most proteases belong to large families with homologies extending from bacteria to humans as is the case for the chymotrypsin family (Rawlings and Barrett, 1993). Based on this observation, one would expect to find homologues of the caspases in lower organisms which would shed light on the origin of the family. One of the most prominent features of the chymotrypsin family of proteases is the dominance of the S_1 pocket which has evolved to accept a large variety of different P_1 substituents on a common structural scaffold. The dominant P_1 specificity is one of the most notable properties shared between the caspases and serine proteases of the chymotrypsin family, and in particular with granzyme B (Odake *et al.*, 1991). There is currently no definite evidence for the existence of other proteases which share the unusual caspase fold and exhibit a distinct and dominant P_1 specificity different from that of the caspases. However, it has been proposed that bacterial proteases clostripain and R-gingipain (dominant P_1 specificity for Arg), K-gingipain (dominant P_1 specificity for Lys), and the plant and animal legumains (dominant P_1 specificity for Asn) may represent such a group of proteases. Though all are cysteine proteases, they may share the caspase fold, and therefore represent very distant homologues (Chen *et al.*, 1998). Thus, we may so far only have found a few of the members of the family sharing the caspase fold and may still find a number of other proteases that will eventually close the apparent evolutionary gap between the bacterial and the mammalian members.

References

Alnemri, E.S., Livingston, D.J., Nicholson, D.W., Salvesen, G., Thornberry, N.A., Wong, W.W., and Yuan, J. (1996) Human ICE/CED-3 protease nomenclature. *Cell* 87: 171.

Bateman, A., Birney, E., Durbin, R., Eddy, S.R., Finn, R.D., and Sonnhammer, E.L. (1999) Pfam 3.1: 1313 multiple alignments and profile HMMs match the majority of proteins. *Nucleic Acids Res.* 27: 260–262.

Bergeron, L., Perez, G.I., Macdonald, G., Shi, L., Sun, Y., Jurisicova, A., Varmuza, S., Latham, K.E., Flaws, J.A., Salter, J.C., Hara, H., Moskowitz, M.A., Li, E., Greenberg, A., Tilly, J.L. and Yuan, J. (1998) Defects in regulation of apoptosis in caspase-2-deficient mice. *Genes Dev.* 12: 1304–1314.

Bertin, J., Mendrysa, S.M., LaCount, D.J., Gaur, S., Krebs, J.F., Armstrong, R.C., Tomaselli, K.J. and Friesen, P.D. (1996) Apoptotic suppression by baculovirus P35 involves cleavage by and inhibition of a virus-induced CED-3/ICE-like protease. *J. Virol.* 70: 6251–6259.

Bian, X., Hughes, F.M., Jr., Huang, Y., Cidlowski, J.A. and Putney, J.W., Jr. (1997) Roles of cytoplasmic Ca^{2+} and intracellular Ca^{2+} stores in induction and suppression of apoptosis in S49 cells. *Am. J. Physiol.* 272: C1241–1249.

Bossy-Wetzel, E. and Green, D.R. (1999) Caspases induce cytochrome c release from mitochondria by activating cytosolic factors. *J. Biol. Chem.* 274: 17484–17490.

Bump, N.J., Hackett, M., Hugunin, M., Seshagiri, S., Brady, K., Chen, P., Ferenz, C., Franklin, S., Ghayur, T., Li, P. *et al.* (1995) Inhibition of ICE family proteases by baculovirus antiapoptotic protein p35. *Science* 269: 1885–1888.

Cardone, M.H., Salvesen, G.S., Widmann, C., Johnson, G. and Frisch, S.M. (1997) The regulation of anoikis: MEKK-1 activation requires cleavage by caspases. *Cell* 90: 315–323.

Casciola-Rosen, L., Nicholson, D.W., Chong, T., Rowan, K.R., Thornberry, N.A., Miller, D.K. and Rosen, A. (1996) Apopain/CPP32 cleaves proteins that are essential for cellular repair: a fundamental principle of apoptotic death. *J. Exp. Med.* 183: 1957–1964.

Caulin, C., Salvesen, G.S. and Oshima, R.G. (1997) Caspase cleavage of keratin 18 and reorganization of intermediate filaments during epithelial apoptosis. *J. Cell. Biol.* 138: 1379–1394.

Chen, J.M., Rawlings, N.D., Stevens, R.A. and Barrett, A.J. (1998) Identification of the active site of legumain links it to caspases, clostripain and gingipains in a new clan of cysteine endopeptidases. *FEBS Lett.* 441: 361–365.

Chinnaiyan, A.M., Chaudhary, D., O'Rourke, K., Koonin, E.V. and Dixit, V.M. (1997) Role of CED-4 in the activation of CED-3. *Nature* 388: 728–729.

Cohen, G.M. (1997) Caspases: the executioners of apoptosis. *Biochem. J.* 326: 1–16.

Deveraux, Q., Takahashi, R., Salvesen, G.S. and Reed, J.C. (1997) X-linked IAP is a direct inhibitor of cell death proteases. *Nature* 388: 300–304.

Deveraux, Q.L., Roy, N., Stennicke, H.R., Zhou, Q., Srinivasula, S.M., Alnemri, E.S., Salvesen, G.S. and Reed, J. (1998) IAPs block apoptotic events induced by caspase 8 and cytochrome C by direct inhibition of distinct caspases. *EMBO J.* 17: 2215–2223.

Fernandes-Alnemri, T., Takahashi, A., Armstrong, R., Krebs, J., Fritz, L., Tomaselli, K.J., Wang, L., Yu, Z., Croce, C.M., Salvesen, G., Ernshaw, W.C., Litwack, G. and Alnemri, E.S. (1995) Mch3, a novel human apoptotic cysteine protease highly related to CPP32. *Cancer Res.* 55: 6045–6052.

Fisher, A.J., Cruz, W., Zoog, S.J., Schneider, C.L. and Friesen, P.D. (1999) Crystal structure of baculovirus P35: role of a novel reactive site loop in apoptotic caspase inhibition. *EMBO J.* 18: 2031–2039.

Garcia-Calvo, M., Peterson, E.P., Leiting, B., Ruel, R., Nicholson, D.W. and Thornberry, N.A. (1998) Inhibition of human caspases by peptide-based and macromolecular inhibitors. *J. Biol. Chem.* **273**: 32608–32613.

Garcia-Calvo, M., Peterson, E.P., Rasper, D.M., Vaillancourt, J.P., Zamboni, R., Nicholson, D.W. and Thornberry, N.A. (1999) Purification and catalytic properties of human caspase family members. *Cell Death Diff.* **6**: 362–369.

Ghayur, T., Hugunin, M., Talanian, R.V., Ratnofsky, S., Quinlan, C., Emoto, Y., Pandey, P., Datta, R., Huang, Y., Kharbanda, S., Allen, H., Kamen, R., Wong, W. and Kufe, D. (1996) Proteolytic activation of protein kinase C delta by an ICE/CED 3-like protease induces characteristics of apoptosis. *J. Exp. Med.* **184**: 2399–2404.

Ghayur, T., Banerjee, S., Hugunin, M., Butler, D., Herzog, L., Carter, A., Quintal, L., Sekut, L., Talanian, R., Paskind, M., Wong, W., Kamen, R., Tracey, D., and Allen, H. (1997) Caspase-1 processes IFN-gamma-inducing factor and regulates LPS-induced IFN-gamma production. *Nature* **386**: 619–623.

Goltsev, Y.V., Kovalenko, A.V., Arnold, E., Varfolomeev, E. E., Brodianskii, V.M. and Wallach, D. (1997) CASH, a novel caspase homologue with death effector domains. *J. Biol. Chem.* **272**: 19641–19644.

Han, D.K.M., Chaudhary, P.M., Wright, M.E., Friedman, C., Trask, B.J., Riedel, R.T., Baskin, D.G., Schwartz, S.M. and Hood, L. (1997) MRIT, a novel death-effector domain-containing protein, interacts with caspases and Bcl-X$_L$ and initiates cell death. *Proc. Natl Acad. Sci. USA* **94**: 11333–11338.

Hofmann, K., Bucher, P. and Tschopp, J. (1997) The CARD domain: a new apoptotic signalling motif. *Trends Biochem. Sci.* **22**: 155–156.

Howard, A.D., Kostura, M.J., Thornberry, N., Ding, G.J.F., Limjuco, G., Weidner, J., Salley, J.P., Hogquist, K.A., Chaplin, D.D., Mumford, R.A., Schmidt, J.A. and Tocci, M.J. (1991) IL-1 converting enzyme requires aspartic acid residues for processing of the IL-1β precursor at two distinct sites and does not cleave 31-kDa IL-1α. *J. Immunol.* **147**: 2964–2969.

Hu, S., Vincenz, C., Ni, J., Gentz, R. and Dixit, V.M. (1997) I-FLICE, a novel inhibitor of tumor necrosis factor receptor-1- and CD-95-induced apoptosis. *J. Biol. Chem.* **272**: 17255–17257.

Humke, E.W., Ni, J. and Dixit, V.M. (1998) ERICE, a novel FLICE-activatable caspase. *J. Biol. Chem.* **273**: 15702–15707.

Inohara, N., Koseki, T., Hu, Y., Chen, S. and Nunez, G. (1997) CLARP, a death effector domain containing protein interacts with caspase-8 and regulates apoptosis. *Proc. Natl Acad. Sci. USA* **94**: 10717–10722.

Irmler, M., Thome, M., Hahne, M., Schneider, P., Hofmann, K., Steiner, V., Bodmer, J.-L., Schroter, M., Burns, K., Mattmann, C., Rimoldi, D., French, L.E. and Tschopp, J. (1997) Inhibition of death receptors signals by cellular FLIP. *Nature* **388**: 190–195.

Jiang, Z.H., Zhang, W.J., Rao, Y. and Wu, J.Y. (1998) Regulation of Ich-1 pre-mRNA alternative splicing and apoptosis by mammalian splicing factors. *Proc. Natl Acad. Sci. USA* **95**: 9155–9160.

Kuida, K., Lippke, J.A., Ku, G., Harding, M.W., Livingston, D.J., Su, M.S.S. and Flavell, R.A. (1995) Altered cytokine export and apoptosis in mice deficient in interleukin-1-beta converting enzyme. *Science* **267**: 2000–2003.

Kuida, K., Zheng, T.S., Na, S., Kuan, C.-Y., Yang, D., Karasuyama, H., Rakic, P. and Flavell, R.A. (1996) Decreased apoptosis in the brain and premature lethality in CPP32-deficient mice. *Nature* **384**: 368–372.

Kuida, K., Haydar, T.F., Kuan, C.Y., Gu, Y., Taya, C., Karasuyama, H., Su, M.S., Rakic, P. and Flavell, R.A. (1998) Reduced apoptosis and cytochrome c-mediated caspase activation in mice lacking caspase 9. *Cell* **94**: 325–337.

Lazebnik, Y.A., Kaufmann, S.H., Desnoyers, S., Poirier, G.G. and Earnshaw, W.C. (1994) Cleavage of poly(ADP-ribose) polymerase by a proteinase with properties like ICE. *Nature* **371**: 346–347.

Lewis, E.R., Johnson, F.A. and Shafer, J.A. (1981) Effect of cystein-25 on the ionization of histidine-159 in papain as determined by proton nuclear magnetic resonance spectroscopy. Evidence for a His-159–Cys-25 ion pair and its possible role in catalysis. *Biochemistry* **20**: 48–51.

Li, H., Zhu, H., Xu, C.J. and Yuan, J. (1998) Cleavage of BID by caspase 8 mediates the mitochondrial damage in the Fas pathway of apoptosis. *Cell* **94**: 491–501.

Li, P., Nijhawan, D., Budihardjo, I., Srinivasula, S.M., Ahmad, M., Alnemri, E.S. and Wang, X. (1997) Cytochrome c and dATP-dependent formation of Apaf-1/Caspase-9 complex initiates an apoptotic protease cascade. *Cell* **91**: 479–489.

Liu, X., Kim, C.N., Yang, J., Jemmerson, R. and Wang, X. (1996) Induction of apoptotic program in cell-free extracts: requirement for dATP and cytochrome c. *Cell* **86**: 147–157.

Liu, X., Zou, H., Slaughter, C. and Wang, X. (1997) DFF, a heterodimeric protein that functions downstream of caspase-3 to trigger DNA fragmentation during apoptosis. *Cell* **89**: 175–184.

Luo, X., Budihardjo, I., Zou, H., Slaughter, C. and Wang, X. (1998) Bid, a Bcl2 interacting protein, mediates cytochrome c release from mitochondria in response to activation of cell surface death receptors. *Cell* **94**: 481–490.

Margolin, N., Raybuck, S.A., Wilson, K.P., Chen, W., Fox, T., Gu, Y. and Livingston, D.J. (1997) Substrate and inhibitor specificity of interleukin-1 beta-converting enzyme and related caspases. *J. Biol. Chem.* **272**: 7223–7228.

Martin, D.A., Siegel, R.M., Zheng, L. and Lenardo, M.J. (1998) Membrane oligomerization and cleavage activates the caspase-8 (FLICE/MACHalpha1) death signal. *J. Biol. Chem.* **273**: 4345–4349.

Mittl, P.R., Di Marco, S., Krebs, J.F., Bai, X., Karanewsky, D.S., Priestle, J.P., Tomaselli, K.J. and Grutter, M.G. (1997) Structure of recombinant human CPP32 in complex with the tetrapeptide acetyl-Asp–Val–Ala–Asp fluoromethyl ketone. *J. Biol. Chem.* **272**: 6539–6547.

Muzio, M., Stockwell, B.R., Stennicke, H.R., Salvesen, G.S. and Dixit, V.M. (1998) An induced proximity model for caspase-8 activation. *J. Biol. Chem.* **273**: 2926–2930.

Nicholson, D.W. and Thornberry, N.A. (1997) Caspases: killer proteases. *Trends Biochem. Sci.* **22**: 299–306.

Nicholson, D.W., Ali, A., Thornberry, N.A., Vaillancourt, J.P., Ding, C.K., Gallant, M., Gareau, Y., Griffin, P.R., Labelle, M., Lazebnik, Y.A., Munday, N.A., Raju, S.M., Smulson, M.E., Yamin, T.T., Yu, V.L. and Miller, D.K. (1995) Identification and inhibition of the ICE/ced-3 protease necessary for mammalian apoptosis. *Nature* **376**: 37–43.

Odake, S., Kam, C.M., Narasimhan, L., Poe, M., Blake, J.T., Krahenbuhl, O., Tschopp, J. and Powers, J.C. (1991) Human and murine cytotoxic T lymphocyte serine proteases: Subsite mapping with peptide thioester substrates and inhibition of enzyme activity and cytolysis by isocoumarins. *Biochemistry* **30**: 2217–2227.

Perry, D.K., Smyth, M.J., Stennicke, H.R., Salvesen, G.S., Duriez, P., Poirier, G.G. and Hannun, Y.A. (1997) Zinc is a potent inhibitor of the apoptotic protease, caspase-3. A novel target for zinc in the inhibition of apoptosis. *J. Biol. Chem.* **272**: 18530–18533.

Prasad, C.V.C., Prouty, C.P., Hoyer, D., Ross, T.M., Salvino, J.M., Awad, M., Graybill, T.L., Schmidt, S.J., Osifo, I.K., Dolle, R.E., Helaszek, C.T., Miller, R.E. and Ator, M.A. (1995) Structural and stereochemical requirements of time-dependent inactivators of the interleukin-1β converting enzyme. *Bioorg. Med. Chem. Lett.* **5**: 315–318.

Rano, T.A., Timkey, T., Peterson, E.P., Rotonda, J., Nicholson, D.W., Becker, J.W., Chapman, K.T. and Thornberry, N.A. (1997) A combinatorial approach for determining protease specificities: application to interleukin-1β converting enzyme (ICE). *Chem. Biol.* **4**: 149–155.

Rawlings, N.D. and Barrett, A.J. (1993) Evolutionary families of peptidases. *Biochem. J.* **290**: 205–218.

Ray, C.A., Black, R.A., Kronheim, S.R., Greenstreet, T.A., Sleath, P.R., Salvesen, G.S. and Pickup, D.J. (1992) Viral inhibition of inflammation: cowpox virus encodes an inhibitor of the interleukin-1β converting enzyme. *Cell* **69**: 597–604.

Rotonda, J., Nicholson, D.W., Fazil, K.M., Gallant, M., Gareau, Y., Labelle, M., Peterson, E.P., Rasper, D.M., Tuel, R., Vaillancourt, J.P., Thornberry, N.A. and Becher, J.W. (1996) The three-dimensional structure of apopain/CPP32, a key mediator of apoptosis. *Nature Struct. Biol.* **3**: 619–625.

Roy, N., Deveraux, Q.L., Takahashi, R., Salvesen, G.S. and Reed, J.C. (1997) The c-IAP-1 and c-IAP-2 proteins are direct inhibitors of specific caspases. *EMBO J.* **16**: 6914–6925.

Salvesen, G.S. and Dixit, V.M. (1997) Caspases: Intracellular signaling by proteolysis. *Cell* **91**: 443–446.

Samali, A., Zhivotovsky, B., Jones, D., Nagata, S. and Orrenius, S. (1999) Apoptosis: cell death defined by caspase activation. *Cell Death Differ.* **6**: 495–496.

Scaffidi, C., Fulda, S., Srinivasan, A., Friesen, C., Li, F., Tomaselli, K.J., Debatin, K.M., Krammer, P.H. and Peter, M.E. (1998) Two CD95 (APO-1/Fas) signaling pathways. *EMBO J.* **17**: 1675–1687.

Shu, H.B., Halpin, D.R. and Goeddel, D.V. (1997) Casper is a FADD- and caspase-related inducer of apoptosis. *Immunity* **6**: 751–763.

Srinivasula, S.M., Fernandes-Alnemri, T., Zangrilli, J., Robertson, N., Armstrong, R.C., Wang, L., Trapani, J.A., Tomaselli, K.J., Litwack, G. and Alnemri, E.S. (1996) The Ced-3/interleukin 1beta converting enzyme-like homolog Mch6 and the lamin-cleaving enzyme Mch2alpha are substrates for the apoptotic mediator CPP32. *J. Biol. Chem.* **271**: 27099–27106.

Srinivasula, S.M., Ahmad, M., Ottile, S., Bullrich, F., Banks, S., Wang, Y., Fernandes-Alnemri, T., Croce, C.M., Litwack, G., Tomaselli, K.J. *et al.* (1997) FLAME-1, a novel FADD-like anti-apoptotic molecule that regulates Fas/TNFR1-induced apoptosis. *J. Biol. Chem.* **272**: 18542–18545.

Stennicke, H.R. and Salvesen, G.S. (1997) Biochemical characteristics of caspases-3, -6, -7, and -8. *J. Biol. Chem.* **272**: 25719–26723.

Stennicke, H.R. and Salvesen, G.S. (1998) Properties of the caspases. *Biochim. Biophys. Acta* **1387**: 17–31.

Stennicke, H.R., Jurgensmeier, J.M., Shin, H., Deveraux, Q., Wolf, B.B., Yang, X., Zhou, Q., Ellerby, H.M., Ellerby, L.M., Bredesen, D. *et al.* (1998) Pro-caspase-3 is a major physiologic target of caspase-8. *J. Biol. Chem.* **273**: 27084–27090.

Stennicke, H.R., Deveraux, Q.L., Humke, E.W., Reed, J.C., Dixit, V.M. and Salvesen, G.S. (1999) Caspase-9 can be activated without proteolytic processing. *J. Biol. Chem.* **274**: 8359–8362.

Sun, J., Bird, C.H., Sutton, V., McDonald, L., Coughlin, P. B., De Jong, T.A., Trapani, J.A. and Bird, P.I. (1996) A cytosolic granzyme B inhibitor related to the viral apoptotic regulator cytokine response modifier A is present in cytotoxic lymphocytes. *J. Biol. Chem.* **271**: 27802–27809.

Szecsi, P.B. (1992) The aspartic proteases. *Scand. J. Clin. Lab. Invest. Suppl.* **210**: 5–22.

Takahashi, A., Alnemri, E.S., Lazebnik, Y.A., Fernandes-Alnemri, T., Litwack, G., Moir, R.D., Goldman, R.D., Poirier, G.G., Kaufmann, S.H. and Earnshaw, W.C. (1996) Cleavage of lamin A by Mch2 alpha but not CPP32: multiple interleukin 1 beta-converting enzyme-related proteases with distinct substrate recognition properties are active in apoptosis. *Proc. Natl Acad. Sci. USA* **93**: 8395–8400.

Talanian, R.V., Quinlan, C., Trautz, S., Hackett, M.C., Mankovich, J.A., Banach, D., Ghayur, T., Brady, K.D. and Wong, W.W. (1997) Substrate specificities of caspase family proteases. *J. Biol. Chem.* **272**: 9677–9682.

Thome, M., Schneider, P., Hofmann, K., Fickenscher, H., Meinl, E., Neipel, F., Mattmann, C., Burns, K., Bodmer, J.-L., Schroter, M. *et al.* (1997) Viral FLICE-inhibitory proteins (FLIPs) prevent apoptosis induced by death receptors. *Nature* **386**: 517–521.

Thornberry, N.A., Bull, H.G., Calaycay, J.R., Chapman, K.T., Howard, A.D., Kostura, M.J., Miller, D.K., Molineaux, S.M., Weidner, J.R., Aunins, J. *et al.* (1992) A novel heterodimeric cysteine protease is required for interleukin-1beta processing in monocytes. *Nature* **356**: 768–774.

Thornberry, N.A., Peterson, E.P., Zhao, J.J., Howard, A.D., Griffin, P.R. and Chapman, K.T. (1994) Inactivation of interleukin-1β converting enzyme by peptide (acyloxy)methyl ketones. *Biochemistry* **33**: 3934–3940.

Thornberry, N.A., Rano, T.A., Peterson, E.P., Rasper, D.M., Timkey, T., Garcia-Calvo, M., Houtzager, V.M., Nordstrom, P.A., Roy, S., Vaillancourt, J.P. *et al.* (1997) A combinatorial approach defines specificities of members of the caspase family and granzyme B. *J. Biol. Chem.* **272**: 17907–17911.

Varfolomeev, E.E., Schuchmann, M., Luria, V., Chiannilkulchai, N., Beckmann, J.S., Mett, I.L., Rebrikov, D., Brodianski, V.M., Kemper, O.C., Kollet, O. *et al.* (1998) Targeted disruption of the mouse Caspase 8 gene ablates cell death induction by the TNF receptors, Fas/Apo1, and DR3 and is lethal prenatally. *Immunity* **9**: 267–276.

Vernet, T., Tessier, D.C., Chatellier, J., Plouffe, C., Lee, T.S., Thomas, D.Y., Storer, A.C. and Menard, R. (1995) Structural and functional roles of asparagine 175 in the cysteine protease papain. *J. Biol. Chem.* **270**: 16645–16652.

Walker, N.P.C., Talanian, R.V., Brady, K.D., Dang, L.C., Bump, N.J., Ferenz, C.R., Franklin, S., Ghayur, T., Hackett, M.C., Hammill, L.D. *et al.* (1994) Crystal structure of the cysteine protease interleukin-1beta-converting enzyme: A (p20/p10)2 homodimer. *Cell* **78**: 343–352.

Wang, L., Miura, M., Bergeron, L., Zhu, H. and Yuan, J. (1994) Ich-1, an Ice/ced-3-related gene, encodes both positive and negative regulators of programmed cell death. *Cell* **78**: 739–750.

Ware, C.F., Santee, S. and Glass, A. (1998) Tumor necrosis factor-related ligands and receptors. In: *The Cytokine Handbook*, 3rd edn (ed A. Thomson). Academic Press, San Diego, pp. 549–592.

Wilson, K.P., Black, J.A., Thomson, J.A., Kim, E.E., Griffith, J.P., Navia, M.A., Murcko, M.A., Chambers, S.P., Aldape, R.A., Raybuck, S.A. and Livingston, D.J. (1994) Structure and mechanism of interleukin-1 beta converting enzyme. *Nature* **370**: 270–275.

Xue, D. and Horvitz, H.R. (1995) Inhibition of the *Caenorhabditis elegans* cell-death protease CED-3 by a CED-3 cleavage site in baculovirus p35 protein. *Nature* **377**: 248–251.

Yang, X., Chang, H.Y. and Baltimore, D. (1998a) Autoproteolytic activation of pro-caspase by oligomerization. *Mol. Cell* **1**: 319–325.

Yang, X., Stennicke, H.R., Wang, B., Green, D.R., Janicke, R.U., Srinivasan, A., Seth, P., Salvesen, G.S., and Froelich, C.J. (1998b) Granzyme B mimics apical caspases. Description of a unified pathway for trans-activation of executioner caspase-3 and -7. *J. Biol. Chem.* **273**: 34278–34283.

Yuan, J. and Horvitz, H.R. (1990) The *Caenorhabditis elegans* genes ced-3 and ced-4 act cell autonomously to cause programed cell death. *Dev. Biol.* **138**: 33–41.

Yuan, J., Shaham, S., Ledoux, S., Ellis, H.M. and Horvitz, H.M. (1993) The *C. elegans* cell death gene ced-3 encodes a protein similar to mammalian interleukin-1β-converting enzyme. *Cell* **75**: 641–652.

Zhou, Q. and Salvesen, G.S. (1997) Activation of pro-caspase-7 by serine proteases includes a non-canonical specificity. *Biochem. J.* **324**: 361–364.

Zhou, Q., Snipas, S., Orth, K., Dixit, V.M. and Salvesen, G.S. (1997) Target protease specificity of the viral serpin CrmA: analysis of five caspases. *J. Biol. Chem.* **273**: 7797–7800.

Zhou, Q., Krebs, J.F., Snipas, S.J., Price, A., Alnemri, E.S., Tomaselli, K.J. and Salvesen, G.S. (1998) Interaction of the baculovirus anti-apoptotic protein p35 with caspases: specificity, kinetics, and characterization of the caspase/p35 complex. *Biochemistry* **37**: 10757–10765.

Zou, H., Henzel, W.J., Liu, X., Lutschg, A. and Wang, X. (1997) Apaf-1, a human protein homologous to *C. elegans* CED-4, participates in cytochrome c-dependent activation of caspase-3. *Cell* **90**: 405–413.

Zou, H., Li, Y., Liu, X. and Wang, X. (1999) An APAF-1.Cytochrome c multimeric complex is a functional apoptosome that activates procaspase-9. *J. Biol. Chem.* **274**: 11549–11556.

3

Modulation of death receptor signalling

Pascal Schneider and Jürg Tschopp

1. Introduction

The death domain-containing members of the TNF receptor family mediate apoptosis upon engagement by their respective ligands. The aggregation of the receptors allows the recruitment of intracellular adaptor proteins (via the death domain) which in turn interact with and activate members of the caspase family by means of distinct protein–protein interaction domains (death effector domain, caspase recruitment domain), ultimately leading to apoptosis. The induction of apoptosis by death receptors is regulated by several extracellular and intracellular factors. The cytotoxic ligand TRAIL, for instance, mediates cell death via two death domain-containing receptors, TRAIL-R1 and TRAIL-R2, but also interacts with soluble or membrane-bound decoy receptors such as the glycosylphosphatidylinositol-anchored TRAIL-R3 whose stable expression protects TRAIL-sensitive cells from TRAIL-mediated effects.

In Fas-mediated apoptosis, the aggregation state of the ligand is critically related to its ability to induce cell death. Naturally processed sFasL occurs in a soluble, trimeric form which normally interacts with Fas, yet is very inefficient at signalling cell death in comparison to membrane-bound FasL. However, cross-linking of soluble FasL fully restores its cytotoxic activity, suggesting that the aggregation of more than three Fas molecules is required to transduce the death signal efficiently. Apoptotic signals are also regulated inside the cell by several anti-apoptotic factors. FLIP is a homologue of caspase 8, containing an inactive protease domain which specifically interferes with recruitment and activation of caspase 8 to several death domain-containing receptors, thereby inhibiting apoptosis.

2. Death receptors and apoptosis

2.1 *Activation of caspases during apoptotic cell death*

The central event of apoptotic cell death is the proteolytic activation of intracellular cysteine proteases, called caspases (Chapters 1 and 2), that pre-exist in the cell as inactive zymogen precursors (Thornberry and Lazebink, 1998). The zymogen is

Programmed Cell Death in Animals and Plants, edited by J.A. Bryant, S.G. Hughes and J.M. Garland.
© 2000 BIOS Scientific Publishers Ltd, Oxford.

composed of a prodomain, which can be very short in effector caspases or consist of a protein–protein interaction domain in regulatory caspases, followed by a large and a small subunit which, after proteolytic processing, form the active caspase. Once activated, caspases specifically cleave a variety of cellular proteins, leading to the apoptotic phenotype via the inactivation of DNA repairing enzymes or structural proteins of the nuclear envelope and of the cytoskeleton, and the activation of DNases.

There are two main pathways leading to the activation of effector caspases: the first one is activated in response to a variety of stress conditions such as UV irradiation, growth factor deprivation, abnormal mitosis or anti-cancer drugs, and proceeds via the recruitment and activation of caspase 9 within a complex containing Apaf-1 and cytochrome c. This pathway is regulated by pro- and anti-apoptotic members of the Bcl-2 family (Chapter 4), which act on the mitochondria (Green and Reed, 1998; see also Chapter 6). The second pathway uses specialized membrane receptors, which are efficiently coupled to caspase activation, especially caspase 8 (Ashkenazi and Dixit, 1998). These death-inducing receptors belong to the tumour necrosis factor receptor (TNF-R) family and are activated upon engagement by their cognate ligand of the TNF family. In addition, the serine protease granzyme B (see Chapter 5), which is secreted by cytotoxic lymphocytes, has the ability to directly activate intracellular caspases, thus acting as a short cut to apoptosis (Medema et al., 1997).

2.2 The TNF and TNF-R families of ligands and receptors

Members of the TNF cytokine family are critically involved in the control of infections and in the regulation of inflammation and of tissue homeostasis (Smith et al., 1994). The family members are type II membrane proteins that can act in a membrane-bound form or as proteolytically processed soluble cytokines in an autocrine, paracrine or endocrine manner (Smith et al., 1994). These proteins exist as homotrimers, which display receptor-binding sites at the boundaries between monomers. One ligand can therefore bind three times its respective receptor at three independent interaction sites, leading to receptor oligomerization and initiation of downstream signalling events.

Signalling pathways stimulated by TNF ligand members are diverse, including the activation of caspases, the translocation of the nuclear factor-κB (NF-κB) or the activation of mitogen-activated protein kinases such as JNK or ERK. Thus, TNF-related ligands can lead to apoptosis, differentiation, or proliferation. Presently 16 ligands and 22 receptors of the TNF/TNF-R families have been described, several amongst them have important regulatory roles in function and development of the immune system (*Figure 1*). For instance, TNF acts as an inflammatory cytokine coordinating host defences in response to aggression by pathogens (Vassalli, 1992). The lymphotoxin system is crucial in the development of peripheral lymphoid organs and the organization of splenic architecture (De Togni et al., 1994; Koni et al., 1997). Fas ligand (FasL), TNF and CD30L are responsible for T-cell receptor-mediated apoptosis of T cells and immature thymocytes (Amakawa et al., 1996; Russell et al., 1993; Zheng et al., 1995). Several of the TNF members and their receptors, in conjunction with T-cell receptor stimulation, enhance T-cell proliferation. Upregulation of TNF-related ligands on T cells is also important for the activation and stimulation of neighbouring cells. CD40L is required for B-cell proliferation and differentiation and for inducing isotype switch (van Kooten and Banchereau, 1997). VEGI appears to play a role in the

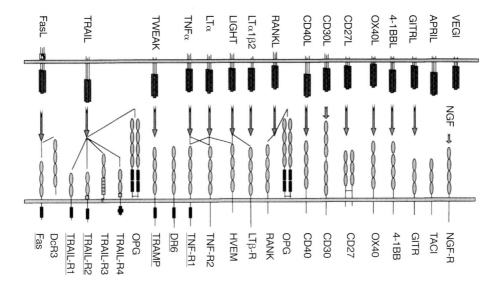

Figure 1. *TNF and TNF receptor family members. Trimeric ligands are shown at the top and receptors at the bottom of the figure. The extracellular portion of the receptors is schematized to indicate the number of cysteine-rich domains characteristic of the family. Interactions that have been reported in the literature are indicated and the six death domain-containing receptors are underlined. The death domain motif is shown as a black box. TRAIL-R4 contains a truncated, inactive death domain and the soluble receptor OPG contains two death domain motives in its C-terminal portion.*

physiology of vascular endothelial cells and RANKL/OPGL is critically involved in osteoclast generation and bone metabolism (Kong *et al.*, 1999; Zhai *et al.*, 1999).

FasL, TNF, TRAIL, and to a lesser extent TWEAK, have all been implicated in the activation of programmed cell death. Their cognate receptors form a subclass of the TNF receptor family that is characterized by the presence of an intracellular domain through which apoptosis is signalled and which has been called the death domain. To date, six death receptors have been described, namely TNF-R1, Fas/CD95/Apo-1, TRAMP/Wsl/DR3, TRAIL-R1/DR4, TRAIL-R2/DR5, and DR6 (Ashkenazi and Dixit, 1998) (*Figure 1*). Fas is a potent activator of caspases and of cell death and its ligand represents one of the major cytolytic weapon of cytotoxic T cells (Kagi *et al.*, 1994; Lowin *et al.*, 1994). FasL is also implicated in the elimination of activated T cells during the down regulation of the immune response in the periphery (Nagata, 1997) and in the maintenance of immune privilege of tissues by preventing the infiltration of Fas-sensitive lymphocytes (Griffith *et al.*, 1995). Although the role of TRAIL and of its receptors in the immune response is not yet fully understood, its rather selective cytotoxicity towards several cancer cell lines makes it an attractive candidate for cancer therapy (Walczak *et al.*, 1999; see also Chapter 20).

2.3 *Signalling through death receptors*

The death domain is an homotypic protein–protein interaction domain of about 90 amino acids, which allows the recruitment of adaptor proteins to the activated receptor. Similarly, two related domains have been identified that promote protein–protein interactions within the apoptotic-signalling pathway, namely the death effector domain and the caspase recruitment domain (CARD). Many of the interactions leading to caspase activation proceed via homotypic interactions of these domains. For example, the death domain of TNF-R1 interacts with the death domain of TRADD and FADD, which in turn binds to caspase 8 via death effector domains, leading to the activation of caspase 8. Similarly, caspase 2 appears to be recruited to activated TNF-R1 via a CARD–CARD interaction with RAIDD. The death domain containing kinase RIP and the TRAFs molecules are involved in parallel pathways leading to NF-κB and JNK activation, respectively. The signalling pathway of TRAMP appears to be similar to that of TNF-R1, which is an efficient inducer of NF-κB and signals cell death only under some circumstances. In the case of Fas, caspase 8 is recruited more directly through the adapter FADD in the absence of TRADD. Fas has also been reported to signal JNK activation via Daxx and the kinase ASK-1 (Yang *et al.*, 1997).

3. Activation of death receptors by death ligands

3.1 *Processing of TNF family members to soluble forms*

TNF-α is a potent pro-inflammatory and immunomodulatory agent. However, elevated concentrations of circulating TNF-α seen during many host responses may be harmful or even fatal to the organism. Membrane-bound TNF-α is synthesized as a type II membrane protein that acts locally through cell to cell contact. Soluble TNF-α (s TNF-α) is released from the cell surface as the result of cleavage by a metallopro-teinase (McGeehan *et al.*, 1994; Mohler *et al.*, 1994). Although both cell surface and secreted TNF-α appear to be biologically active, deleterious physiological responses such as cachexia and endotoxic shock are mediated by the latter.

FasL is also processed and shed from the surface of human cells. Until recently, it was not clear whether the activities described for FasL were primarily due to its cell surface or secreted form. Elevated levels (up to 10 ng ml^{-1}) of sFasL were found in sera from patients with large granular lymphocytic leukaemias and natural killer cell lymphomas (Tanaka *et al.*, 1996). Since the administration of recombinant FasL or agonistic Fas antibodies into mice leads to liver failure and to rapid death of the animals (Ogasawara *et al.*, 1993; Rensing-Ehl *et al.*, 1995), it has been proposed that sFasL is implicated in the pathogenesis of various diseases such as hepatitis or AIDS (Ameisen *et al.*, 1995; Galle *et al.*, 1995; Nagata and Golstein, 1995). The possible implication of sFasL in pathological processes prompted us to investigate the structure and activity of naturally processed sFasL in more detail.

3.2 *Determination of the cleavage site of human FasL*

The 40 kDa membrane-bound FasL is cleaved by an uncharacterized metalloproteinase to generate a 26–29 kDa soluble fragment (Mariani *et al.*, 1995; Tanaka *et al.*, 1996). In agreement with these published results, we detected processed sFasL in supernatants of 293 cells transiently or stably transfected with full-length human FasL, and this release was inhibited upon addition of the metalloproteinase inhibitor KB8301. Gel filtration chromatography indicated an apparent molecular size of approximately 70 kDa for native sFasL, in agreement with its predicted trimeric structure. sFasL was isolated by affinity chromatography on recombinant Fas receptor. N-terminal sequence analysis revealed that sFasL begins with Leu127, showing that the cleavage occurred in the sequence TASS–LEKQ located in the stalk region that links the TNF-homology domain to the transmembrane domain. Cleavage in TNF-α occurs in the equivalent region between Ala76 and Val77, and is mediated by at least two zinc metalloproteases (TACE, AD10) (Black *et al.*, 1997; Rosendahl *et al.*, 1997). As mutations of the flanking amino acids Ala76 and Val77 to acidic residues negatively influences the processing of TNF-α (Rosendahl *et al.*, 1997), similar mutations were introduced at the cleavage site of FasL. Mutations at the P1 (Ser126Glu) and P1′ (Leu127Glu) sites partially prevented the release of sFasL whereas the double mutation (Ser126Glu; Leu127Glu) resulted in the loss of detectable sFasL, indicating that the metalloproteinase present in 293 cells uses this cleavage site.

3.3 *sFasL is poorly cytotoxic, in contrast to membrane bound FasL*

Unexpectedly, the specific activity of sFasL after removal of residual membrane-bound FasL by Triton X-114 extraction and affinity purification was marginal. Human Jurkat T cells, which are highly sensitive to apoptosis induced by anti-Fas antibodies (Dhein *et al.*, 1995), remained viable at sFasL concentrations exceeding 1 μg ml^{-1}. Recombinant sFasL corresponding in length to the naturally processed form was equally inactive, although it was perfectly able to bind to Fas positive cells or to recombinant Fas. In contrast, supernatants of Neuro-2a cells transfected with murine FasL, known to be highly cytotoxic (Rensing-Ehl *et al.*, 1995), readily induced apoptosis in Jurkat and other cells. These supernatants contain unprocessed, membrane-bound form of FasL displaying high molecular weight under native conditions, suggesting that Neuro-2a cells express little or no processing proteases, and that membrane-fragments or vesicles containing the membrane-bound FasL are released into the supernatant. These results suggested that high apoptotic activity was associated with aggregated forms of FasL.

3.4 *Cross-linking of FasL promotes its cytotoxic activity* in vitro *and* in vivo

We wondered whether controlled cross-linking of sFasL, mimicking membrane-bound FasL, could lead to an increase in activity. We took advantage of the presence of a Flag-tag on the recombinant sFasL to induce sFasL aggregation. The mere addition of cross-linking anti-FasL antibodies resulted in a more than 1000-fold increase of recombinant sFasL activity on Jurkat and many other cell lines, including murine cells, with IC$_{50}$ as low as 1–10 ng ml^{-1} being obtained with aggregated sFasL.

These results suggested that sFasL should either be inactive or only poorly active *in vivo*. Indeed, mice injected with 12.5 μg of sFasL survived, and showed no signs of sickness. In contrast, injection of sFasL followed by cross-linking anti-Flag antibodies was fatal and mice died within 3 hours, similar to mice treated with agonistic anti-Fas antibodies (Ogasawara *et al.*, 1993) or with FasL contained in supernatants of transfected Neuro-2a cells (Rensing-Ehl *et al.*, 1995). The cross-linked FasL induced lethal liver haemorrhages and hepatocyte apoptosis, whereas in the absence of the anti-Flag antibody, hepatocytes had a normal morphology (Schneider *et al.*, 1998).

Our experiments suggest that the mere trimerization of Fas by FasL may not suffice to transmit death signals. It is thought that the trimeric FasL induces Fas trimer formation, followed by the binding of FADD to the death domain of Fas (Boldin *et al.*, 1995; Chinnaiyan *et al.*, 1995). FADD then recruits caspase 8 (Boldin *et al.*, 1996; Muzio *et al.*, 1996), providing the connection of death receptors to caspases. The exact stoichiometry of adapters mediating caspase activation has, however, never been determined, and it is possible that multiple Fas trimers are required to allow the assembly of multiple FADD adapters bringing several caspase 8 proteins together, and allowing reciprocal cleavage and subsequent activation of the caspase (*Figure 2*). The requirement of the formation of large aggregates of Fas for signal transduction is also suggested by the efficacy of agonistic antibodies of the IgM and IgG3 type which are known to be multimeric or to form aggregates.

Local cell to cell contact appears to be critical for T-cell specific immunosurveillance, and indeed, FasL-mediated killing of virus-infected or tumor target cells is a highly specific process which assures that neighbouring cells or tissues are not affected. This specificity can only be guaranteed if the cytotoxic ligand remains associated with the

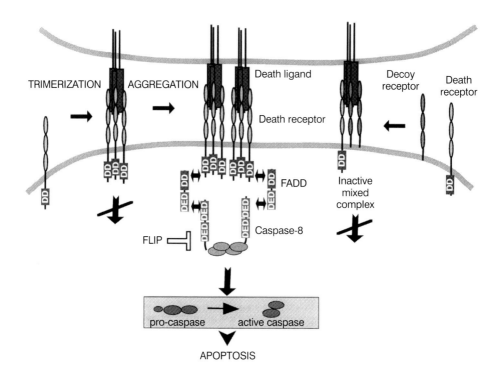

Figure 2. Model showing the effects of ligand aggregation, decoy receptor and intracellular FLIP on the function of death receptors. Ligands of the TNF family are trimeric and can recruit three receptors, which in some instances provide a full signal (e.g. TNF-α acting on TNF-R1). This is however insufficient in other systems (e.g. FasL on Fas or TRAIL on TRAIL-R1/R2) where additional aggregation is required to allow death signalling. This would ensure that only membrane-bound ligand could provide the death signal during a cell to cell contact. In the case of TRAIL, overexpression of TRAIL-R3 (decoy receptor) provides specific protection against TRAIL. One hypothesis is that the decoy receptor forms mixed complexes with the death receptor and therefore acts as a dominant negative inhibitor. In Fas-mediated apoptosis, caspase 8 is recruited to the activated Fas via the adapter molecule FADD. FLIP interacts with and probably hinders proper activation of caspase 8, resulting in death signalling blockade.

lymphocytes. However, FasL is rapidly released from cells and detectable levels of sFasL have been found in sera from patients with LGL-leukaemia, NK-lymphomas (Tanaka *et al.*, 1996), or melanoma (Hahne *et al.*, 1996). Considering the hepatic toxicity of FasL, circulating sFasL, if active, would be pathological. Neither patients with tumours nor individuals with high numbers of activated T cells (for example after a viral infection) suffer from hepatitis, implying that sFasL *in vivo* is at best poorly active, in agreement with our results. We cannot exclude the possibility, however, that high local concentrations of sFasL, in conjunction with other cytotoxic agents, may lead to cell death.

3.5 *Cross-linking enhances the activity of some, but not all members of the TNF family*

We tested whether other members of the TNF-ligand family known to induce apoptosis also required cross-linking for their activity. Similar to FasL, TRAIL (Pitti *et al.*, 1996; Wiley *et al.*, 1995) was active at low concentrations only in the presence of cross-linking antibodies. No apoptosis was observed with sTRAIL at concentrations as high as 5 μg ml^{-1}. In contrast, cross-linking of sTWEAK (Chicheportiche *et al.*, 1997) did not increase its cytotoxic activity on HT-29 cells, suggesting that this ligand can exert its apoptotic activity in the soluble form. Interestingly, the activity of sTNF-α was differentially influenced by the presence of cross-linking antibodies depending on the receptor triggered. To distinguish between TNF-R1 and TNF-R2-mediated effects, we took advantage of the fact that human TNF-α only binds to murine TNF-R1 but not to murine TNF-R2 (Lewis *et al.*, 1991). When TNF-α was added to WEHI164 cells, which are known to undergo apoptosis upon stimulation of TNF-R1, the activity was independent of the presence of cross-linking antibodies. In contrast, the proliferative effect of TNF-α on mouse CT6 cells, which is mediated through TNF-R2-dependent signalling pathways and is therefore only observed in response to murine TNF-α, was highly increased in the presence of cross-linking antibodies. This is in line with results demonstrating that membrane-bound TNF-α is the prime activating ligand for TNF-R2-mediated responses (Grell *et al.*, 1995).

4. Decoy receptors

4.1 *The TRAIL receptor family*

The ligand TRAIL shows a broad tissue distribution (Wiley *et al.*, 1995) and like FasL, induces rapid apoptosis of various cell lines. To date, five receptors interacting with TRAIL have been described (*Figure 1*). Two of them (TRAIL-R1/DR4 and TRAIL-R2/DR5) contain a death domain and are cytotoxic. The other three may act as decoy receptors. TRAIL-R3/DcR1 and TRAIL-R4/DcR2 are membrane-bound but lack a functional death domain, whereas OPG is a soluble receptor with dual specificity for TRAIL and RANKL/OPGL. In contrast to FasL, TRAIL is not cytotoxic to tissues despite the wide expression of TRAIL-R1 and -R2, suggesting that TRAIL receptor signalling is under the control of strong regulatory mechanisms.

4.2 *TRAIL-R3 is a glycosylphosphatidylinositol (GPI) anchored membrane protein*

Several groups, including ours, cloned the cDNA of TRAIL-R3. The predicted amino acid sequence of this short receptor is devoid of an intracellular domain and ends immediately after a carboxy-terminal stretch of 15 hydrophobic amino acids, which is reminiscent of a signal for the addition of a glycosyl-phosphatidylinositol (GPI) anchor (Udenfriend and Kodukula, 1995). In GPI-anchored proteins, the carboxy-terminal portion of the protein is exchanged for a pre-formed glycolipid by the action of a GPI-transamidase. A functional signal for GPI addition fulfils several sequence requirements (Udenfriend and Kodukula, 1995): first, the GPI acceptor amino acid residue (at the ω site) and the amino acid at position $\omega + 2$ are small. Second, the amino acid at position $\omega + 1$ is different from Trp or Pro, and third, the presence of a hinge region located between position $\omega + 3$ and approximately $\omega + 8$, followed by a stretch of 8–20 highly hydrophobic amino acids is observed. According to this rule, Ala^{236} would be the predicted attachment site for a GPI-anchor in TRAIL-R3.

Several lines of evidence indicate that TRAIL-R3 is indeed a GPI-anchored protein. First, TRAIL-R3 expressed in 293 cells partitioned quantitatively into the detergent phase of a Triton X-114 phase separation, as expected for a GPI-anchored protein. Second, TRAIL-R3 was biosynthetically labelled with [^{3}H]inositol, which, in a protein, is highly specific of a GPI. The third line of evidence comes from the radiochemical analysis of TRAIL-R3. The hallmark of GPI-anchors is a non-acetylated glucosamine that can be deaminated with nitrous acid and reduced with sodium borotritide to yield a radioactive anhydromannitol residue, with concomitant loss of the phosphatidylinositol. Subsequent dephosphorylation with hydrofluoric acid releases the neutral carbohydrate fraction of the GPI which, in the case of TRAIL-R3, co-migrated with authentic Man$_3$-anhydromannitol, the minimal conserved carbohydrate structure present in all GPI anchors analysed so far. Acid hydrolysis of this product yielded the expected and very characteristic anhydromannitol residue. Taken together, these data demonstrate the presence of a GPI in TRAIL-R3. Finally, a number of GPI-anchored proteins can be solubilized by the action of phosphatidylinositol-specific phospholipase C (PI-PLC). This reaction is however not possible if the hydroxyl group on position 2 of the inositol is acylated. This is most likely the case of TRAIL-R3 expressed in 293 cells, which was resistant to the action of PI-PLC.

4.3 TRAIL-R3 can act as a dominant negative inhibitor of death-signalling TRAIL receptors

Because TRAIL-R3 can bind to TRAIL, yet is unable to transduce a death signal, it was expected that it could counteract the action of TRAIL in sensitive cells either by scavenging TRAIL or by forming inactive mixed receptors with TRAIL-R1/R2. MCF-7 cells respond to TRAIL or TNF by stopping proliferation and displaying morphological features of apoptotic cells. However, MCF-7 cells stably transfected with a dominant negative version of FADD became entirely resistant to the effects of TRAIL, TNF and agonistic anti-TRAIL-R2 antibodies, indicating that FADD-DN efficiently interacts with death signalling as previously described (Chinnaiyan et al., 1996; Schneider et al., 1997). MCF-7 cells stably transfected with TRAIL-R3 bound approximately 20 times more TRAIL on their surface than non-transfected cells, but only became susceptible to TRAIL with 100 times greater doses than those necessary to kill wild type cells. These TRAIL-R3 transfectants however still had functional apoptotic signalling pathways as demonstrated by their unaltered susceptibility to TNF and agonistic anti-TRAIL-R2 antibodies.

These results point to the fact that TRAIL-R3 can indeed act as a dominant negative inhibitor of TRAIL action. Whether this reflects the role of TRAIL-R3 *in vivo* remains to be determined. However, experiments performed on tumour cell lines revealed no correlation between mRNA expression of the different TRAIL-receptors and their sensitivity towards TRAIL (Griffith *et al.*, 1998). Clearly, intracellular regulators of the pathway also play an essential role. In addition, northern blot analysis indicated that TRAIL-R3 expression is less widespread than those of TRAIL-R1 and TRAIL-R2, suggesting that it may play a role in a restricted number of cell types.

5. Inhibition of death receptor by FLIP

5.1 *Viral and cellular FLIPs inhibit death receptor-mediated apoptosis*

Caspase 8 is recruited to FADD via interaction between death effector domains. We found that ORF E8 of the equine herpes virus 2 encoded a protein with two predicted death effector domains. This protein, called v-FLIP, is also present in several other viruses including human herpes virus 8. It interferes with the apoptotic pathway of FasL by binding to FADD and potently inhibits TRAIL-mediated cell death (Thome *et al.*, 1997).

The sequence information from v-FLIP led to the discovery of a mammalian homologue, cellular FLIP (c-FLIP, also called CASPER/I-FLICE/Flame/CASH/CLARP/MRIT or usurpin) which is expressed in several splice variants (reviewed by Tschopp *et al.*, 1998). The short forms contain two death effector domains and therefore resemble v-FLIP, whereas the long form contains an additional caspase-like domain with significant homology to caspase 8 (and caspase 10). In contrast to caspase 8, however, c-FLIP/long lacks essential features that are required for substrate catalysis and is therefore devoid of proteolytic activity. c-FLIP/long binds to both FADD and caspase 8 via its death effector domains. The caspase-like domain also binds to caspase 8, suggesting that the intimate contact occurring between c-FLIP and caspase 8 may interfere with the activity of the latter. Indeed, both c-FLIP/short and c-FLIP/long inhibited cell death induced by several death receptors. c-FLIP/long appeared to be a better inhibitor than c-FLIP/short and both were especially effective against TRAIL-mediated apoptosis (Irmler *et al.*, 1997).

It is predicted that Fas-recruited pro-caspase 8 is able to bind to and cleave neighbouring c-FLIP. However, pro-caspase 8 itself cannot be processed by the proteolytically inactive c-FLIP, resulting in the blockade of the death signal. FLIP and other viral or cellular proteins involved in the intracellular regulation of death pathways, such as the IAPs, CrmA and Bcl-2 family members may contribute to the resistance of death receptor expressing cells to the action of death ligands (Tschopp *et al.*, 1998). This could be the case of T cells during the early phase of activation, which are resistant to FasL despite abundant Fas expression.

6. Conclusion

The activation of apoptosis via death receptors is a tightly regulated event, which can be modulated at several levels (*Figure 2*). Soluble or membrane-bound decoy receptors can interfere with the death pathway by either scavenging the death ligand or by forming inactive mixed receptor complexes, thereby acting as dominant negative inhibitors. The aggregation state of the ligand is probably a crucial factor. Some receptors can efficiently signal cell death only if they are sufficiently aggregated in a complex comprising more than three receptors. This requirement counteracts unde-sired activation of the receptor by systemic soluble trimeric ligands but still allows specific activation of the receptor during a cell to cell contact. In this context, processing of a membrane-bound ligand to a soluble form may correspond to an inac-tivation step. Finally, intracellular proteins that interfere with the apoptotic machinery also modulate cell death, and may dictate the outcome of death receptor triggering.

References

Amakawa, R., Hakem, A., Kundig, T.M., Matsuyama, T., Simard, J.J., Timms, E., Wakeham, A., Mittruecker, H.W., Griesser, H., Takimoto, H. *et al.* (1996) Impaired negative selection of T cells in Hodgkin's disease antigen CD30-deficient mice. *Cell* 84: 551–562.

Ameisen, J.C., Estaquier, J., Idziorek, T. and De Bels, F. (1995) The relevance of apoptosis to AIDS pathogenesis. *Trends Cell Biol.* 5: 27–32.

Ashkenazi, A. and Dixit, V.M. (1998) Death receptors: signaling and modulation. *Science* 281: 1305–1308.

Black, R.A., Rauch, C.T., Kozlosky, C.J., Peschon, J.J., Slack, J.L., Wolfson, M.F., Castner, B.J., Stocking, K.L., Reddy, P., Srinivasan, S. *et al.* (1997) A metalloproteinase disintegrin that releases tumour-necrosis factor-alpha from cells. *Nature* 385: 729–733.

Boldin, M.P., Varfolomeev, E.E., Pancer, Z., Mett, I.L., Camonis, J.H. and Wallach, D. (1995) A novel protein that interacts with the death domain of Fas/APO1 contains a sequence motif related to the death domain. *J. Biol. Chem.* 270: 7795–7798.

Boldin, M.P., Goncharov, T.M., Goltsev, Y.V. and Wallach, D. (1996) Involvement of MACH, a novel MORT1/FADD-interacting protease, in Fas/APO-1- and TNF receptor-induced cell death. *Cell* 85: 803–815.

Chicheportiche, Y., Bourdon, P.R., Xu, H., Hsu, Y.M., Scott, H., Hession, C., Garcia, I. and Browning, J.L. (1997) TWEAK, a new secreted ligand in the tumor necrosis factor family that weakly induces apoptosis. *J. Biol. Chem.* 272: 32401–32410.

Chinnaiyan, A.M., O'Rourke, K., Tewari, M. and Dixit, V.M. (1995) FADD, a novel death domain-containing protein, interacts with the death domain of Fas and initiates apoptosis. *Cell* 81: 505–512.

Chinnaiyan, A.M., Tepper, C.G., Seldin, M.F., O'Rourke, K., Kischkel, F.C., Hellbardt, S., Krammer, P.H., Peter, M.E. and Dixit, V.M. (1996) FADD/MORT1 is a common mediator of CD95 (Fas/APO-1) and tumor necrosis factor receptor-induced apoptosis. *J. Biol. Chem.* 271: 4961–4965.

De Togni, P., Goellner, J., Ruddle, N.H., Streeter, P.R., Fick, A., Mariathasan, S., Smith, S.C., Carlson, R., Shornick, L.P., Strauss-Schoenberger, J. *et al.* (1994) Abnormal development of peripheral lymphoid organs in mice deficient in lymphotoxin. *Science* 264: 703–707.

Dhein, J., Walczak, H., Bäumler, C., Debatin, K.-M. and Krammer, P.H. (1995) Autocrine T-cell suicide mediated by APO-1/(Fas/CD95). *Nature* 373: 438–441.

Galle, P.R., Hofmann, W.J., Walzak, H., Schaller, H., Otto, G., Stremmel, W., Krammer, H.P. and Runkel, L. (1995) Involvement of the CD95 (Apo-1/Fas) receptor and ligand in liver damage. *J. Exp. Med.* 182: 1223–1230.

Green, D.R. and Reed, J.C. (1998) Mitochondria and apoptosis. *Science* 281: 1309–1312.

Grell, M., Douni, E., Wajant, H., Lohden, M., Clauss, M., Maxeiner, B., Georgopoulos, S., Lesslauer, W., Kollias, G., Pfizenmaier, K. and Scheurich, P. (1995) The transmembrane form of tumor necrosis factor is the prime activating ligand of the 80 kDa tumor necrosis factor receptor. *Cell* 83: 793–802.

Griffith, T.S., Brunner, T., Fletcher, S.M., Green, D.R. and Ferguson, T.A. (1995) Fas ligand-induced apoptosis as a mechanism of immune privilege. *Science* 270: 1189–1192.

Griffith, T.S., Chin, W.A., Jackson, G.C., Lynch, D.H. and Kubin, M.Z. (1998) Intracellular regulation of TRAIL-induced apoptosis in human melanoma cells. *J. Immunol.* 1998: 2833–2840.

Hahne, M., Rimoldi, D., Schroter, M., Romero, P., Schreier, M., French, L.E., Schneider, P., Bornand, T., Fontana, A., Lienard, D. *et al.* (1996) Melanoma cell expression of Fas(Apo-1/CD95) ligand: implications for tumor immune escape. *Science* 274: 1363–1366.

Irmler, M., Thome, M., Hahne, M., Schneider, P., Hofmann, K., Steiner, V., Bodmer, J.L., Schroter, M., Burns, K., Mattmann, C. *et al.* (1997) Inhibition of death receptor signals by cellular FLIP. *Nature* 388: 190–195.

Kagi, D., Vignaux, F., Ledermann, B., Burki, K., Depraetere, V., Nagata, S., Hengartner, H. and Golstein, P. (1994) Fas and perforin pathways as major mechanisms of T cell-mediated cytotoxicity. *Science* 265: 528–530.

Kong, Y.-Y., Yoshida, H., Sarosi, I., Timms, E., Capparelli, C., Morony, S., Oliviera-dos-Santos, A.J., Van, G., Itie, A., Khoo, W. *et al.* (1999) OPGL is a key regulator of osteoclastogenesis, lymphocyte development and lymph-node organogenesis. *Nature* 397: 315–323.

Koni, P.A., Sacca, R., Lawton, P., Browning, J.L., Ruddle, N.H. and Flavell, R.A. (1997) Distinct roles in lymphoid organogenesis for lymphotoxins alpha and beta revealed in lymphotoxin beta-deficient mice. *Immunity* 6: 491–500.

Lewis, M., Tartaglia, L.A., Lee, A., Bennett, G.L., Rice, G.C., Wong, G.H., Chen, E.Y. and Goeddel, D.V. (1991) Cloning and expression of cDNAs for two distinct murine tumor necrosis factor receptors demonstrate one receptor is species specific. *Proc. Natl Acad. Sci. USA* 88: 2830–2834.

Lowin, B., Hahne, M., Mattmann, C. and Tschopp, J. (1994) Cytolytic T-cell cytotoxicity is mediated through perforin and Fas lytic pathways. *Nature* 370: 650–652.

McGeehan, G.M., Becherer, J.D., Bast, R.C., Jr., Boyer, C.M., Champion, B., Connolly, K.M., Conway, J.G., Furdon, P., Karp, S., Kidao, S. *et al.* (1994) Regulation of tumour necrosis factor-alpha processing by a metalloproteinase inhibitor. *Nature* 370: 558–561.

Mariani, S.M., Matiba, B., Baumler, C. and Krammer, P.H. (1995) Regulation of cell surface APO-1/Fas (CD95) ligand expression by metalloproteases. *Eur. J. Immunol.* 25: 2303–2307.

Medema, J.P., Toes, R.E., Scaffidi, C., Zheng, T.S., Flavell, R.A., Melief, C.J., Peter, M.E., Offringa, R. and Krammer, P.H. (1997) Cleavage of FLICE (caspase 8) by granzyme B during cytotoxic T lymphocyte-induced apoptosis. *Eur. J. Immunol.* 27: 3492–3498.

Mohler, K.M., Sleath, P.R., Fitzner, J.N., Cerretti, D.P., Alderson, M., Kerwar, S.S., Torrance, D.S., Otten-Evans, C., Greenstreet, T., Weerawarna, K. *et al.* (1994) Protection against a lethal dose of endotoxin by an inhibitor of tumour necrosis factor processing. *Nature* 370: 218–220.

Muzio, M., Chinnaiyan, A.M., Kischkel, F.C., O'Rourke, K., Shevchenko, A., Ni, J., Scaffidi, C., Bretz, J.D., Zhang, M., Gentz, R. *et al.* (1996) FLICE, a novel FADD-homologous ICE/CED-3-like protease, is recruited to the CD95 (Fas/APO-1) death-inducing signaling complex. *Cell* 85: 817–827.

Nagata, S. (1997) Apoptosis by death factor. *Cell* 88: 355–365.

Nagata, S. and Golstein, P. (1995) The Fas death factor. *Science* 267: 1449–1456.

Ogasawara, J., Watanabe, F.R., Adachi, M., Matsuzawa, A., Kasugai, T., Kitamura, Y., Itoh, N., Suda, T. and Nagata, S. (1993) Lethal effect of the anti-Fas antibody in mice. *Nature* 364: 806–809.

Pitti, R.M., Marsters, S.A., Ruppert, S., Donahue, C.J., Moore, A. and Ashkenazi, A. (1996) Induction of apoptosis by Apo-2 ligand, a new member of the tumor necrosis factor cytokine family. *J. Biol. Chem.* 271: 12687–12690.

Rensing-Ehl, A., Frei, K., Flury, R., Matiba, B., Mariani, S.M., Weller, M., Aebischer, P., Krammer, P.H. and Fontana, A. (1995) Local Fas/APO-1 (CD95) ligand-mediated tumor cell killing in vivo. *Eur. J. Immunol.* 25: 2253–2258.

Rosendahl, M.S., Ko, S.C., Long, D.L., Brewer, M.T., Rosenzweig, B., Hedl, E., Anderson, L., Pyle, S.M., Moreland, J., Meyers, M.A. *et al.* (1997) Identification and characterization of a pro-tumor necrosis factor-alpha-processing enzyme from the ADAM family of zinc metalloproteases. *J. Biol. Chem.* 272: 24588–24593.

Russell, J.H., Rush, B., Weaver, C. and Wang, R. (1993) Mature T cells of the autoimmune lpr/lpr mice have a defect in antigen-stimulated suicide. *Proc. Natl Acad. Sci. USA* **90**: 4409–4413.

Schneider, P., Thome, M., Burns, K., Bodmer, J.L., Hofmann, K., Kataoka, T., Holler, N. and Tschopp, J. (1997) TRAIL receptors 1 (DR4) and 2 (DR5) signal FADD-dependent apoptosis and activate NF-kappaB. *Immunity* **7**: 831–836.

Schneider, P., Holler, N., Bodmer, J.L., Hahne, M., Frei, K., Fontana, A. and Tschopp, J. (1998) Conversion of membrane-bound Fas(CD95) ligand to its soluble form is associated with downregulation of its proapoptotic activity and loss of liver toxicity. *J. Exp. Med.* **187**: 1205–1213.

Smith, C.A., Farrah, T. and Goodwin, R.G. (1994) The TNF-receptor superfamily of cellular and viral proteins: activation, costimulation, and death. *Cell* **76**: 959–962.

Tanaka, M., Suda, T., Haze, K., Nakamura, N., Sato, K., Kimura, F., Motoyoshi, K., Mizuki, M., Tagawa, S., Ohga, S. *et al.* (1996) Fas ligand in human serum. *Nature Med.* **2**: 317–322.

Thome, M., Schneider, P., Hofmann, K., Fickenscher, H., Meinl, E., Neipel, F., Mattmann, C., Burns, K., Bodmer, J.L., Schroter, M. *et al.* (1997) Viral FLICE-inhibitory proteins (FLIPs) prevent apoptosis induced by death receptors. *Nature* **386**: 517–521.

Thornberry, N.A. and Lazebink, Y. (1998) Caspases: enemies within. *Science* **281**: 1312–1316.

Tschopp, J., Irmler, M. and Thome, M. (1998) Inhibition of Fas death signals by FLIPs. *Curr. Opin. Immunol.* **10**: 552–558.

Udenfriend, S. and Kodukula, K. (1995) Prediction of ω site in nascent precursor of glycosylphosphatidylinositol protein. *Meth. Enzymol.* **250**: 571–583.

van Kooten, C. and Banchereau, J. (1997) Functions of CD40 on B cells, dendritic cells and other cells. *Curr. Opin. Immunol.* **9**: 330–337.

Vassalli, P. (1992) The pathophysiology of tumor necrosis factors. *Annu. Rev. Immunol.* **10**: 411–452.

Walczak, H., Miller, R.E., Kiley, A., Gliniak, B., Griffith, T.S., Kubin, M., Chin, W., Jones, J., Woodward, A., Le, T. *et al.* (1999) Tumoricidal activity of tumor necrosis factor-related apoptosis-inducing ligand *in vivo*. *Nature Med.* **5**: 157–163.

Wiley, S.R., Schooley, K., Smolak, P.J., Din, W.S., Huang, C.P., Nicholl, J.K., Sutherland, G.R., Smith, T.D., Rauch, C., Smith, C.A. and Goodwin, R.G. (1995) Identification and characterization of a new member of the TNF family that induces apoptosis. *Immunity* **3**: 673–682.

Yang, X., Khosravi-Far, R., Chang, H.Y. and Baltimore, D. (1997) Daxx, a novel Fas-binding protein that activates JNK and apoptosis. *Cell* **89**: 1067–1076.

Zhai, Y., Ni, J., Jiang, G.-W., Lu, J., Xing, L., Lincoln, C., Carter, K.C., Janat, F., Kozak, D., Xu, S. *et al.* (1999) VEGI, a novel cytokine of the tumor necrosis factor family, is an angiogenesis inhibitor that suppresses the growth of colon carcinomas *in vivo*. *FASEB J.* **13**: 181–189.

Zheng, L., Fisher, G., Miller, R.E., Peschon, J., Lynch, D.H. and Lenardo, M.J. (1995) Induction of apoptosis in mature T cells by tumor necrosis factor. *Nature* **377**: 348–351.

4

The role of the Bcl-2 family in the modulation of apoptosis

Christopher Pepper and Paul Bentley

1. Introduction

Programmed cell death in animals is a mechanism of cell suicide that typically culminates in a clearly defined set of morphological changes known as apoptosis. The changes include chromatin condensation, nuclear fragmentation and cytoplasmic shrinkage. Apoptosis is the most common form of cell death in both physiology and pathology and is essential for the development and maintenance of cellular homeostasis in renewable tissues as well as control of the immune response within all multicellular animals. Alterations in the regulation of this cell death programme have been implicated in many human disease states including neurodegenerative conditions, acquired immune deficiency syndrome (AIDS) and cancer (Kerr *et al.*, 1994; Oyaizu and Pahwa, 1995; Thompson, 1995). Indeed, the common involvement of apoptosis in human disease, and its implication in disease aetiology or pathogenesis, has led to a huge explosion in research into the molecular mechanisms by which it is induced and regulated.

Although many genes participate in the regulation, initiation or execution of apoptosis, the most prominent family of apoptosis regulators is represented by Bcl-2 and its homologues (Reed, 1994; Chao and Korsmeyer, 1998). The Bcl-2 family presides over a critical intracellular checkpoint of apoptosis within a distal common cell death pathway. These proteins exert their apoptosis-regulatory effects primarily by regulating mitochondrial alterations that precede the activation of cysteine proteases (caspases) and nucleases (Zamzami *et al.*, 1998). Modification of the expression of this family of proteins is a common observation in human cancers, contributing to neoplastic cell expansion by the suppression of programmed cell death (see Chapters 20 and 21). Moreover, since most chemotherapeutic drugs ultimately exert their cell killing effect through the induction of apoptosis, it would seem logical that responsiveness to therapy would, at least in part, be determined by Bcl-2 family expression. This chapter discusses how Bcl-2 family proteins function, and the impact that these

Programmed Cell Death in Animals and Plants, edited by J.A. Bryant, S.G. Hughes and J.M. Garland.
© 2000 BIOS Scientific Publishers Ltd, Oxford.

proteins have on cell survival. In addition, the specific example of Bcl-2 family dysregulation in B-cell chronic lymphocytic leukaemia (B-CLL) will be examined and research data will be presented.

2. Bcl-2 family: life or death promoters

Sixteen members of the human Bcl-2 family of proteins have so far been identified, all of which have either pro- or anti-apoptotic functions. Bcl-2 is the proteotypic family member and is homologous to the nematode *Caenorhabditis elegans* ced-9 product (Hengartner and Horvitz, 1994 and Chapter 1), suggesting that not only has this protein been conserved through evolution but also that its role in regulating a similar survival pathway may also be conserved. The *BCL-2* gene was first discovered through its involvement in the t(14;18) chromosomal translocation commonly found in low grade non-Hodgkin's lymphoma (NHL) (Korsmeyer, 1992). The translocation places *BCL-2* under the control of transcriptional enhancers associated with the immunoglobulin heavy chain gene locus on chromosome 14, leading to hyperexpression of the anti-apoptotic gene product, Bcl-2. Since its discovery Bcl-2 has been shown to be over-expressed in many other malignant conditions in the absence of detectable gene rearrangements.

While many of the Bcl-2 family proteins have similar function to Bcl-2 (e.g. Bcl-X_L, Bcl-W, Mcl-1, A1), others have diametrically opposed activity (e.g. Bax, Bad, Bcl-X_S, Bik) (Cory, 1995; Kroemer, 1997). The relative expression of these pro- and anti-apoptotic proteins is thought to determine, to a large extent, how readily an individual cell will succumb to apoptotic signals. Bcl-2 family proteins appear to be widely expressed in human tissues, but their expression patterns are cell-type specific and these patterns are modulated during differentiation and activation (Krajewski *et al.*, 1994).

3. Function of the Bcl-2 family proteins

Three general functions for Bcl-2 have been identified: dimerization with other Bcl-2 family proteins; interaction with non-homologous proteins; and formation of ion channels/pores. Most of the Bcl-2 family of proteins contain a C-terminal hydrophobic transmembrane domain that causes them to post-translationally insert into membranes, primarily the outer mitochondrial membrane, nuclear envelope and the endoplasmic reticulum (Jacobson *et al.*, 1993; Krajewski *et al.*, 1993). The intracellular distribution of different members of the Bcl-2 family is heterogeneous and can depend on the cell-type as well as on the family member. For example Bcl-2, Bcl-X_L and Bax tend to be particularly abundant in the outer mitochondrial membrane (Yang and Korsmeyer, 1996), whereas two pro-apoptotic proteins, Bad and Bid, which lack membrane-anchoring domains, appear to be translocated between the cytosol and the surface of intracellular organelles, depending on their association with other Bcl-2 family proteins (Li *et al.*, 1998; Luo *et al.*, 1998; Wang *et al.*, 1996).

3.1 *Dimerization*

The life/death rheostat that governs susceptibility to apoptotic signals is mediated, at least in part, by competitive dimerization between selective pairs of antagonists and agonists (Oltvai *et al.*, 1993). Sequence comparisons and mutational analysis have revealed the existence of four Bcl-2 homology (BH) domains. All of the anti-apoptotic and several of the pro-apoptotic members (Bax, Bak and Bok) possess three such domains: BH1, BH2 and BH3 which are involved in the formation of homo-dimers and hetero-dimers. A fourth BH domain (BH4) is found only in some of the anti-apoptotic family members and is involved in the interaction with proteins not belonging to the Bcl-2 family such as Bag-1 and Raf-1. Mutagenic assays have revealed that the BH1, BH2 and BH3 domains of Bcl-2 and Bcl-X_L are required for effective hetero-dimerization with the BH3 domain of Bax or Bak (Sattler *et al.*, 1997; Yin *et al.*, 1994). From nuclear magnetic resonance (NMR) experiments it appears that the BH3 domain of one Bcl-2 family protein forms an α-helix which inserts into a hydrophobic pocket formed by the BH1, BH2 and BH3 domains of its dimerizing partner (Muchmore *et al.*, 1996). Therefore a classic ligand/receptor interaction can be envisaged.

Most experimental evidence suggests that each cell type has a set point for the ratio of Bcl-2 family members, such as Bcl-2/Bax, Bcl-X_L/Bak, or Mcl-1/Bok, which gauges the sensitivity of cells towards apoptotic signals. The cumulative effect of dimerization on cell fate not only depends on which Bcl-2 family members are expressed in the respective cell type but also on the level, regulation and context of that expression. For example, Bax and Bak can either promote or suppress apoptosis depending on cell type and tissue distribution (Kiefer *et al.*, 1995; White, 1996).

3.2 *Adapter/docking protein function*

The BH4 binding domain is conserved among anti-apoptotic Bcl-2 homologues, yet is missing from all pro-apoptotic Bcl-2 family members apart from Bcl-X_S. As has been outlined previously, this domain is not required for dimerization with Bcl-2 family proteins but its deletion leads to loss of function or the acquisition of a pro-apoptotic phenotype (Borner, 1996; Hunter and Parslow, 1996). The BH4 domain is, however, necessary for the interaction of Bcl-2 with Bag-1 (Takayama *et al.*, 1995) and the serine threonine kinase Raf-1 (Wang *et al.*, 1996a). These two molecules link signal transduction pathways to the Bcl-2 complex and influence its anti-apoptotic potential. For example, Bcl-2 may localize the predominantly cytosolic Raf-1 molecule to the mitochondrial membrane where it phosphorylates Bad (a pro-apoptotic member of the Bcl-2 family which is inactivated by phosphorylation) (Wang *et al.*, 1996b; Zha *et al.*, 1996). Bcl-2 is also capable of interacting with other proteins such as p53 binding protein BP2, the GTPase Ras and calcineurin. The functional significance of these interactions has yet to be fully elucidated.

Recently, a link between Bcl-2 and caspase activation has been described. Apaf-1 which is the structural analogue of the nematode death regulatory protein CED-4 (Zou *et al.*, 1997) has been shown to interact with cytochrome c and Bcl-2. Apaf-1 contains a CARD domain which interacts with a similar domain within the N-terminus of pro-caspase 9 (and possibly pro-caspase 2). This leads to the formation of a tri-molecular complex containing Bcl-2/Bcl-X_L, Apaf-1 and pro-caspase 9. It has been proposed that the dissociation of Bcl-2/Bcl-X_L from this complex might initiate a reaction that, in the presence of dATP and cytochrome c, would lead to proteolytic activation of caspase 9 and the subsequent activation of other caspase molecules.

3.3 *Pore-forming function of Bcl-2 family proteins*

Structural analysis of Bcl-X$_L$ using NMR and X-ray crystallography has revealed similarities between this protein and the pore-forming domains of diphtheria toxin and the colicins A and E1, indicating that Bcl-X$_L$ and other family members may regulate apoptosis as ion channels (Muchmore *et al.*, 1996). The structure of Bcl-X$_L$ contains a pair of core hydrophobic helices that are long enough to penetrate the lipid bilayer which suggests that if these proteins were incorporated as dimers or higher-order structures effective pores might be formed. Indeed, as predicted by their structures, Bcl-2, Bcl-X$_L$ and Bax have been shown to form ion-channels, at least *in vitro*, when added to synthetic membranes (Minn *et al.*, 1997; Schendel *et al.*, 1997). Though no direct evidence has been presented to confirm that Bcl-2 family proteins form channels *in vivo*, evidence is building to suggest that proteins such as Bax do insert into membranes and exist in two conformational states that might represent active and latent versions of the protein. For example, in some cell types Bax is largely distributed in the cytosol until the cell is exposed to apoptotic signals; Bax then translocates to the mitochondrial membrane where it forms pores which might facilitate the loss of cytochrome c to the cytosol (Gross *et al.*, 1998; Hsu *et al.*, 1997; see also Chapters 6 and 7). The mechanism of cytochrome c release from the mitochondria remains controversial, however, and more than one mechanism might be envisaged. For example, an alternative mechanism might be facilitated through a marked change in the permeability of the inner mitochondrial membrane, leading to osmotic swelling and rupture of the outer membrane (Petit *et al.*, 1996). During apoptosis the membrane potential across the inner mitochondrial membrane is dissipated, probably as a result of the opening of a large conductance channel commonly referred to as the mitochondrial permeability transition (PT) pore (Chapter 6). This channel is poorly defined but is thought to be made up of both inner membrane proteins such as the adenine nucleotide translocator (ANT) and outer membrane proteins such as the voltage-dependent anion channel (VDAC) (Bernardi *et al.*, 1994). It might therefore be proposed that cytochrome c is released from the mitochondria following outer mitochondrial membrane rupture as a direct consequence of PT pore opening.

Reports that cytochrome c release from mitochondria can precede dissipation of the membrane potential across the inner membrane would argue in favour of a Bax-specific pore hypothesis (Bossy-Wetzel *et al.*, 1998); this evidence is not entirely convincing, however. Moreover, Bax has been reported to co-precipitate with the multiprotein PT pore complex from mitochondria and regulates its activity when reconstituted in synthetic liposomes (Hirsch *et al.*, 1997; Marzo *et al.*, 1998). In addition, direct interactions between Bax and the ANT have been reported.

It may well be that both hypotheses are correct. Bax may be capable of forming large pores in the outer membrane, possibly assisted by interactions with the ANT. On the other hand, sustained PT pore opening may be aided by Bax either by translocation of Bax to the inner mitochondrial membrane or interactions with the PT pore complex. In any event it could be envisaged that Bcl-2 or Bcl-X$_L$ might protect mitochondrial membranes by preventing Bax dimerization with either itself or other PT pore proteins.

4. Cellular control of Bcl-2 family proteins

One way in which Bcl-2 family proteins can be regulated is through phosphorylation by certain protein kinases and dephosphorylation by protein phosphatases. Bcl-2 appears to be phosphorylated in response to a variety of stimuli, including chemotherapeutic agents and the phosphatase inhibitor okadaic acid (Gajewski and Thompson, 1996). This phosphorylation most often results in inhibition of the anti-apoptotic function of the protein and this appears to be independent of its ability to form dimers.

One of the Bcl-2 antagonists, Bad, is also regulated by phosphorylation. When phosphorylated by either the kinase Akt (protein kinase B) or by Raf-1, Bad can no longer heterodimerize with anti-apoptotic proteins such as Bcl-X_L (Blume-Jensen *et al.*, 1998). Thus phosphorylation of Bad represents a mechanism for inactivating this pro-apoptotic transdominant suppressor of Bcl-2 and Bcl-X_L, whereas dephosphorylation of Bad can reactivate the protein by allowing it to bind to the apoptosis inhibitors. The universal relevance of this mechanism of apoptosis regulation is questionable since not all cell-types express Bad (Kitada *et al.*, 1998). However, it is possible that other Bcl-2 family proteins may be influenced by this phosphorylation mechanism in a similar way.

An alternative mechanism for the regulation of Bcl-2 family proteins may be through proteolytic cleavage. Bid, a Bcl-2 family member which only contains one homology domain (BH3) is controlled in this fashion. Bid can be cleaved by caspase 8 and upon removal of its N-terminus, is translocated to the mitochondrial membrane from the cytosol. Here, it dimerizes with Bax which appears to be critical for its pro-apoptotic function (Desagher *et al.*, 1998).

5. Over-expression of Bcl-2 family proteins and chemoresistance

Bcl-2 over-expression can render tumour cells markedly more resistant to the cytotoxic effects of essentially all currently available anti-cancer drugs (Reed, 1995). Although cytotoxic drugs damage tumour cells by several different mechanisms (e.g. DNA cleavage, alkylation, microtubule aggregation, nucleotide precursor inhibition), ultimately their cell killing effect is mediated through the induction of apoptosis. In this regard, over-expression of Bcl-2 or Bcl-X_L has been shown to prevent proteolytic processing and activation of caspases in response to chemotherapy-induced damage (Datta *et al.*, 1996; Ibrado *et al.*, 1996). The cytoprotection provided by Bcl-2 allows the damaged malignant cells to remain viable. This may either be facilitated by repair of the DNA damage or by adaptation of the cell in order that it might survive despite the additional mutation. In this regard, over-expression of Bcl-2 and Bcl-X_L has been shown to create a cellular environment that is permissive for accumulation of mutations and aberrant chromosomal segregation (Minn *et al.*, 1997). This presumably occurs because cells with genetic defects and chromosomal damage would normally be eliminated by apoptosis. Thus, damaged tumour cells that over-express Bcl-2 or Bcl-X_L may survive and acquire additional secondary mutations that can lead to the emergence of other drug resistance mechanisms.

6. Drug resistance in B-CLL

B-CLL is characterized by the progressive expansion of a slowly dividing subpopulation of monoclonal B-lymphocytes which are functionally ineffective but have a long lifespan *in vivo*. Although most patients with B-CLL show initial responsiveness to treatment, this ultimately diminishes, with most treated patients becoming resistant to chemotherapeutic drugs. Indeed, the development of drug resistance in B-CLL is probably the single biggest obstacle to the successful management of this condition. B-CLL is a quintessential example of a human malignancy which is caused primarily by defects in the programmed cell death machinery rather than by dysregulation of the cell cycle. It therefore represents an excellent model for investigations into the molecular mechanisms that regulate apoptosis and drug resistance.

Constitutively high levels of Bcl-2 mRNA and protein, equivalent to those seen in t(14;18)-containing NHLs, are found in the majority of cases of B-CLL but without any evidence of gene rearrangements (Hanada *et al.*, 1993; Robertson *et al.*, 1996). Regardless of the underlying reason for high Bcl-2 expression, the expected outcome of over-expression would be the inhibition of programmed cell death; this has consistently been found to be the case in B-CLL. However, the balance of pro- and anti-apoptotic proteins within B-CLL cells appears to determine their fate. Experimental evidence from our own laboratory and others has shown that the ratio of Bcl-2 to Bax is a critical factor in determining the relative sensitivity of B-CLL cells to chemotherapeutic drugs *in vitro* (Aguilar-Santelises *et al.*, 1996; Pepper *et al.*, 1996). Interestingly, the Bcl-2/Bax ratios were found to be highest in those patients who had been previously treated suggesting that the development of resistance may be mediated through selection of high Bcl-2/Bax expressing subclones (Pepper *et al.*, 1999).

Several studies have attempted to assess the usefulness of Bcl-2 family expression as prognostic indicators in B-CLL (Kitada *et al.*, 1998; Robertson *et al.*, 1996). Whereas the levels of Bcl-2 by themselves have not always proven useful in terms of segregating patients into different prognostic groups, higher Bcl-2/Bax mRNA or protein ratios have been correlated with progressive disease and with resistance to treatment (Aguilar-Santelises *et al.*, 1996; Pepper *et al.*, 1997). In addition, the anti-apoptotic protein, Mcl-1 has also been correlated with failure to achieve complete remission after single-agent therapy (Kitada *et al.*, 1998). Although larger prospective studies are undoubtedly required, it seems that consideration of Bcl-2 family protein expression may be useful in predicting clinical outcome in the future (see also Chapter 21).

7. Strategies for inhibiting Bcl-2 function in B-CLL

Finding strategies for overcoming the cytoprotective effects of Bcl-2 and Bcl-X$_L$ is a major goal of research in this area. Several alternatives can be envisaged for abrogating the effects of Bcl-2 in human cancers; these include small molecule inhibitors that block the homodimerization or heterodimerization of Bcl-2 in order to prevent Bcl-2 interacting with accessory proteins such as Bag-1 or Raf-1 that collaborate with Bcl-2 in order to promote survival. Alternatively, therapeutic agents may be developed as the pore-like functions of Bcl-2 family proteins are further elucidated and their biological significance with regard to apoptosis becomes better understood. Another approach to inhibit the activity of Bcl-2 would be to increase the phosphorylation of Bcl-2 and thereby potentially inactivate the protein. A more complete understanding of the kinases and phosphatases that control phosphorylation of Bcl-2 may offer additional strategies for controlling Bcl-2 activity. In this regard, while Paclitaxel and other anti-microtubule drugs can induce phosphorylation of Bcl-2, typically less than half of the Bcl-2 protein in the cell becomes phosphorylated and, moreover, the phosphorylation takes place only during M-phase arrest of the cell cycle. Therefore, since the proliferative index of most solid tumours and lymphoproliferative diseases is relatively low these drugs presumably have limited usefulness in these conditions.

7.1 *Antisense oligonucleotides*

A strategy that has been increasingly explored in recent years for the silencing of gene expression is the use of antisense oligonucleotides. These short, synthetic single-stranded DNA molecules present the opportunity to modify gene expression in a sequence specific manner and have been used to inhibit several target gene-products. Antisense molecules targeted to Bcl-2 have been used successfully to decrease Bcl-2 protein expression in a human pre-B-cell leukaemia cell line (Reed *et al.*, 1990), in a lymphoma cell line containing the t(14;18) translocation (Kitada *et al.*, 1993) and in a SCID (severe combined immunodeficiency disease) mouse model containing human B-cell lymphoma cells with dysregulated Bcl-2 expression (Cotter *et al.*, 1994). In addition, this methodology has been applied to clinical therapy for non-Hodgkin's lymphoma with effective down-regulation of Bcl-2 in tumour tissue (Webb *et al.*, 1997).

In a recent study in our laboratory a Bcl-2 antisense molecule was employed in order to establish the importance of Bcl-2 in the B-CLL disease process and to examine the potential for therapy incorporating silencing of the BCL-2 gene. The antisense molecule used in this study was designed to hybridize to the coding region of the Bcl-2 mRNA. Bcl-2 mRNA was down-regulated in a sequence-specific manner and this led to a reduction in Bcl-2 protein expression in cultured B-CLL cells (*Figure 1*). This reduction was accompanied by a significant increase in apoptotic cell death over the 48-hour culture period when compared to those cells either treated with a scrambled oligonucleotide sequence or those cells incubated without oligonucleotide (*Figure 2*).

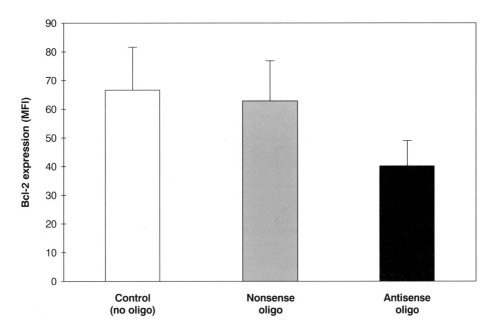

Figure 1. *Mean Bcl-2 protein expression (± SD) in B-CLL cells following a 48-hour incubation with either 5 μM antisense oligonucleotide, 5 μM of a scrambled (nonsense) oligonucleotide sequence or without oligonucleotide (n = 18). Bcl-2 protein expression was measured using a fluorescence-labelled monoclonal antibody, was quantified using flow cytometry and expressed as units of mean fluorescent intensity (MFI).*

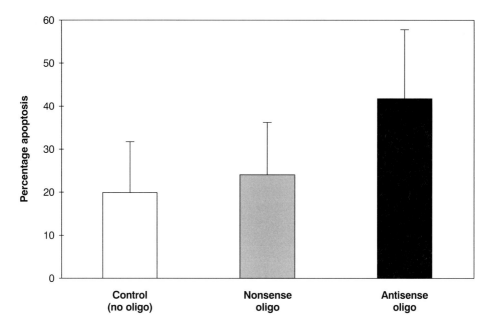

Figure 2. Mean percentage apoptosis (± SD) in B-CLL cells following a 48-hour incubation with either 5 μM antisense oligonucleotide, 5 μM of a scrambled (nonsense) oligonucleotide sequence or without oligonucleotide (n = 18). Apoptosis was measured using a fluorescence-labelled Annexin V assay and was quantified using flow cytometry.

These findings suggest that Bcl-2 plays a significant role in the B-CLL disease process and may contribute to the essentially incurable nature of the disease. Potentially, chemosensitization of B-CLL cells by antisense-mediated down-regulation of Bcl-2 protein could provide a therapeutic benefit and possibly a more durable response to treatment.

8. Conclusions

Bcl-2 family proteins play a key role in the regulation of apoptosis and therefore impact heavily on clinical effectiveness of essentially all chemotherapeutic agents. Although much work remains to be done, it seems likely that strategies designed to modulate the balance between pro- and anti-apoptotic proteins in many human diseases may offer the prospect of improving response rates and reversing chemo-resistance to existing therapeutic drugs.

References

Aguilar-Santelises, M., Rottenberg, M.E., Lewin, N., Mellstedt, H. and Jondal, M. (1996) Bcl-2, bax and p53 expression in B-CLL in relation to in vitro survival and clinical progression. *Int. J. Cancer* **69**: 114–119.

Bernardi, P., Broekemeier, K.M. and Pfieffer, D.R. (1994) Recent progress on regulation of the mitochondrial permeability transition pore; a cyclosporin-sensitive pore in the inner mitochondrial membrane. *J. Bioenerg. Biomembr.* **26**: 509–517.

Blume-Jensen, P., Janknecht, R. and Hunter, T. (1998) The Kit receptor promotes cell survival via activation of PI-3-kinase and subsequent Akt-mediated phosphorylation of Bad on ser136. *Curr. Biol.* **8**: 779–782.

Borner, C. (1996) Diminished cell proliferation associated with the death protective activity of Bcl-2. *J. Biol. Chem.* **271**: 695–698.

Bossy-Wetzel, E., Newmeyer, D.D. and Green, D.R. (1998) Mitochondrial cytochrome c release in apoptosis occurs upstream of DEVD-specific caspase activation and independently of mitochondrial transmembrane polarization. *EMBO J.* **17**: 37–49.

Chao, D.T. and Korsmeyer, S.J. (1998) Bcl-2 family: regulators of cell death. *Annu. Rev. Immunol.* **16**: 395–419.

Cory, S. (1995) Regulation of lymphocyte survival by the bcl-2 gene family. *Annu. Rev. Immunol.* **13**: 513–524.

Cotter, F.E., Johnson, P., Hall, P., Pocock, C., Al Mahdi, N. and Cowell, J.K. (1994) Antisense oligonucleotides suppress B-cell lymphoma growth in SCID-hu mouse model. *Oncogene* **9**: 3049–3055.

Datta, R., Banach, D., Kojima, H., Talanian, R.V., Alnemri, E.S., Wong, W.W. and Kufe, D.W. (1996) Activation of the CPP32 protease in apoptosis induced by 1-beta-D-arabinofuranosyl-cytosine and other DNA damaging agents. *Blood* **88**: 1936–1943.

Desagher, S., Osen-Sand, A., Nichols, A., Eskes, R., Montessuit, S., Lauper, S., Maundrell, K., Antonsson, B. and Martinou, J-C. (1998) Bid-induced conformational change of Bax is responsible for mitochondrial cytochrome c release during apoptosis. *Proc. SEB Symp: Cell Death in Health and Disease Exeter*, p 6 (abstract).

Gajewski, T.F. and Thompson, C.B. (1996) Apoptosis meets signal transduction – elimination of a bad influence. *Cell* **87**: 589–592.

Gross, A., Jockel, J., Wei, M. and Korsmeyer, S.J. (1998) Enforced dimerization of BAX results in its translocation, mitochondrial dysfunction and apoptosis. *EMBO J.* **17**: 3878–3885.

Hanada, M., Delia, D., Aiello, A., Stadtmauer, E. and Reed, J.C. (1993) bcl-2 gene hypomethylation and high level expression in B-cell chronic lymphocytic leukemia. *Blood* **82**: 1820–1828.

Hengartner, M.O. and Horvitz, H.R. (1994) *C. elegans* cell survival gene ced-9 encodes a functional homolog of the mammalian proto-oncogene *bcl-2*. *Cell* **76**: 665–676.

Hirsch, T., Marchetti, P., Susin, S.A., Dallaporta, B., Zamzami, N., Marzo, I., Geuskens, M. and Kroemer, G. (1997) The apoptosis-necrosis paradox. Apoptogenic proteases activated after mitochondrial permeability transition determine the mode of cell death. *Oncogene* **15**: 1573–1582.

Hsu, Y.T., Wolter, K.G. and Youle, R.J. (1997) Cytosol-to-membrane redistribution of Bax and Bcl-X(L) during apoptosis. *Proc. Natl Acad. Sci. USA* **94**: 3668–3672.

Hunter, J.J. and Parslow, T.G. (1996) A peptide sequence from Bax that converts Bcl-2 into an activator of apoptosis. *J. Biol. Chem.* **271**: 8521–8524.

Ibrado, A.M., Huang, Y., Fang, G. and Bhalla, K. (1996) Bcl-XL overexpression inhibits Taxol-induced Yama protease activity and apoptosis. *Cell Growth Differ.* **7**: 1087–1094.

Jacobson, M.D., Burne, J.F., King, M.P., Miyashita, T., Reed, J.C. and Raff, M.C. (1993) bcl-2 blocks apoptosis in cells lacking mitochondrial DNA. *Nature* **361**: 365–368.

Kerr, J.F.R., Winterford, C.M. and Harmon, B.V. (1994) Apoptosis: its significance in cancer and cancer therapy. *Cancer* **73**: 2013–2026.

Kiefer, M.C., Brauer, M.J., Powers, V.C., Wu, J.J., Umansky, S.R., Tomei, L.D. and Barr, P.J. (1995) Modulation of apoptosis by the widely distributed Bcl-2 homologue Bak. *Nature* 374: 736–739.

Kitada, S., Miyashita, T., Tanaka, S. and Reed, J.C. (1993) Investigations of antisense oligonucleotides targeted against bcl-2 RNAs. *Antisense Res. Dev.* **3**: 157–169.

Kitada, S., Andersen, J., Akar, S., Zapata, J.M., Takayama, S., Krajewski, S., Wang, H-G., Zhang, X., Bullrich, F., Croce, C.M. *et al.* (1998) Expression of apoptosis-regulating proteins in chronic lymphocytic leukemia: correlations with in vitro and in vivo chemoresponses. *Blood* **91**: 3379–3389.

Korsmeyer, S. (1992) bcl-2 initiates a new category of oncogenes: regulators of cell death. *Blood* 80: 879–886.

Krajewski, S., Tanaka, S., Takayama, S., Schibler, M.J., Fenton, W. and Reed, J.C. (1993) Investigation of the subcellular distribution of the bcl-2 oncoprotein: residence in the nuclear envelope, endoplasmic reticulum, and outer mitochondrial membranes. *Cancer Res.* 53: 4701–4714.

Krajewski, S., Krajewska, M., Shabaik, A., Wang, H.G., Irie, S., Fong, L. and Reed, J.C. (1994) Immunohistochemical analysis of in vivo patterns of Bcl-X expression. *Cancer Res.* 54: 5501–5507.

Kroemer, G. (1997) The proto-oncogene Bcl-2 and its role in regulating apoptosis. *Nature Med.* 3: 614–620.

Li, H., Zhu, H., Xu, C. and Yuan, J. (1998) Cleavage of BID by caspase-8 mediates the mito-chondrial damage in the Fas pathway of apoptosis. *Cell* 94: 491–501.

Luo, X., Budihardjo, I., Zou, H., Slaughter, C. and Wang, X. (1998) Bid, a Bcl-2 interacting protein, mediates cytochrome c release from mitochondria in response to activation of cell surface death receptors. *Cell* 94: 481–490.

Marzo, I., Brenner, C., Zamzami, N., Jurgensmeier, J.M., Susin, S.A., Vieira, H.L.A., Prevost, M-C., Xie, Z., Matsuyama, S., Reed, J.C. and Kroemer, G. (1998) Bax and adenine nucleotide translocator cooperate in the mitochondrial control of apoptosis. *Science* 281: 2027.

Minn, A.J., Velez, P., Schendel, S.L., Liang, H., Muchmore, S.W., Fesik, S.W., Fill, M. and Thompson, C.B. (1997) Bcl-XL forms an ion channel in synthetic lipid membranes. *Nature* 385: 353–357.

Muchmore, S.W., Sattler, M., Liang, H., Meadows, R.P., Harlan, J.E., Yoon, H.S., Nettesheim, D., Chang, B.S., Thompson, C.B., Wong, S. *et al.* (1996) X-ray and NMR structure of human Bcl-XL, an inhibitor of programmed cell death. *Nature* 381: 335–341.

Oltvai, Z.N., Milliman, C.L. and Korsmeyer, S.J. (1993) bcl-2 heterodimerizes in vitro with a conserved homolog, bax, that accelerates programmed cell death. *Cell* 74: 609–619.

Oyaizu, N. and Pahwa, S. (1995) Role of apoptosis in HIV disease pathogenesis. *J. Clin. Immunol.* 15: 217–231.

Pepper, C., Bentley, P. and Hoy, T. (1996) Regulation of clinical chemoresistance by bcl-2 and bax oncoproteins in B-cell chronic lymphocytic leukaemia. *Br. J. Haematol.* 95: 513–517.

Pepper, C., Hoy, T. and Bentley, P. (1997) Bcl-2/Bax ratios in chronic lymphocytic leukaemia and their correlation with in vitro apoptosis and clinical resistance. *Br. J. Cancer* 76: 935–938.

Pepper, C., Thomas, A., Hoy, T. and Bentley, P. (1999) Chlorambucil resistance in B-cell chronic lymphocytic leukaemia is mediated through failed Bax induction and selection of high Bcl-2-expressing subclones. *Br. J. Haematol.* 104: 581–588.

Petit, P.X., Susin, S-A., Zamzami, N., Mignotte, B. and Kroemer, G. (1996) Mitochondria and programmed cell death: back to the future. *FEBS Lett.* 396: 7–13.

Reed, J.C. (1994) Bcl-2 and the regulation of programmed cell death. *J. Cell Biol.* 124: 1–6.

Reed, J.C. (1995) Bcl-2 family proteins: regulators of chemoresistance in cancer. *Toxicol. Lett.* 82/83: 155–158.

Reed, J.C., Stein, C., Subasinghe, C., Haldar, S., Croce, C.M., Yum, S. and Cohen, J. (1990) Antisense-mediated inhibition of BCL2 protooncogene expression and leukemic cell growth and survival: comparisons of phosphodiester and phosphorothioate oligodeoxynucleotides. *Cancer Res.* 50: 6565–6570.

Robertson, L.E., Plunkett, W., McConnell, K., Keating, M.J. and McDonnell, T.J. (1996) Bcl-2 expression in chronic lymphocytic leukemia and its correlation with the induction of apoptosis and clinical outcome. *Leukemia* 10: 456–459.

Sattler, M., Liang, H., Nettesheim, D., Meadows, R.P., Harlan, J.E., Eberstadt, M., Yoon, H.S., Shuker, S.B., Chang, B.S., Minn, A.J. *et al.* (1997) Structure of Bcl-X(L)-Bak peptide complex-recognition between regulators of apoptosis. *Science* 275: 983–986.

Schendel, S.L., Xie, Z., Oblatt-Montal, M., Matsuyama, S., Montal, M. and Reed, J.C. (1997) Channel formation by anti-apoptotic protein Bcl-2. *Proc. Natl Acad. Sci. USA* **94**: 5113–5118.

Takayama, S., Sato, T., Krajewski, S., Kochel, K., Irie, S., Millan, J.A. and Reed, J.C. (1995) Cloning and functional analysis of BAG-1: a novel Bcl-2-binding protein with anti-cell death activity. *Cell* **80**: 279–284.

Thompson, C.B. (1995) Apoptosis in the pathogenesis and treatment of disease. *Science* **267**: 1456–1462.

Wang, H-G., Takayama, S., Rapp, U.R. and Reed, J.C. (1996a) Bcl-2 interacting protein, BAG-1, binds to and inactivates the kinase Raf-1. *Proc. Natl Acad. Sci. USA* **93**: 7063–7068.

Wang, H-G., Rapp, U.R. and Reed, J.C. (1996b) Bcl-2 targets the protein kinase Raf-1 to mitochondria. *Cell* **87**: 629–638.

Wang, K., Yin, X-M., Chao, D.T., Milliman, C.L. and Korsmeyer, S.J. (1996) BID: a novel BH3 domain-only death agonist. *Genes Dev.* **10**: 2859–2869.

Webb, A., Cunningham, D., Cotter, F., Clarke, P.A., di Stafano, F., Ross, P., Corbo, M. and Dziewanowska, Z. (1997) Bcl-2 antisense therapy in patients with non-Hodgkin lymphoma. *Lancet* **349**: 1137–1141.

White, E. (1996) Life, death and the pursuit of apoptosis. *Genes Dev.* **10**: 1–15.

Yang, E. and Korsmeyer, S.J. (1996) Molecular thanatopsis: a discourse on the Bcl-2 family and cell death. *Blood* **88**: 386–401.

Yin, X.M., Oltvai, Z.N. and Korsmeyer, S.J. (1994) BH1 and BH2 domains of Bcl-2 are required for inhibition of apoptosis and heterodimerization with Bax. *Nature* **369**: 321–323.

Zamzami, N., Brenner, C., Marzo, I., Susin, S.A. and Kroemer, G. (1998) Subcellular and submitochondrial mode of action of Bcl-2-like oncoproteins. *Oncogene* **16**: 2265–2282.

Zha, J., Harada, H., Yang, E., Jockel, J. and Korsmeyer, S.J. (1996) Serine phosphorylation of death agonist BAD in response to survival factor results in binding to 14–3–3 not Bcl-X(L). *Cell* **87**: 619–628.

Zou, H., Henzel, W.J., Liu, X., Lutschg, A. and Wang, X. (1997) Apaf-1, a human protein homologous to C. elegans CED-4, participates in cytochrome c-dependent activation of caspase-3. *Cell* **90**: 405–413.

5

The role of granzymes and serpins in regulating cell growth and death

Paul Coughlin, Emma Morris and Lynne Hampson

1. Introduction

Whilst the cell is the fundamental functional unit of higher organisms it is, under most circumstances, individually expendable. Nature has taken advantage of this fact by developing a sophisticated cell death process which can be precisely activated to the overall benefit of the organism. Cells undergoing programmed cell death, or apoptosis, are recognized by their healthy neighbours or by specialized phagocytes, engulfed and destroyed silently without activating broadly damaging inflammatory pathways. The apoptotic process has been further refined and specialized by the mammalian immune system to allow the elimination of virally infected, foreign or transformed cells. This presents a special problem in that the effector cells of the immune system must be resistant to their own cytotoxic mechanisms. On the other hand cellular components of the immune system which are dangerous or are no longer required need to be eliminated.

This review outlines the dual killing mechanism used by the immune system and suggests a way in which effector cells may act with impunity while remaining susceptible to regulation by apoptosis. In particular we will deal with cell death caused by granular serine proteases, granzymes, which gain access to the interior of target cells and activate the apoptotic process. We will also examine evidence showing that a member of the intracellular serpin family called granzyme B inhibitor (or PI-9) is expressed in immune cells and may protect them from endogenous granzymes. The role of other intracellular serpins in the regulation of apoptosis, cell growth and differentiation will also be discussed.

2. Induction of apoptosis by cytotoxic T cells and natural killer cells

The principal effectors of the mammalian immune system are cytotoxic T cells and natural killer cells. In response to signals from immune regulatory cells and target cells

Programmed Cell Death in Animals and Plants, edited by J.A. Bryant, S.G. Hughes and J.M. Garland.
© 2000 BIOS Scientific Publishers Ltd, Oxford.

they express Fas ligand and TNF which bind receptors on targets activating a cascade of intracellular events leading to apoptosis. Recent evidence shows that immune effector cells direct Fas ligand to cytolytic granules as transmembrane proteins (Bossi and Griffiths, 1999). Upon activation Fas ligand is transported to the cell surface where it is cleaved and released in soluble form. These cytolytic granules, whose contents are released into the intercellular space, also contain several cytotoxic serine proteases called granzymes together with the pore-forming protein perforin. The most abundant of these proteases are granzyme A, which cleaves substrates at basic residues, and granzyme B, which cleaves at acidic residues. These enzymes are well conserved between species while other less abundant granular proteases vary considerably between rodents and man (Caughey, 1994). Granzymes are produced as proenzymes which are activated by dipeptidyl protease prior to secretion (Kummer et al., 1996; Smyth et al., 1995). The importance of granzymes in apoptosis has been highlighted by targeted disruption studies. Mice lacking granzyme B are morphologically normal but killer cell cytotoxicity is dramatically reduced, although not eliminated (Heusel et al., 1994). The granzyme A knockout mouse demonstrates no significant loss of cytotoxic activity although it may still play an adjunctive role (Ebnet et al., 1995). The perforin knockout mouse also shows loss of killer cell activity, confirming the importance of this protein as well as granzyme B in granular cytotoxicity (Lowin et al., 1994; Walsh et al., 1994).

Although these and other studies have clearly established granzyme B and perforin as the key players in effecting granule-mediated cytotoxicity, the mechanism of this activity at a cellular and molecular level has not been clearly defined. Several recent papers have addressed the question of how granzyme B gains access to the interior of the target cell. There is also accumulating evidence indicating that granzyme B directly interacts with and activates the caspase pathway within cells.

3. How does granzyme B gain access to the interior of target cells?

Until recently the prevailing view was that granzyme B was secreted into the intercellular space and, in the presence of sublytic concentrations of perforin, entered the target cell through pores in the plasma membrane. It was therefore surprising when Froelich et al. (1996) and Shi et al. (1997) showed that granzyme B bound to the surface of cells and was internalized in the absence of perforin. However, the subsequent addition of perforin to the cells was required in order to induce apoptosis. The effect of perforin could be mimicked by replication-defective adenovirus type 2 particles which have previously been shown to cause membrane disruption. Shi et al. (1997), using HeLa cells as targets showed by immuno-electron microscopy that granzyme B entered the cytosol and that apoptosis occurring after perforin exposure coincided with nuclear translocation of granzyme B. Froelich et al. (1996) on the other hand suggested that granzyme B was internalized into a membrane-bound compartment initially and that perforin allowed egress of the enzyme into the cytosol. In a refinement of these studies, Pinkowski et al. (1996) showed granzyme B trafficking through early endosomal vesicles and subsequent localization in a novel membrane-bound compartment. When taken together, these data show that granzyme B enters cells autonomously but that perforin is required to initiate apoptosis and permits nuclear translocation of the granzyme.

4. Molecular targets of granzyme B

Granzyme B shares with the caspase family of cysteine proteases a specificity for cleavage at aspartate residues. Caspases are activated during apoptosis and cleave key cellular proteins which render cell death irreversible. They are produced as inactive pro-enzymes and are activated by cleavage at sites which are themselves substrates for caspases (Salvesen and Dixit, 1997). This gave rise to the concept of an enzyme cascade analogous to the coagulation and complement systems in which so-called apical caspases are activated in response to apoptotic stimuli. Subsequently, downstream, or executioner, caspases would be activated by the apical enzymes and cleave key cellular proteins. Support for involvement of granzyme B in this system was suggested by Darmon *et al.* (1995, 1996) who showed that granzyme B was able to cleave and activate caspase 3. Detailed studies using combinatorial libraries of peptide have defined the substrate specificity of granzyme B (Harris *et al.*, 1998; Thornberry *et al.*, 1997). These show that granzyme B has extended substrate specificity, cleaving short peptides inefficiently. Four amino acids are required on the N-terminal side of the cleavage site for efficient catalysis and, in keeping with its regulatory role, the amino acids C-terminal to the cleavage site are also important.

The preferred substrate for granzyme B is $(I/L/V)EXD^vXG$ which is identical to caspases 6, 8, 9 and 10 supports its role as an activator of the downstream caspase 3 whose activation sequence is $IETD^vSG$. However, *in vitro* studies using recombinant proteins show that granzyme B cleaves and activates 7 and 10 more efficiently than caspase 3 (Talanian *et al.*, 1997). This dilemma may have been resolved by Yang *et al.* (1998), who showed that *in vitro* caspase 3 efficiently cleaved the caspase 7 propeptide fragment but did not cleave at the activation site between light and heavy chains. *In vivo* studies have previously demonstrated that caspase 7 is first cleaved at the propeptide site during apoptosis suggesting prior activation of caspase 3 (Froelich *et al.*, 1996; Orth *et al.*, 1996). Using MCF-7 cells deficient in caspases 3 and 10, Yang *et al.* (1998) showed that although granzyme B efficiently activates caspases 7 and 10 *in vitro*, in the absence of caspase 3 this does not occur *in vivo*, suggesting that these caspases are sequestered or inaccessible to the granzyme. When caspase 3 was re-expressed in MCF-7 cells exposed to granzyme B, early cleavage of caspase 7 pro-peptide and subsequent full activation was seen.

These data suggest a two-step model in which granzyme B activates caspase 3 which then cleaves the caspase 7 propeptide, releasing it from a sequestered site, and making it available for full activation by granzyme B. This model suggests that granzyme B activates the caspase cascade downstream of the cell surface-associated signalling complex used by TNF and Fas, allowing it to bypass inhibitors of apical components of the pathway.

Further interdependence between granzyme B and caspase activation is suggested by Trapani *et al.* (1998) who showed that caspase activation is essential for transport of granzyme B into the nucleus. Cells treated with YVAD-fmk, which efficiently inhibits caspases but has no effect on granzyme B, then exposed to granzyme B/perforin, did not show nuclear localization of the granzyme B. Surprisingly the cells still underwent cell death as evidenced by markedly reduced clonogenicity. They suggested that granzyme B may have cytosolic non-caspase targets which activate other death pathways.

5. Regulation of granzyme B activity

It is axiomatic that biochemical pathways leading to important intra- and extracellular events are controlled by negative regulators. Apoptosis is a critical process in cellular function and is also tightly controlled. The activation of caspases 8 and 10 at the apex of the signalling pathway is modulated by proteins which show selective recruitment to the cell membrane signalling complex. IAP proteins which competitively inhibit caspases are known to exist and viral homologues demonstrate similar activities. Members of the serpin (**ser**ine **p**roteinase **in**hibitor) superfamily of proteins may also regulate apoptotic proteases as there is now unequivocal evidence that some members of the family are retained within the cell (Coughlin *et al.*, 1993; Scott *et al.*, 1996; Sun *et al.*, 1996).

Most serpins exert their biological effects by causing irreversible inhibition of target proteinases (Huber and Carrell, 1989). Upon binding a target protease the serpin undergoes a dramatic stabilizing conformational change, trapping the enzyme in an irreversible complex. This metastable state has been further exploited in nature to modulate serpin activity, allowing the bait region of the molecule to be more or less available (Carrell and Stein, 1996). This mechanism is illustrated by the interaction of antithrombin with heparans in which the serpin is localized to the endothelial surface and its activity against target proteases is increased several hundred fold (Carrell *et al.*, 1991). This capacity of serpins to be localized and modulated makes intracellular members of the family good candidate regulators of apoptotic proteases.

The first serpin to be described with anti-apoptotic activity was the poxvirus protein *crmA* which inhibits proteases cleaving at aspartate residues (Komiyama *et al.*, 1994). *In vitro crmA* displays cross-class inhibition, efficiently inhibiting caspases 1 and 8 as well as granzyme B (Zhou *et al.*, 1997). When transfected into cells it also blocks Fas and TNF mediated apoptosis (Tewari and Dixit, 1995). Physiologically, in the setting of viral infection, it appears to function principally as a caspase 1 inhibitor, delaying host inflammatory response.

Recently, a novel intracellular serpin, PI-9, has been described, expressed in the cytosol of activated T lymphocytes and natural killer cells, with the acidic residue glutamate at the predicted P_1 site suggesting that it may inhibit either caspases or granzyme B (Sun *et al.*, 1996) (the P_1 site is the bait amino acid on the N-terminal side of the bond cleavable by the protease). Recombinant PI-9 was found to be a good inhibitor of granzyme B with an association rate constant of $1.7 \times 10^6 \, M^{-1} s^{-1}$ but was a poor inhibitor of caspases. The presence of a glutamate in the P_1 position of PI-9 was consistent with granzyme B inhibition but it might be expected that aspartate would be preferred. Therefore Bird *et al.* (1998) made a mutant in which the P_1 Glu was changed to Asp and found improved caspase inhibitory activity but paradoxically, decreased granzyme B inhibition. Analysis of complex formation between the serpin and enzyme on SDS-PAGE showed that the P_1 Asp mutant was preferentially cleaved by the granzyme B rather than forming an inhibitory complex. Therefore the P_1 Asp mutant displayed substrate-like properties. These data may be explained by the known mechanism of action of serpins in which rapid conformational change in the reactive site loop causes distortion of the protease active site thereby preventing completion of the cleavage reaction and locking the enzyme into an inhibitory complex (Stein and Carrell, 1995). The slightly less favoured P_1 Glu in PI-9 may well slow down the otherwise efficient cleavage of the loop allowing time for inhibitory conformational change to occur. It was also shown that the presence of a P_1 Glu makes PI-9 a specific granzyme B inhibitor with poor activity against caspases (Bird *et al.*, 1998).

6. The intracellular regulation of granzyme B by PI-9

The *in vivo* function of PI-9 was assessed by treating FDCP-1 cells, transfected with PI-9, to exogenous granzyme B and perforin (Bird *et al.*, 1998). Transfected cells were protected from apoptosis and the degree of protection correlated with level of PI-9 expression. Consistent with the results of *in vitro* studies, the P_1 Asp mutant was less effective than wild type while a predictably dysfunctional mutant (in the proximal hinge region) was completely ineffective. The ability of PI-9 to protect cells from cellular cytotoxic attack was tested using MCF-7 cells (which do not express Fas) transfected with PI-9 and exposed to activated T cells. Using this model, MCF-7 cells expressing PI-9 were protected from cytotoxic cells expressing low levels of granzyme B but not from cells expressing high levels. It is possible that T cells expressing high levels of granzyme B also possess other cytotoxic activities against which PI-9 is ineffective. As expected PI-9 was ineffective in preventing Fas-mediated apoptosis.

Further studies are required to clarify the functions of PI-9 *in vivo*. It is, however, clear that PI-9 effects only the granzyme B-mediated arm of cell-mediated cytotoxicity. Therefore immune effector cells are susceptible to elimination by Fas-mediated apoptosis while being relatively protected from inadvertent exposure to granzyme B. This could occur when small amounts of granzyme B leak from secretory vesicles into the cytoplasm or alternatively when secreted granzyme B re-enters the immune effector cell rather than the target cell. It is possible that dysfunction or deletion of PI-9 could lead to unusual sensitivity to immune attack and immune deficiency states. Similarly ectopic expression of PI-9 may allow cells to evade immune surveillance leading to autoimmunity or clonal expansion as a precursor to malignant change. These possibilities are currently being investigated.

7. The role of intracellular serpins in cell growth and death

Several other intracellular serpins have been proposed to play a role in cell growth, death and differentiation. Most of these belong to the ovalbumin-like serpins and are closely related in sequence to PI-9. Furthermore, the human intracellular serpins are clustered on two chromosomal loci at 6p25 and 18q21–23 and have probably evolved through gene duplication (Sun *et al.*, 1998). At least three members of this family appear to share with PI-9 the ability to target proteases which are normally sequestered in membrane-bound organelles. PI-6 and squamous cell carcinoma antigen 2 have recently been shown to be effective inhibitors of lysosomal cathepsin G (Schick *et al.*, 1997); squamous cell carcinoma antigen 1 (SCCA1) is a good cross-class inhibitor of lysosomal cysteine proteases (Schick *et al.*, 1998) and the monocyte/neutrophil elastase inhibitor targets granular elastase (Remold-O'Donnell *et al.*, 1992). Although there are no data on the extent to which lysosomal proteases escape into the cytosol, it is well described that uncontrolled proteolytic activity can lead to caspase activation (Williams and Henkart, 1994). It seems likely that there would be multiple adverse effects on the cell if sequestered proteases leaked into other compartments.

Another serpin, plasminogen activator inhibitor 2 (PAI-2), exists in both an intracellular and glycosylated extracellular form. The intracellular expression of PAI-2 has been shown to inhibit TNF induced apoptosis and this activity is dependent on the presence of an unusually long loop between the serpin C and D helices (Dickinson *et al.*, 1998). Although the mechanism of this anti-apoptotic activity is unclear, the C–D interhelical loop has been shown to be a substrate for intracellular transglutaminases (Jensen *et al.*, 1994).

Our interest in this field was further stimulated by the discovery of the first of a newly-discovered family of intracellular serpins expressed in a primitive, karyotypically normal, haemopoietic cell line (Hampson *et al.*, 1997). This murine multipotent cell line, FDCP-MixA4, can be induced to differentiate down different lineages under the influence of growth factors and the 'new' gene, known as serpin2A, was identified by virtue of its differential expression between immature and differentiated cells.

Northern blot studies confirmed that serpin2A was down-regulated upon differentiation and this pattern of expression was also seen in haemopoietic progenitors isolated directly from murine bone marrow. Northern analysis also showed that the principal tissues expressing the serpin were thymus, spleen, lung and testis. Expression in bone marrow was low, presumably because the serpin is restricted to rare primitive cells. Further studies have shown that when T cells were activated in primary splenocyte culture there was a rapid increase in the level of expression (Hampson *et al.*, 1997). A similar pattern is seen in transgenic mice with a monoclonal T-cell population. Upon activation of the clonal T cells, serpin2A expression rises dramatically within hours (D. Kioussis, personal communication). A truncated serpin2A cDNA had previously been isolated from a murine teratocarcinoma cell line, EB22, and it had been shown to belong to Spi2 cluster of genes on chromosome 12. These genes are most closely related to human antichymotrypsin and distantly related to the ovalbumin serpins (Inglis and Hill, 1991; Inglis *et al.*, 1991). It is therefore interesting that an intracellular expression of human antichymotrypsin has been reported several times.

Immunofluorescence studies of transiently transfected COS-7 cells and undifferentiated FDCP-MixA4 cells visualized with laser confocal scanning microscopy demonstrate diffuse cytoplasmic distribution of the recombinant protein similar to other intracellular serpins (Scott *et al.*, 1996). Surprisingly, a significant proportion of cells show serpin2A principally localized in the nucleus. The protein does not have a recognizable nuclear localization signal but this is unlikely to be a passive event as other intracellular serpins of a similar molecular weight are clearly excluded from the cell nucleus (Scott *et al.*, 1996). The fact that serpin2A is both nuclear and cytosolic, together with its predicted redox sensitivity, suggests a role in the modulation of apoptosis but it has been impossible to produce cell lines constitutively expressing the serpin to test this possibility. Transfected cells in selection media are only able to express detectable protein for short periods (2 weeks). This may be because cells expressing the protein die, or because of selection of non-expressing cells, or because of silencing of the construct. We therefore suspect that serpin2A is either growth inhibitory or pro-apoptotic and may require co-expression of a 'cofactor' to prevent cell death.

The amino acid sequence of serpin2A shows the following features: (1) there is no conventional N-terminal signal sequence, (2) the predicted reactive site scissile bond (P_1–P_1') is Cys–Cys which had not been previously seen in the serpin family, and (3) there is an unusual C-terminal extension of some 30 amino acids containing two further cysteine residues. The presence of oxidizable residues in the reactive site loop and C-terminal extension suggests that the activity of the serpin will be sensitive to prevailing redox conditions. The fact that the reactive site sequence does not correspond to any protease substrate makes it impossible to predict an enzyme target and the recombinant serpin does not form inhibitory complexes when incubated with a panel of proteases. However, the proximal hinge region of the reactive site loop is typical of inhibitory serpins and recombinant serpin2A undergoes the stressed → relaxed conformation change upon cleavage of the reactive site loop. It is therefore uncertain whether serpin2A behaves as an *in vivo* protease inhibitor but it may act as a redox sensor whose molecular interactions are modulated by conformational change.

In order to address the possibility that serpin 2A interacts with unknown proteases or alternatively that it takes part in non-protease-inhibitor interactions, we have screened yeast two-hybrid libraries derived from mouse testis and human T cells. These screens have yielded several possible interactors including (1) mov34, a nuclear protein which promotes G_2–M phase cell cycle progression (Mahalingham *et al.*, 1998), (2) thioredoxin, which acts as a cellular redox sensor and modulates the activity of transcription factors including NF kappaB (Hirota *et al.*, 1997), (3) minichromosome maintenance protein 2 (MCM2) which belongs to a group of MCM proteins required for G_1–S phase cell cycle progression (Ritzi *et al.*, 1998), (4) a B-cell antigen receptor binding protein, and (5) Tat interactive protein, which interacts with a cysteine-rich region of the HIV protein Tat (Kamine *et al.*, 1996). Although unverified, the Y2H results suggest that serpin2A may act as a redox-sensitive regulator of cell cycle progression. Further analysis of potential interactors will focus on demonstrating physiological relevance as shown by co-expression and co-immunoprecipitation. A feature of the Y2H results so far is the absence of possible target proteases which may be caused by the C-terminal extension of the protein obstructing the reactive site loop. We will therefore repeat the Y2H screens with a truncated bait protein lacking the C-terminal cysteines.

Further evidence to suggest a role of antichymotrypsin-like serpins in cell growth came from a report showing that antisense expression of a Spi2 serpin in NIH3T3 cells gave rise to a transformed phenotype (Whitehead *et al.*, 1995). By using RT-PCR we have demonstrated the presence of serpin2A message in NIH3T3 and Swiss3T3 cells and also identified a closely related novel serpin sequence which possesses a different reactive site sequence (Leu–Cys) which we have called 3T3Spi2. We have further preliminary evidence that the expression of these genes is increased in G_0 arrested NIH3T3 cells.

8. Conclusions

Recent work has led to the description of intracellular proteolytic cascades, involving caspases and granzyme B, which are of comparable complexity to the extracellular haemostatic and complement pathways. Within the cell these biochemical events are further constrained and modified by separation of molecules into compartments. The movement of proteases and substrates between membrane-bound compartments or other sites of sequestration allows strict control over critical events in the apoptotic mechanism.

The role of serpins in control of intracellular proteolysis is only now beginning to be appreciated. It is likely that proteins such as PI-9, PI-6 and SCCA1 play an important role in restricting protease activity to specific cellular organelles. Work on serpin2A hints at the way in which the serpin structure may be adapted to link modulation of protease activity with changes in the intracellular environment, such as redox conditions. These molecular events may be further integrated with cell cycle progression or cell death by interaction with non-protease cellular targets. While this speculation may seem far fetched, the unique structure and mechanism of action of serpins has caused them to be recognized as important players in processes as diverse as deposition of amyloid in Alzheimer's disease, modulation of cell interactions with tissue matrix and the control of haemostasis (Carrell and Gooptu, 1998; Stefansson and Lawrence, 1996).

References

Bird, C.H., Sutton, V.R., Sun, J., Hirst, C.E., Novak, A., Kumar, S., Trapani, J.A. and Bird, P.I. (1998) Selective regulation of apoptosis: the cytotoxic lymphocyte serpin proteinase inhibitor 9 protects against granzyme B-mediated apoptosis without perturbing the Fas cell death pathway. *Mol. Cell. Biol.* **18**: 6387–6398.

Bossi, G. and Griffiths, G.M. (1999) Degranulation plays an essential part in regulating cell surface expression of Fas ligand in T cells and natural killer cells. *Nat. Med.* **5**: 90–96.

Carrell, R. and Gooptu, B. (1998) Conformational changes and disease – serpins, prions and Alzheimer's. *Curr. Opin. Struct. Biol.* **8**: 799–809.

Carrell, R.W. and Stein, P.E. (1996) The biostructural pathology of the serpins – critical function of sheet opening mechanism. *Biol. Chem. Hoppe Seyler* **377**: 1–17.

Carrell, R.W., Evans, D. and Stein, P. (1991) Structure and function of antithrombin. *Thromb. Haemost.* **65**: 691–691.

Caughey, G.H. (1994) Serine proteinases of mast cell and leukocyte granules. A league of their own. *Am. J. Resp. Crit. Care. Med.* **150**: S138–S142.

Coughlin, P., Sun, J.R., Cerruti, L., Salem, H.H. and Bird, P. (1993) Cloning and molecular characterization of a human intracellular serine proteinase-inhibitor. *Proc. Natl Acad. Sci. USA* **90**: 9417–9421.

Darmon, A.J., Nicholson, D.W. and Bleackley, R.C. (1995) Activation of the apoptotic protease CPP32 by cytotoxic T-cell-derived granzyme B. *Nature* **377**: 446–448.

Darmon, A.J., Ley, T.J., Nicholson, D.W. and Bleackley, R.C. (1996) Cleavage of CPP32 by granzyme B represents a critical role for granzyme B in the induction of target cell DNA fragmentation. *J. Biol. Chem.* **271**: 21709–21712.

Dickinson, J., Norris, B., Jensen, P. and Antalis, T. (1998) The C–D interhelical domain of the serpin plasminogen activator inhibitor-type 2 is required for protection from TNF-alpha induced apoptosis. *Cell Death Differ.* **5**: 163–171.

Ebnet, K., Hausmann, M., Lehmann-Grube, F., Mullbacher, A., Kopf, M., Lamers, M. and Simon, M.M. (1995) Granzyme A-deficient mice retain potent cell-mediated cytotoxicity. *EMBO J.* **14**: 4230–4239.

Froelich, C.J., Orth, K., Turbov, J., Seth, P., Gottlieb, R., Babior, B., Shah, G.M., Bleackley, R.C., Dixit, V.M. and Hanna, W. (1996) New paradigm for lymphocyte granule-mediated cytotoxicity. Target cells bind and internalise granzyme B, but an endosmolytic agent is necessary for cytosolic delivery and subsequent apoptosis. *J. Biol. Chem.* **271**: 29073–29079.

Hampson, I.N., Hampson, L., Pinkoski, M., Cross, M., Heyworth, C.M., Bleackley, R.C., Atkinson, E. and Dexter, T.M. (1997) Identification of a serpin specifically expressed in multipotent and bipotent hematopoietic progenitor cells and in activated T cells. *Blood* **89**: 108–118.

Harris, J.L., Peterson, E.P., Hudig, D., Thornberry, N.A. and Craik, C.S. (1998) Definition and redesign of the extended substrate specificity of granzyme B. *J. Biol. Chem.* **273**: 27364–27373.

Heusel, J.W., Wesselschmidt, R.L., Shresta, S., Russel, J.H. and Ley, T.J. (1994) Cytotoxic lymphocytes require granzyme B for the rapid induction of DNA fragmentation and apoptosis in allogeneic target cells. *Cell* **76**: 977–987.

Hirota, K., Matsui, M., Iwata, S., Nishiyama, A., Mori, K. and Yodoi, J. (1997) AP-1 transcriptional activity is regulated by a direct association between thioredoxin and Ref-1. *Proc. Natl Acad. Sci. USA* **94**: 3633–3638.

Huber, R. and Carrell, R.W. (1989) Implications of the three-dimensional structure of α1-antitrypsin for structure and function of serpins. *Biochemistry* **28**: 8951–8966.

Inglis, J.D. and Hill, R.E. (1991) The murine Spi-2 proteinase inhibitor locus: a multigene family with a hypervariable reactive site domain. *EMBO J.* **10**: 255–261.

Inglis, J.D., Lee, M., Davidson, D.R. and Hill, R.E. (1991) Isolation of two cDNAs encoding novel antichymotrypsin-like proteins in a murine chondrocytic cell line. *Gene* **106**: 213–220.

Jensen, P., Schuler, E., Woodrow, G., Richardson, M., Goss, N., Hojrup, P., Petersen, T. and Rasmussen, L. (1994) A unique interhelical insertion in plasminogen activator inhibitor-2 contains three glutamines, Gln83, Gln84, Gln86, essential for transglutaminase-mediated cross-linking. *J. Biol. Chem.* **269**: 15394–15398.

Kamine, J., Elongovan, B., Subramanian, T., Coleman, D. and Chinnadurai, G. (1996) Identification of a cellular protein that specifically interacts with the essential cysteine region of the HIV-1 Tat transactivator. *Virology* **216**: 367–366.

Komiyama, T., Ray, C.A., Pickup, D.J., Howard, A.D., Thornberry, N.A., Peterson, E.P. and Salveson, G. (1994) Inhibition of interleukin-1β converting enzyme by the cowpox virus serpin *crmA*. *J. Biol. Chem.* **269**: 19331–19337.

Kummer, J., Kamp, A., Citarella, F., Horrevoets, A. and Hack, C. (1996) Expression of human recombinant granzyme A zymogen and its activation by the cysteine proteinase cathepsin C. *J. Biol. Chem.* **271**: 9281–9286.

Lowin, B., Beermann, F., Schmidt, A. and Tschopp, J. (1994) A null mutation in the perforin gene impairs cytolytic T lymphocyte- and natural killer cell-mediated cytotoxicity. *Proc. Natl Acad. Sci. USA* **91**: 11571–11575.

Mahalingham, S., Ayyavoo, V., Patel, M., Kieber-Emmons, T., Kao, G.D., Muschel, R.J. and Weiner, D.B. (1998) HIV-1 Vpr interacts with a human 34-kDa mov34 homologue, a cellular factor linked to the G2/M phase transition of the mammalian cell cycle. *Proc. Natl Acad. Sci. USA* **95**: 3419–3424.

Orth, K., Orourke, K., Salvesen, G.S. and Dixit, V.M. (1996) Molecular ordering of apoptotic mammalian CED-3/ICE-like proteases. *J. Biol. Chem.* **271**: 20977–20980.

Pinkoski, M.J., Winkler, U., Hudig, D. and Bleackley, R.C. (1996) Binding of granzyme B in the nucleus of target cells. Recognition of an 80-kilodalton protein. *J. Biol. Chem.* **271**: 10225–10229.

Remold-O'Donnell, E., Chin, J. and Alberts, M. (1992) Sequence and molecular characterization of human monocyte/neutrophil elastase inhibitor. *Proc. Natl Acad. Sci. USA* **15**: 5635–5639.

Ritzi, M., Baack, M., Musahl, C., Romanowski, P., Laskey, R.A. and Knippers, R. (1998) Human minichromosome maintenance proteins and human origin recognition complex 2 protein on chromatin. *J. Biol. Chem.* **273**: 24543–24549.

Salvesen, G.S. and Dixit, V.M. (1997) Caspases: Intracellular signalling by proteolysis. *Cell* **91**: 443–446.

Schick, C., Kamachi, Y., Bartuski, A.J., Cataltepe, S., Schechter, N.M., Pemberton, P.A. and Silverman, G.A. (1997) Squamous cell carcinoma antigen 2 is a novel serpin that inhibits the chymotrypsin-like proteinases cathepsin G and mast cell chymase. *J. Biol. Chem.* **272**: 1849–1855.

Schick, C., Pemberton, P.A., Shi, G.P., Kamachi, Y., Cataltepe, S., Bartuski, A.J., Gornstein, E.R., Bromme, D., Chapman, H.A. and Silverman, G.A. (1998) Cross-class inhibition of the cysteine proteinases cathepsins K, L, and S by the serpin squamous cell carcinoma antigen 1: A kinetic analysis. *Biochemistry* **37**: 5258–5266.

Scott, F.L., Coughlin, P.B., Bird, C., Cerruti, L., Hayman, J.A. and Bird, P. (1996) Proteinase-inhibitor-6 cannot be secreted, which suggests it is a new-type of cellular serpin. *J. Biol. Chem.* **271**: 1605–1612.

Shi, L., Mai, S., Israels, S., Browne, K., Trapani, J.A. and Greenberg, A.H. (1997) Granzyme B (GraB) autonomously crosses the cell membrane and perforin initiates apoptosis and GraB nuclear localisation. *J. Exp. Med.* **185**: 855–866.

Smyth, M., McGuire, M. and Thia, K. (1995) Expression of recombinant human granzyme B. A processing and activation role for dipeptidyl peptidase I. *J. Immunol.* **154**: 6299–6305.

Stefansson, S. and Lawrence, D. (1996) The serpin PAI-1 inhibits cell migration by blocking integrin alpha V beta 3 binding to vitronectin. *Nature* **383**: 441–443.

Stein, P.E. and Carrell, R.W. (1995) What do dysfunctional serpins tell us about molecular mobility and disease. *Nat. Struct. Biol.* **2**: 96–113.

Sun, J.R., Bird, C.H., Sutton, V., Mcdonald, L., Coughlin, P.B., Dejong, T.A., Trapani, J.A. and Bird, P.I. (1996) A cytosolic granzyme-B inhibitor related to the viral apoptotic regulator cytokine response modifier A is present in cytotoxic lymphocytes. *J. Biol. Chem.* **271**: 27802–27809.

Sun, J., Stephens, R., Mirza, G., Kanai, H., Ragoussis, J. and Bird, P. (1998) A serpin gene cluster on human chromosome 6p25 contains PI6, PI9 and ELANH2 which have a common structure almost identical to the 18q21 ovalbumin serpin genes. *Cytogenet. Cell Genet.* **82**: 273–277.

Talanian, R.V., Yang, X.H., Turbov, J., Seth, P., Ghayur, T., Casiano, C.A., Orth, K. and Froelich, C.J. (1997) Granule-mediated killing: Pathways for granzyme B-initiated apoptosis. *J. Exp. Med.* **186**: 1323–1331.

Tewari, M. and Dixit, V.M. (1995) Fas-induced and tumor necrosis factor-induced apoptosis is inhibited by the poxvirus *crmA* gene-product. *J. Biol. Chem.* **270**: 3255–3260.

Thornberry, N.A., Rano, T.A., Peterson, E.P., Rasper, D.M., Timkey, T., Garcia-Calvo, M., Houtzager, V.M., Nordstrom, P.A., Roy, S., Vaillancourt, J.P., Chapman, K.T. and Nicholson, D.W. (1997) A combinatorial approach defines specificities of members of the caspase family and granzyme B. *J. Biol. Chem.* **272**: 17907–17911.

Trapani, J.A., Jans, D.A., Jans, P.J., Smyth, M.J., Browne, K.A. and Sutton, V.R. (1998) Efficient nuclear targeting of granzyme B and the nuclear consequences of apoptosis induced by granzyme B and perforin are caspase-dependent, but cell death is caspase independent. *J. Biol. Chem.* **273**: 27934–27938.

Walsh, C.M., Matloubian, M., Liu, C.C., Ueda, R., Kurahara, C.G., Christensen, J.L., Huang, M.T., Young, J.D., Ahmed, R. and Clark, W.R. (1994) Immune function in mice lacking the perforin gene. *Proc. Natl Acad. Sci. USA* **92**: 10854–10858.

Whitehead, I., Kirk, H. and Kay, R. (1995) Expression cloning of oncogenes by retroviral transfer of cDNA libraries. *Mol. Cell Biol.* **15**: 704–710.

Williams, M. and Henkart, P. (1994) Apoptotic cell death induced by intracellular proteolysis. *J. Immunol.* **153**: 4247–4255.

Yang, X., Stennicke, H.R., Wang, B., Green, D.R., Janicke, R.U., Srinivasan, A., Seth, P., Salvesen, G.S. and Froelich, C.J. (1998) Granzyme B mimics apical caspases. Description of a unified pathway for the trans-activation of executioner caspase-3 and -7. *J. Biol. Chem.* **273**: 34278–34283.

Zhou, Q., Snipas, S., Orth, K., Muzio, M., Dixit, V.M. and Salvesen, G.S. (1997) Target protease specificity of the viral serpin CrmA – Analysis of five caspases. *J. Biol. Chem.* **272**: 7797–7800.

6

Mitochondria and cell death: a pore way to die?

Andrew P. Halestrap, J. P. Gillespie, Alison O'Toole and Elena Doran

1. Introduction

Mitochondria provide most of the ATP used by the many energy-requiring processes of the cell. In addition, mitochondria are responsible for some key reactions in biosynthetic pathways, including gluconeogenesis and fatty acid synthesis. To fulfil these roles, the mitochondrial inner membrane must present a permeability barrier to protons to maintain a membrane potential and pH gradient that provides the driving force for ATP synthesis through oxidative phosphorylation. Movement of metabolites across the inner membrane is also restricted and occurs only by means of specific transport mechanisms. These allow gradients of metabolites to be maintained between the cytosol and mitochondrial matrix which are essential for metabolic regulation.

If the permeability barrier of the inner membrane is disrupted in some way, mitochondria become uncoupled with dire consequences for the cell. It can neither separate its cytosolic and mitochondrial pools of metabolites nor synthesize ATP by oxidative phosphorylation. Even glycolytically derived ATP is hydrolysed by uncoupled mitochondria, as the ATPase reverses in the absence of a membrane potential or pH gradient. A cell presented with this scenario could no longer maintain its normal function, nor repair any damage that it might suffer: it would be doomed to die. Unlikely as it may seem, mitochondria possess a latent pore within their inner membrane that, when open, does in fact produce such a non-specific increase in permeability. Known as the mitochondrial permeability transition pore (MPTP) this is now thought to play a critical role in both necrotic and apoptotic cell death (Bernardi, 1996; Halestrap *et al.*, 1998; Lemasters *et al.*, 1998).

2. The mitochondrial permeability transition

2.1 *Molecular mechanism*

The phenomenon of the mitochondrial permeability transition (MPT) was first described several decades ago, and refers to the massive swelling of mitochondria that

Programmed Cell Death in Animals and Plants, edited by J.A. Bryant, S.G. Hughes and J.M. Garland.
© 2000 BIOS Scientific Publishers Ltd, Oxford.

occurs when mitochondria are exposed to high calcium concentrations, especially when this is associated with adenine nucleotide depletion, membrane depolarization and oxidative stress (Gunter and Pfeiffer, 1990). Although initially thought to be a result of non-specific damage to the inner membrane by phospholipases, elegant experiments by Haworth and Hunter implied the opening of a non-specific pore that transports any molecule of molecular weight up to 1500 (Haworth and Hunter, 1979; Hunter and Haworth, 1979). The extensive swelling of the mitochondria is the result of the equilibration of all small solutes across the membrane leaving a colloidal osmotic pressure exerted by the matrix proteins.

A major breakthrough elucidating the molecular mechanism of the MPT came in 1988 when Crompton and colleagues (1988) demonstrated that the process could be inhibited specifically by submicromolar concentrations of the immunosuppressive drug, cyclosporin A (CsA). In this laboratory we demonstrated that this effect of CsA was exerted through inhibition of a mitochondrial enzyme, peptidyl-prolyl *cis–trans* isomerase (PPIase), otherwise known as cyclophilin (CyP) because of its CsA binding properties (Connern and Halestrap, 1992; Griffiths and Halestrap, 1991; Halestrap and Davidson, 1990). We have since purified the protein, cloned the corresponding cDNA, and shown it to be a unique nuclear encoded cyclophilin (CyP-D) that is translocated into the mitochondria followed by cleavage of the N-terminal signal sequence (Connern and Halestrap, 1992; Woodfield *et al.*, 1997).

PPIases are involved in changing the shape of proteins by altering the conformation of the peptide bond around proline residues. This provides a clue as to how the MPT pore could be formed; an existing membrane protein might undergo a PPIase-induced conformational change, converting it into a channel. Work in this and other laboratories suggested that this membrane protein might be the adenine nucleotide translocase (ANT) whose normal function is to catalyse the transport of ADP into and ATP out of the mitochondria (Halestrap and Davidson, 1990; Halestrap *et al.*, 1997a). We suggested that under conditions favouring the opening of the MPT, CyP binds to the ANT and when triggered by calcium this causes a conformational change converting the specific transporter into a non-specific pore.

Since we first proposed this molecular mechanism in 1990, the evidence in support of it has accumulated (reviewed in Halestrap *et al.*, 1998) and led to the refined model illustrated in *Figure 1*. The most recent data are perhaps the most convincing, since purified proteins have been used to confirm the specific interaction of CyP-D with the ANT and their ability to reconstitute an MPTP-like pore in proteoliposomes. Brustovetsky and Klingenberg (1996) reported that the purified ANT, reconstituted into proteoliposomes, could become a non-specific high conductance pathway at high (>1 mM) concentrations of calcium (Ca^{2+}) ions but this was not inhibited by CsA. Brdiczka's laboratory demonstrated that such proteoliposomes containing purified ANT also become permeable to oxaloacetate at high (> 1 mM) calcium ion concentrations [Ca^{2+}] (Rück *et al.*, 1998). In earlier experiments these workers had also shown that detergent-solubilized brain extracts could be reconstituted into proteoliposomes to produce MPTP-like pores that opened at micromolar [Ca^{2+}] in a CsA-inhibitable manner (Beutner *et al.*, 1996). These proteoliposomes contained many proteins, including ANT, porin, hexokinase and CyP-D.

Data from this laboratory had demonstrated that CyP-D binds to the inner mitochondrial membrane in a CsA-inhibitable manner and that this binding is increased by oxidative stress (Connern and Halestrap, 1994, 1996). In order to confirm that it was

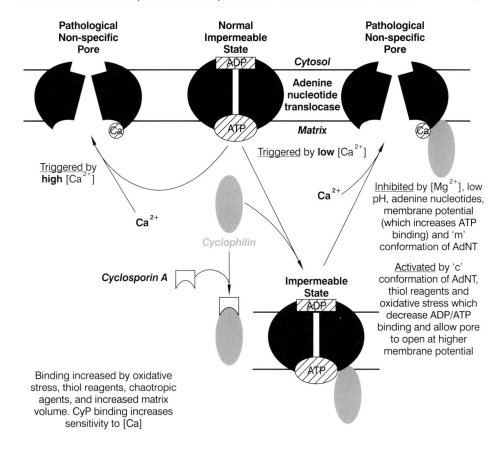

Figure 1. Proposed model for the mechanism of action of the mitochondrial permeability transition pore.

the ANT in the membrane that was responsible for the CyP-D binding, we over-expressed CyP-D as a glutathione-S-transferase (GST) fusion protein and immobilized this on a glutathione sepharose matrix (Woodfield *et al.*, 1998). Triton-solubilized inner mitochondrial membranes or purified ANT were added to this affinity matrix and all but the most tightly bound proteins removed by extensive washing. The remaining bound protein was eluted using glutathione to displace the GST-CyP-D and proteins analysed by SDS-PAGE (sodium dodecyl sulphate polyacrylamide gel electrophorosis) and western blotting with anti-ANT antibodies. Only one membrane protein was found to be bound under these conditions, and this was the ANT. Binding was greatly reduced by prior treatment of the affinity column with CsA. Crompton and colleagues have reported similar results (Crompton *et al.*, 1998), although their use of a different detergent led to the binding of the outer membrane protein porin as well as the ANT. It is well established that the ANT and porin can form a very tight complex under some conditions. Surprisingly, under these conditions, CsA did not appear able to prevent the ANT binding to the GST–CyP-D. However, these workers were able to reconstitute their eluted protein into proteoliposomes and demonstrate a calcium activated, CsA-inhibited increase in permeability to fluorescein.

These data all confirm a critical role for the ANT and CyP-D in the formation of the MPTP, but do not exclude the possibility of porin also being involved. Arguing against this, opening of the MPTP can be demonstrated in mitoplasts stripped of their outer membrane (Halestrap *et al.*, 1997a) which suggests that porin may not be essential for formation of the MPTP. We have very recently confirmed this directly by reconstituting the purified ANT (totally devoid of porin) in proteoliposomes and adding purified CyP-D. Whereas in the absence of CyP-D, 100 μM Ca^{2+} caused little increase in permeability to solutes of 100–700 daltons, in the presence of CyP-D opening permeability was induced at calcium ion concentrations below 1.0 μM. The effects of CyP-D were abolished by prior incubations with CsA. It seems clear, therefore, that the MPTP can be reconstituted from the ANT and CyP-D alone as originally proposed. However, this does not exclude a regulatory role for porin (and possibly other outer mitochondrial membrane components) through an interaction with the ANT at the contact sites between the inner and outer membrane.

2.2 *Potential role in necrotic cell death*

When a cell is exposed to some form of insult such as ischaemia, osmotic shock or a chemical toxin, damage occurs as a result of the activation of a range of degradative enzymes including phospholipases, nucleases and proteases. If left unopposed these enzymes would cause breakdown of intracellular components and ultimately the rupture of the plasma membrane that is characteristic of necrosis. Repair processes may be capable of reversing this damage provided that ATP concentrations can be maintained to drive the necessary biosynthetic and salvage pathways involved. It is mitochondria that supply this ATP and provided they continue to do so, necrotic cell death may be prevented. However, if the MPTP opens this will not be possible and necrotic cell death is inevitable. There is now convincing evidence that this sequence of events does occur in tissues subject to ischaemia followed by reperfusion and when cells are exposed to chemical toxins or other stimuli that cause calcium overload and oxidative stress. In this laboratory we have made extensive studies on reperfusion injury in the heart (Halestrap *et al.*, 1998). For information on the role of the MPTP in cells subjected to other stresses the reader is referred to Lemasters *et al.* (1998).

2.3 *Measurement of MPTP opening in situ*

Isolated cells. In order to prove the prediction that necrotic cell death involves the MPT, a direct measurement of the MPT within intact cells is required. With isolated cells (freshly prepared from tissues or maintained in culture) some workers have used dissipation of mitochondrial membrane potential measured with fluorescent dyes as an indicator of pore opening (Kroemer *et al.*, 1998; Lemasters *et al.*, 1998). However, this technique does not discriminate between opening of the MPTP and other means of mitochondrial uncoupling; as such it must be treated with caution. It is more convincing if it can also be shown that specific inhibitors of the MPTP prevent dissipation of the membrane potential.

A more rigorous measurement of MPTP opening can be obtained using confocal fluorescence microscopy with a fluorescent molecule normally unable to permeate the inner mitochondrial membrane, but which can permeate through the MPTP. Lemasters and colleagues have employed the green fluorescent dye, calcein, for this purpose (Lemasters et al., 1997, 1998; Nieminen et al., 1995; Qian et al., 1997). Confocal microscopy reveals that with appropriate loading conditions, the dye remains confined to the cytosol with mitochondria showing up as dark spheres against a green fluorescent background (Lemasters et al., 1997, 1998). However, under conditions which stimulate opening of the MPTP the dye enters mitochondria and the 'black holes' are lost. An advantage of the green fluorescence of calcein is that it can be used in conjunction with tetramethylrhodamine (TMRH), enabling coincident measurement of mitochondrial membrane potential. As expected, upon opening of the MPTP, calcein is found to enter the mitochondria at the same time as TMRH is lost.

Experiments such as these have confirmed that the MPT does occur in isolated cells subjected to a variety of stresses that induce necrosis. Furthermore, low pH and drugs such as CsA that inhibit the MPT, protect cells from necrotic death under these conditions. Elegant though this technique is, it is highly dependent on calcein being excluded from the mitochondria during dye loading. Other workers have found this is often not possible. Indeed, it has been possible to visualize mitochondrial calcein preferentially, by quenching cytosolic dye with cobalt, and then demonstrating opening of the MPTP by loss of calcein from the mitochondria (Petronilli et al., 1999).

Perfused tissues. Direct measurement of the MPT in whole tissues is more difficult, but we have devised a technique, sometimes known as the 'hot-dog' technique, that can be applied to perfused organs (Griffiths and Halestrap, 1995; Halestrap et al., 1997b; Kerr et al., 1999). The basis of this technique is that the radioactive glucose analogue, [³H]-2-deoxyglucose ([³H]-DOG), is readily taken up into cells and metabolized to DOG-6-phosphate (DOG-6P) which remains trapped in the cytosol. It cannot enter the mitochondria unless the MPTP opens. Thus isolation of mitochondria from tissues treated in this way and determination of their [³H]-DOG-6P content provides an indication of how many mitochondria have undergone the MPT.

2.4 *Reperfusion injury of the heart*

Reperfusion of a tissue following a period of ischaemia (total or partial loss of blood flow) often causes more damage than the ischaemia itself. This is reflected in loss of intracellular enzymes, and is a phenomenon with serious clinical implications (Halestrap et al., 1998; Lemasters and Thurman, 1995; Reimer and Jennings, 1992). For example, after a heart attack, attempts may be made to re-establish blood flow to an area of the heart blocked by a clot or atherosclerotic plaque, whilst in open heart surgery blood flow to the heart is totally stopped and must be restarted after surgery. There is good reason to suspect that the opening of the MPTP during reperfusion may be a major factor in reperfusion injury (reviewed in Crompton, 1990; Halestrap et al., 1998). During ischaemia the heart attempts to maintain its ATP level using glycolysis. Although this may be sufficient to maintain ionic homeostasis, it cannot keep pace with the ATP demands of contraction and the heart rapidly stops beating as ATP concentrations fall. Although glycolysis remains stimulated initially, it eventually becomes inhibited as intracellular pH (pH_i) decreases in response to the accumulation of lactic acid that cannot be removed from the tissue. Thus intracellular ATP concentrations dwindle further and the heart goes into contracture.

In parallel with this, the Na^+/H^+ antiporter is activated in an attempt to restore pH_i (Lazdunski et al., 1985) and this increases cellular sodium ion concentration $[Na^+]$.

However, without sufficient ATP supply, this Na^+ cannot be pumped out of the cell by the Na^+/K^+ ATPase and it accumulates, preventing the Na^+/Ca^{2+} antiporter from pumping Ca^{2+} out of the cell. Indeed, this process may actually reverse and allow additional Ca^{2+} to enter the cytosol from the plasma, some of which may enter the mitochondria by reversal of its Na^+/Ca^{2+} antiporter. Upon reperfusion further Ca^{2+} is rapidly taken up into the mitochondria from the cytosol by means of the uniporter. This overloads the mitochondrial matrix with large amounts of Ca^{2+} (Miyata et al., 1992) which alone might not be sufficient to open the MPTP. However, the depletion of adenine nucleotides that occurs during ischaemia will sensitize MPTP-opening to $[Ca^{2+}]$, as will the burst of oxygen free radical production that accompanies the sudden influx of oxygen into an anoxic cell. The latter is caused by an interaction of oxygen with ubisemiquinone that is formed during anoxia as a result of respiratory chain inhibition. Even these stresses might not open the MPTP if the low pH of ischaemia were maintained, but during the first 2 minutes of reperfusion the pH_i returns to preischaemic values, enabling the other effectors of the MPTP to exert their full effect.

We have used the mitochondrial DOG-entrapment technique to confirm that the MPT does not occur during the ischaemic period itself, but opens between 2 and 5 minutes after the start of reperfusion (Griffiths and Halestrap, 1995; Halestrap et al., 1997b). Furthermore, we were able to show that the degree to which the heart recovers from ischaemia correlates well with the amount of pore opening (Kerr et al., 1999). Thus a critical role of the MPTP in reperfusion injury would appear likely.

Protection of hearts from reperfusion injury using inhibitors of the MPT. If the MPT is a key event in reperfusion injury of the heart, it would be predicted that inhibition of this process should be protective and this is found to be the case. We and others have demonstrated that CsA protects hearts from reperfusion injury (Duchen et al., 1993; Griffiths and Halestrap, 1993; Nazareth et al., 1991), and others have shown similar results in a range of different tissues (Konukoglu et al., 1998; LeDucq et al., 1998; Shiga et al., 1992; Uchino et al., 1998; Ysebaert et al., 1997). Furthermore, only analogues of CsA that inhibit the MPTP are effective at protecting the heart (Griffiths and Halestrap, 1995). However, CsA shows a narrow concentration range over which it is effective, and has side-effects which make it unsuitable for routine use. These probably relate to the action of CsA (via its complex with cytosolic cyclophilin-A) to inhibit calcineurin, a calcium-dependent protein phosphatase (Schreiber and Crabtree, 1992). For this reason we have investigated other means of inhibiting the MPT. Most successful has been the addition of pyruvate to the perfusion medium prior to ischaemia and during reperfusion; this greatly inhibits opening of the MPTP (measured using DOG entrapment) and improves heart recovery (Kerr et al., 1999).

In these studies an important observation was made – the MPTP can close again after it has opened. This was demonstrated by comparing the amount of mitochondrial DOG entrapment determined after loading hearts with DOG before ischaemia, with that determined when loading was performed after the hearts had reached maximal recovery during reperfusion. In the absence of pyruvate, hearts subjected to 40 minutes of ischaemia recovered about 50% of their left ventricular developed pressure (LVDP) during reperfusion, and their mitochondrial DOG entrapment (loaded after 25 minutes of reperfusion) remained elevated at about 50% of that observed during the initial phase of reperfusion (loaded with DOG before ischaemia). However, in the presence of 10 mM pyruvate, hearts recovered 100% of their function, with DOG entrapment in the initial phase of reperfusion being only 50% of that found in the absence of pyruvate and returning to control values when loaded after 25 minutes of reperfusion. Thus it appears that full recovery of the heart can occur even if some MPTP opening occurs initially, but then resealing follows rapidly. Three mechanisms may be responsible for the protective effect of pyruvate: (1) it increases lactic acid production and thus maintains a lower intracellular pH, (2) it acts as a scavenger of oxygen free radicals, and (3) it is an excellent respiratory fuel to energize mitochondria and so fuel ATP production (Kerr et al., 1999).

Propofol is an anaesthetic that is frequently used during cardiac surgery and in post-operative sedation (Bryson et al., 1995). It has been shown to protect hearts from the injury caused by hydrogen peroxide-induced oxidative stress (Kokita and Hara, 1996) or reperfusion injury (Ko et al., 1997; Kokita et al., 1998) perhaps by acting as a free radical scavenger (Eriksson et al., 1992; Green et al., 1994; Murphy et al., 1992) or through inhibition of plasma membrane calcium channels (Li et al., 1997). These two effects would lead to a decrease in oxidative stress and cytosolic $[Ca^{2+}]$ both of which should protect mitochondria from MPTP opening. Furthermore, there are reports that propofol can inhibit the MPT in isolated mitochondria (Eriksson, 1991; Sztark et al., 1995), although the concentrations used in these studies were considerably greater than those employed clinically. We have confirmed the protective effects of propofol against reperfusion injury in the Langendorff perfused heart at concentrations of 2 µg ml^{-1}, lower than that employed by others and more typical of concentrations employed in clinical anaesthesia (Bryson et al., 1995; Cockshott, 1985; Servin et al., 1988). The recovery of propofol-treated hearts after 30 minutes of ischaemia was significantly improved with the LVDP expressed as a percentage of the pre-ischaemic value (mean ± SEM) increasing from 36 ± 8% ($n = 10$) in the absence of propofol to 70 ± 11% ($n = 8; p < 0.05$) in its presence. This improvement was accompanied by a 25% decrease in mitochondrial entrapment of preloaded [^{3}H]-DOG (Javador et al., 1999) Whatever the mechanism, propofol provides another example of a reagent whose protection of the heart from reperfusion injury is accompanied by a decrease in MPTP opening in vivo. These data suggest that propofol and pyruvate may be useful adjuncts to the cardioplegic solutions used in cardiac surgery.

3. Mitochondria and apoptosis

The central role of mitochondria in necrotic cell death seems well established, but a surprising and very exciting development in the last few years is the discovery of their role in apoptosis (reviewed in Green and Reed, 1998; Kroemer et al., 1998; see also Chapters 7 and 8). The literature in this field is increasing at an exponential rate and some of it is conflicting. However, what is now firmly established is that mitochondria release cytochrome c, and possibly other factors (Susin et al., 1999), into the cytosol from the intermembrane space in response to a wide range of apoptotic stimuli. This process is inhibited by the anti-apoptotic protein Bcl-2 and related family members such as Bcl-X_L that are associated with the outer mitochondrial membrane (Adams and Cory, 1998). Other Bcl-2 family members such as BAD and BAX are pro-apoptotic and enhance cytochrome c release. In the presence of dATP, the released cytochrome c binds with APAF-1 and pro-caspase 9 to form a complex that leads to the proteolytic activation of pro-caspase 9 to active caspase 9 (see Chapters 1 and 2). This then cleaves pro-caspase 3 to caspase 3, which directly or indirectly activates a range of enzymes critical for the rearrangement of the nucleus, cytoskeleton and plasma membrane that are characteristic of apoptosis (Thornberry and Lazebnik, 1998).

3.1 Mechanisms involved in cytochrome c release

The MPTP. Since the MPT produces swelling of mitochondria and rupture of the outer mitochondrial membrane, it will cause cytochrome c release. Although recently redis-covered (Petit *et al.*, 1998; Yang and Cortopassi, 1998) this is actually an old obser-vation (see Halestrap, 1982). It may seem an attractive hypothesis that outer membrane rupture in response to the MPT is the normal mechanism for cytochrome c release in apoptosis. The group of Kroemer (1998), in particular, has proposed that the MPT is an early and critical event in apoptosis induced by a wide range of stimuli. The main evidence in favour of this is the detection of an early drop in mitochondrial membrane potential that precedes release of mitochondrial cytochrome c. Furthermore, some workers have reported that mitochondria in cells that are expressing increased levels of the anti-apoptotic protein Bcl-2 are less prone to opening of the MPTP (Kroemer *et al.*, 1998). In addition, other studies have shown that inhibitors of the MPT such as CsA and bongkrekic acid can inhibit apoptosis whilst activators of the MPT such as oxidative stress and atractyloside may induce apoptosis. However, these reagents cannot be regarded as specific modulators of the MPT. For example, atractyloside is used at concentrations three orders of magnitude greater than required to activate the MPT of isolated mitochondria. Furthermore, bongkrekic acid will block mitochondrial ATP/ADP transport irrespective of any effect on the MPT. This will prevent hydrolysis of glycolytically derived ATP by damaged mitochondria which may act as an alternative mechanism for protecting cells from death. Yet another contrary observation is that CsA does not inhibit apoptosis in many cells; indeed, it may even induce apoptosis (e.g. McDonald *et al.*, 1996).

   We have investigated the effects of CsA on the induction of apoptosis by TNF-α in the murine fibrosarcoma cell line WEHI-164. Whilst a degree of protection was observed in response to 0.2 μM CsA in combination with trifluoperazine (5 μM), the immunosuppressant FK506 (0.02 μM) also conferred protection. This suggests that a calmodulin/calcineurin-dependent step may be the locus of action of CsA rather than the MPT. Lemasters has demonstrated that TNF-α induces the MPT in rat hepato-cytes expressing an IκB suppressor. However, the effect is relatively late (nine hours) and appears to occur upstream of cytochrome c release and caspase 3 activation (Bradham *et al.*, 1998). In contrast, our own studies suggest that caspase 3 activation and changes in cell surface are occurring much earlier than this implying that the MPT is a late downstream event. In fact many studies (e.g. Garland and Halestrap, 1997; Garland *et al.*, 1997; Kluck *et al.*, 1997; Yang *et al.*, 1997) have failed to show any decrease in mitochondrial membrane potential or mitochondrial swelling associated with cytochrome c release in response to a range of apoptotic stimuli. It seems possible that opening of the MPTP and consequent decrease in membrane potential may reflect secondary necrosis which would not normally occur *in vivo*. Here changes in the cell surface would enable macrophages to engulf apoptotic cells/bodies before mitochon-drial damage and secondary necrosis occur. Indeed, if the MPT were to occur early in apoptosis, mitochondria would actively hydrolyse ATP rather than synthesize it. Yet apoptosis is an active process that requires cellular ATP levels to be maintained; if ATP levels fall the cell death pathway may change from apoptotic to necrotic (Leist and Nicotera, 1997).

Nevertheless, circumstances can be envisaged where apoptosis might be initiated by opening of the MPTP. A severe insult will cause widespread opening of the MPTP in mitochondria with substantial swelling, outer membrane rupture and cytochrome c release. Although this will initiate the apoptotic cascade, death will still be necrotic because the mitochondria remain 'open' causing inhibition of ATP synthesis and activation of the hydrolysis of glycolytically derived ATP. However, if the initial insult to a cell is not too severe and relatively few mitochondria undergo the MPT, or those that do reseal, cellular ATP concentrations can be maintained. Under these conditions, cytochrome c release will cause activation of the caspase cascade and induction of apoptosis. In this scenario, illustrated in *Figure 2*, mitochondria determine not only whether a cell lives or dies, but the route of cell death taken. This provides an explanation for the co-existence of apoptotic and necrotic cell death within the same tissue. For example, in the ischaemic area that is produced in a myocardial infarction, the centre of the infarct shows necrotic cell death (severe insult), whereas the peripheral area exhibits apoptosis (Bartling *et al.*, 1998; Fliss and Gattinger, 1996). Another example of transient MPTP opening that leads to apoptosis is the death of hippocampal neurons that occurs several days after a 30-minute insulin-induced hypoglycaemic insult. Protection is afforded by

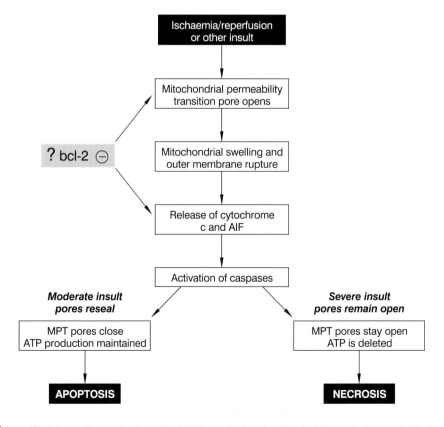

Figure 2. Scheme illustrating how the MPT may be involved in deciding whether a cell dies by necrosis or apoptosis

treatment with CsA (but not FK506) during the hypoglycaemia and this also prevents the swelling of mitochondria observed in apoptotic cells (Friberg *et al.*, 1998). Interestingly, mitochondria isolated from the hippocampus are more sensitive to calcium-induced opening of the MPTP than mitochondria from cortex or cerebellum, and this correlated with a lower adenine nucleotide content (Friberg *et al.*, 1999).

MPT-independent mechanisms. There is now increasing evidence that the pro-apoptotic members of the Bcl-2 family, such as BAX, BAD and most recently BID can stimulate cytochrome c release from mitochondria without any involvement of the MPT (i.e. no uncoupling or mitochondrial swelling) (Jurgensmeier *et al.*, 1998; Li *et al.*, 1998; Luo *et al.*, 1998; Manon *et al.*, 1997). It is unclear how these proteins interact with the anti-apoptic members of the family such as Bcl-2 and Bcl-X$_L$ in order to maintain the required balance between cell survival and proliferation on the one hand and apoptosis on the other. Proteolytic modifications and phosphorylations of the different Bcl-2 family members have been suggested to provide the link between cell surface receptors and stimulation of cytochrome c release (Adams and Cory, 1998). Very recently a more coherent picture has begun to emerge. It is well established that activation of members of the TNF receptor family leads to death domain-dependent activation of caspase 8 (Ashkenazi and Dixit, 1998). It has now been shown that caspase 8 can proteolytically cleave a BID precursor to form active BID which migrates to the mitochondria and causes cytochrome c release. This then activates caspase 9 and hence caspase 3 which then sets in motion the apoptotic cascade (Li *et al.*, 1998; Luo *et al.*, 1998).

These data explain why caspase 8 activation of caspase 3 in cell extracts is greatly enhanced by the presence of mitochondria (Kuwana *et al.*, 1998). What is still missing from the picture is how BID (and BAX and BAD) causes the release of cytochrome c from the intermembrane space (Green and Reed, 1998). Two possibilities have been proposed; one is outer membrane rupture (Marzo *et al.*, 1998a, 1998b; VanderHeiden *et al.*, 1997) and the other is that BAX, alone or in combination with other proteins, may form a specific pathway (channel) for cytochrome c transport (Eskes *et al.*, 1998). The former mechanism would be associated with release of other intermembrane proteins such as adenylate kinase (AK), but no data have been published that address this possibility.

In this laboratory we have demonstrated that isolated rat liver mitochondria contain some BAX dimer. When these mitochondria were incubated in KCl medium under energized conditions, a loss of cytochrome c to the medium was observed over several minutes. This release was not associated with mitochondrial swelling nor with release of AK (measured enzymically). It was not prevented by CsA, confirming that the MPT was not involved, but could be greatly reduced by increasing the colloidal osmotic pressure of the incubation medium with 5% dextran. This contrasts with the effects of increasing the matrix volume using hypo-osmotic media, where a 50% increase in matrix volume caused substantial cytochrome c release, associated with extensive AK release whether or not dextran was present. These data are illustrated in *Figure 3*. The effect of increasing the colloidal osmotic pressure with Dextran is to bring the inner and outer mitochondrial membrane in closer proximity, and to increase the number of contact sites (Wicker *et al.*, 1993). This suggests that disruption of an interaction between the inner and outer membrane, perhaps at the contact sites, may open up a permeability pathway to cytochrome c. In this context it is of interest that Bcl-2 is preferentially located at contact sites where porin and the outer membrane benzodiazepine receptor (Benzd-R) are also concentrated (Dejong *et al.*, 1994;

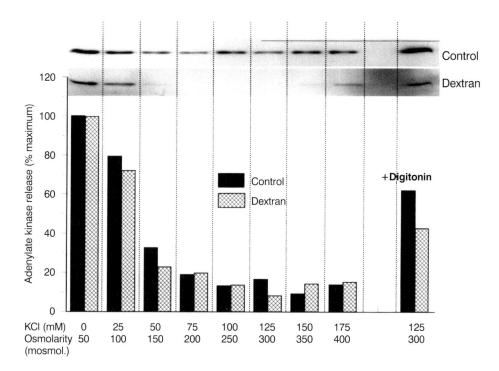

Figure 3. The release of cytochrome c by isolated rat liver mitochondria. Isolated rat liver mitochondria (1 mg protein ml^{-1}) were incubated for 1 min at 25°C in medium containing 20 mM Tris, 10 mM MOPS, 2 mM EGTA, 0.2 µM rotenone, 1 mM succinate, pH 7.4 with additional KCl at the concentration indicated. Where indicated dextran was present at 5% (w/v). After incubation mitochondria were placed on ice and rapidly sedimented by centrifugation. Cytochrome c content of the supernatant was determined using SDS-PAGE and western blotting whilst the adenylate kinase activity was assayed spectrophotometrically.

Kroemer *et al.*, 1998). We have found that polyanions (which bind to porin and force it into its closed state) increase cytochrome c release, whilst neither protoporphyrin IX and diazepam, which bind to the Benzd-R, have any effect. Thus, we propose that it is at the contact sites that pro- and anti-apoptotic members of the Bcl-2 family exert their effects, perhaps through some interaction with porin.

Acknowledgements

This work was supported by project grants from the Medical Research Council and the British Heart Foundation, a Wellcome Trust Fellowship to Alison O'Toole and a Royal Society/NATO Fellowship to Elena Doran.

References

Adams, J.M. and Cory, S. (1998) The Bcl-2 protein family: Arbiters of cell survival. *Science* **281**: 1322–1326.

Ashkenazi, A. and Dixit, V.M. (1998) Death receptors: Signaling and modulation. *Science* **281**: 1305–1308.

Bartling, B., Holtz, J. and Darmer, D. (1998) Contribution of myocyte apoptosis to myocardial infarction? *Basic Res. Cardiol.* **93**: 71–84.

Bernardi, P. (1996) The permeability transition pore. Control points of a cyclosporin A-sensitive mitochondrial channel involved in cell death. *Biochim. Biophys. Acta* **1275**: 5–9.

Beutner, G., Rück, A., Riede, B., Welte, W. and Brdiczka, D. (1996) Complexes between kinases, mitochondrial porin and adenylate translocator in rat brain resemble the permeability transition pore. *FEBS Lett.* **396**: 189–195.

Bradham, C.A., Qian, T., Streetz, K., Trautwein, C., Brenner, D.A. and Lemasters, J.J. (1998) The mitochondrial permeability transition is required for tumor necrosis factor alpha-mediated apoptosis and cytochrome c release. *Mol. Cell. Biol.* **18**: 6353–6364.

Brustovetsky, N. and Klingenberg, M. (1996) Mitochondrial ADP/ATP carrier can be reversibly converted into a large channel by Ca^{2+}. *Biochemistry* **35**: 8483–8488.

Bryson, H.M., Fulton, B.R. and Faulds, D. (1995) Propofol – an update of its use in anesthesia and conscious sedation. *Drugs* **50**: 513–559.

Cockshott, I.D. (1985) Propofol ('Diprivan') pharmacokinetics and metabolism – an overview. *Postgrad. Med. J.* **61**(suppl. 3): 45–50.

Connern, C.P. and Halestrap, A.P. (1992) Purification and N-terminal sequencing of peptidyl-prolyl cis-trans-isomerase from rat liver mitochondrial matrix reveals the existence of a distinct mitochondrial cyclophilin. *Biochem. J.* **284**: 381–385.

Connern, C.P. and Halestrap, A.P. (1994) Recruitment of mitochondrial cyclophilin to the mitochondrial inner membrane under conditions of oxidative stress that enhance the opening of a calcium-sensitive non-specific channel. *Biochem. J.* **302**: 321–324.

Connern, C.P. and Halestrap, A.P. (1996) Chaotropic agents and increased matrix volume enhance binding of mitochondrial cyclophilin to the inner mitochondrial membrane and sensitize the mitochondrial permeability transition to $[Ca^{2+}]$. *Biochemistry* **35**: 8172–8180.

Crompton, M. (1990) The role of Ca^{2+} in the function and dysfunction of heart mitochondria. In: *Calcium and the Heart* (ed. G.A. Langer), Raven Press, New York. pp. 167–198.

Crompton, M., Ellinger, H. and Costi, A. (1988) Inhibition by cyclosporin A of a Ca^{2+}-dependent pore in heart mitochondria activated by inorganic phosphate and oxidative stress. *Biochem. J.* **255**: 357–360.

Crompton, M., Virji, S., Ward, J.M. (1998) Cyclophilin-D binds strongly to complexes of the voltage-dependent anion channel and the adenine nucleotide translocase to form the permeability transition pore. *Eur. J. Biochem.* **258**: 729–735.

Dejong, D., Prins, F.A., Mason, D.Y., Reed, J.C., Vanommen, G.B. and Kluin, P.M. (1994) Subcellular localization of the Bcl-2 protein in malignant and normal lymphoid cells. *Cancer Res.* **54**: 256–260.

Duchen, M.R., McGuinness, O., Brown, L.A. and Crompton, M. (1993) On the involvement of a cyclosporin-a sensitive mitochondrial pore in myocardial reperfusion injury. *Cardiovasc. Res.* **27**: 1790–1794.

Eriksson, O. (1991) Effects of the general anaesthetic propofol on the Ca^{2+}-induced permeabilization of rat liver mitochondria. *FEBS Lett.* **279**: 45–48.

Eriksson, O., Pollesello, P. and Saris, N.E. (1992) Inhibition of lipid peroxidation in isolated rat liver mitochondria by the general anaesthetic propofol. *Biochem. Pharmacol.* **44**: 391–393.

Eskes, R., Antonsson, B., OsenSand, A., Montessuit, S., Richter, C., Sadoul, R., Mazzei, G., Nichols, A. and Martinou, J.C. (1998) Bax-induced cytochrome C release from mitochondria is independent of the permeability transition pore but highly dependent on Mg^{2+} ions. *J. Cell Biol.* **143**: 217–224.

Fliss, H. and Gattinger, D. (1996) Apoptosis in ischemic and reperfused rat myocardium. *Circ. Res.* **79**: 949–956.

Friberg, H., FerrandDrake, M., Bengtsson, F., Halestrap, A.P. and Wieloch, T. (1998) Cyclosporin A, but not FK 506, protects mitochondria and neurons against hypoglycemic damage and implicates the mitochondrial permeability transition in cell death. *J. Neurosci.* **18**: 5151–5159.

Friberg, H., Connern, C., Halestrap, A.P. and Wieloch, T. (1999) Differences in the activation of the mitochondrial permeability transition among brain regions in the rat correlates with selective vulnerability. *J. Neurochem.*, **72**: 2488–2497.

Garland, J.M. and Halestrap, A. (1997) Energy metabolism during apoptosis – bcl-2 promotes survival in hematopoietic cells induced to apoptose by growth factor withdrawal by stabilizing a form of metabolic arrest. *J. Biol. Chem.* **272**: 4680–4688.

Garland, J.M., Sondergaard, K.L. and Jolly, J. (1997) Redox regulation of apoptosis in interleukin-3-dependent haemopoietic cells: absence of alteration in both mitochondrial membrane potential (Delta psi(m)) and free radical production during apoptosis induced by IL3 withdrawal. *Br. J. Haematol.* **99**: 756–765.

Green, D.R. and Reed, J.C. (1998) Mitochondria and apoptosis. *Science* **281**: 1309–1312.

Green, T.R., Bennett, S.R. and Nelson, V.M. (1994) Specificity and properties of propofol as an antioxidant free radical scavenger. *Toxicol. Appl. Pharmacol.* **129**: 163–169.

Griffiths, E.J. and Halestrap, A.P. (1991) Further evidence that cyclosporin-a protects mitochondria from calcium overload by inhibiting a matrix peptidyl-prolyl cis-trans isomerase – implications for the immunosuppressive and toxic effects of cyclosporin. *Biochem. J.* **274**: 611–614.

Griffiths, E.J. and Halestrap, A.P. (1993) Protection by cyclosporin A of ischemia reperfusion-induced damage in isolated rat hearts. *J. Mol. Cell. Cardiol.* **25**: 1461–1469.

Griffiths, E.J. and Halestrap, A.P. (1995) Mitochondrial non-specific pores remain closed during cardiac ischaemia, but open upon reperfusion. *Biochem. J.* **307**: 93–98.

Gunter, T.E. and Pfeiffer, D.R. (1990) Mechanisms by which mitochondria transport calcium. *Am. J. Physiol.* **258**: C755–C786.

Halestrap, A.P. (1982) The nature of the stimulation of the respiratory chain of rat liver mitochondria by glucagon pretreatment of animals. *Biochem. J.* **204**: 37–47.

Halestrap, A.P. and Davidson, A.M. (1990) Inhibition of Ca^{2+}-induced large amplitude swelling of liver and heart mitochondria by Cyclosporin A is probably caused by the inhibitor binding to mitochondrial matrix peptidyl-prolyl cis-trans isomerase and preventing it interacting with the adenine nucleotide translocase. *Biochem. J.* **268**: 153–160.

Halestrap, A.P., Woodfield, K.Y. and Connern, C.P. (1997a) Oxidative stress, thiol reagents, and membrane potential modulate the mitochondrial permeability transition by affecting nucleotide binding to the adenine nucleotide translocase. *J. Biol. Chem.* **272**: 3346–3354.

Halestrap, A.P., Connern, C.P., Griffiths, E.J. and Kerr, P.M. (1997b) Cyclosporin A binding to mitochondrial cyclophilin inhibits the permeability transition pore and protects hearts from ischaemia/reperfusion injury. *Mol. Cell. Biochem.* **174**: 167–172.

Halestrap, A.P., Kerr, P.M., Javadov, S. and Woodfield, K.Y. (1998) Elucidating the molecular mechanism of the permeability transition pore and its role in reperfusion injury of the heart. *Biochim. Biophys. Acta* **1366**: 79–94.

Haworth, R.A. and Hunter, D.S. (1979) The Ca^{2+}-induced membrane transition in mitochondria. II. Nature of the Ca^{2+} trigger site. *Arch. Biochem. Biophys.* **195**: 460–467.

Hunter, D.R. and Haworth, R.A. (1979) The Ca^{2+}-induced membrane transition in mitochondria. I. The protective mechanisms. *Arch. Biochem. Biophys.* **195**: 453–459.

Javadov, S.A., Lim, K.H.H., Kerr, P.M., Suleiman, S-M., Angelini, G.D. and Halestrap, A.P. (1999) Protection of hearts from reperfusion injury by propofol is associated with inhibition of the mitochondrial permeability transition. *Cardiovasc. Res.*, in press.

Jurgensmeier, J.M., Xie, Z.H., Deveraux, Q., Ellerby, L., Bredesen, D. and Reed, J.C. (1998) Bax directly induces release of cytochrome c from isolated mitochondria. *Proc. Natl Acad. Sci. USA* **95**: 4997–5002.

Kerr, P.M., Suleiman, M.-S. and Halestrap, A.P. (1999) Recovery of rat hearts reperfused after a period of ischemia is accompanied by reversal of the mitochondrial permeability transition and is enhanced by pyruvate. *Am. J. Physiol.* **276**: H496–H502.

Kluck, R.M., BossyWetzel, E., Green, D.R. and Newmeyer, D.D. (1997) The release of cytochrome c from mitochondria: A primary site for Bcl-2 regulation of apoptosis. *Science* **275**: 1132–1136.

Ko, S.H., Yu, C.W., Choe, H., Chung, M.J., Kwak, Y.G., Chae, S.W. and Song, H.S. (1997) Propofol attenuates ischaemic-reperfusion injury in the isolated rat heart. *Anesth. Analg.* **85**: 719–724.

Kokita, N. and Hara, A. (1996) Propofol attenuates hydrogen-peroxide induced mechanical and metabolic derangements in the isolated rat heart. *Anesthesiol.* **84**: 117–127.

Kokita, N., Hara, A., Abiko, Y., Arakawa, J., Hashizume, H. and Namiki, A. (1998) Propofol improves functional and metabolic recovery in ischemic reperfused isolated rat hearts. *Anesth. Analg.* **86**: 252–258.

Konukoglu, D., Tasci, I. and Cetinkale, O. (1998) Effects of cyclosporin A and ibuprofen on liver ischemia-reperfusion injury in the rat. *Clin. Chim. Acta.* **275**: 1–8.

Kroemer, G., Dallaporta, B. and RescheRigon, M. (1998) The mitochondrial death/life regulator in apoptosis and necrosis. *Annu. Rev. Physiol.* **60**: 619–642.

Kuwana, T., Smith, J.J., Muzio, M., Dixit, V., Newmeyer, D.D. and Kornbluth, S. (1998) Apoptosis induction by caspase-8 is amplified through the mitochondrial release of cytochrome c. *J. Biol. Chem.* **273**: 16589–16594.

Lazdunski, M., Frelin, C. and Vigne, P. (1985) The sodium/hydrogen exchange system in cardiac cells: its biochemical and pharmacological properties and its role in regulating internal concentrations of sodium and internal pH. *J. Mol. Cell. Cardiol.* **17**: 1029–1042.

LeDucq, N., Delmas Beauvieux, M.C., Bourdel Marchasson, I., Dufour S., Gallis, J.L., Canioni, P. and Diolez, P. (1998) Mitochondrial permeability transition during hypothermic to normothermic reperfusion in rat liver demonstrated by the protective effect of cyclosporin A. *Biochem. J.* **336**: 501–506.

Leist, M. and Nicotera, P. (1997) The shape of cell death. *Biochem. Biophys. Res. Commun.* **236**: 1–9.

Lemasters, J.J. and Thurman, R.G. (1995) The many facets of reperfusion injury. *Gastroenterology* **108**: 1317–1320.

Lemasters, J.J., Nieminen, A.L., Qian, T., Trost, L.C. and Herman, B. (1997) The mitochondrial permeability transition in toxic, hypoxic and reperfusion injury. *Mol. Cell. Biochem.* **174**: 159–165.

Lemasters, J.J., Nieminen, A.L., Qian, T., Trost, L.C., Elmore, S.P., Nishimura, Y., Crowe, R.A., Cascio, W.E., Bradham, C.A., Brenner, D.A. and Herman, B. (1998) The mitochondrial permeability transition in cell death: a common mechanism in necrosis, apoptosis and autophagy. *Biochim. Biophys. Acta* **1366**: 177–196.

Li, H.L., Zhu, H., Xu, C.J. and Yuan, J.Y. (1998) Cleavage of BID by caspase 8 mediates the mitochondrial damage in the Fas pathway of apoptosis. *Cell* **94**: 491–501.

Li, Y.C., Ridefelt, P., Wiklund, L. and Bjerneroth, G. (1997) Propofol induces a lowering of free cytosolic calcium in myocardial cells. *Acta Anaesthesiol. Scand.* **41**: 633–638.

Luo, X., Budihardjo, I., Zou, H., Slaughter, C. and Wang, X.D. (1998) Bid, a Bcl2 interacting protein, mediates cytochrome c release from mitochondria in response to activation of cell surface death receptors. *Cell* **94**: 481–490.

McDonald, J.W., Goldberg, M.P., Gwag, B.J., Chi, S.I. and Choi, D.W. (1996) Cyclosporine induces neuronal apoptosis and selective oligodendrocyte death in cortical cultures. *Ann. Neurol.* **40**: 750–758.

Manon, S., Chaudhuri, B. and Guerin, M. (1997) Release of cytochrome c and decrease of cytochrome c oxidase in Bax-expressing yeast cells, and prevention of these effects by coexpression of Bcl-x(L). *FEBS Lett.* **415**: 29–32.

Marzo, I., Brenner, C., Zamzami, N., Jurgensmeier, J.M., Susin, S.A., Vieira, H.L.A., Prevost, M.C., Xie, Z.H., Matsuyama, S., Reed, J.C. and Kroemer, G. (1998a) Bax and adenine nucleotide translocator cooperate in the mitochondrial control of apoptosis. *Science* **281**: 2027–2031.

Marzo, I., Brenner, C., Zamzami, N., Susin, S.A., Beutner, G., Brdiczka, D., Remy, R., Xie, Z.H., Reed, J.C. and Kroemer, G. (1998b) The permeability transition pore complex: A target for apoptosis regulation by caspases and Bcl-2-related proteins. *J. Exp. Med.* **187**: 1261–1271.

Miyata, H., Lakatta, E.G., Stern, M.D. and Silverman, H.S. (1992) Relation of mitochondrial and cytosolic free calcium to cardiac myocyte recovery after exposure to anoxia. *Circ. Res.* **71**: 605–613.

Murphy, P.G., Myers, D.S., Davies, W.J. and Webster, N.R.J.J.G. (1992) The antioxidant potential of propofol (2,6-diisopropylphenol). *Br. J. Anaesth.* **68**: 616–618.

Nazareth, W., Yafei, N. and Crompton, M. (1991) Inhibition of anoxia-induced injury in heart myocytes by cyclosporin-A. *J. Mol. Cell. Cardiol.* **23**: 1351–1354.

Nieminen, A.L., Saylor, A.K., Tesfai, S.A., Herman, B. and Lemasters, J.J. (1995) Contribution of the mitochondrial permeability transition to lethal injury after exposure of hepatocytes to t-butylhydroperoxide. *Biochem. J.* **307**: 99–106.

Petit, P.X., Goubern, M., Diolez, P., Susin, S.A., Zamzami, N. and Kroemer, G. (1998) Disruption of the outer mitochondrial membrane as a result of large amplitude swelling: the impact of irreversible permeability transition. *FEBS Lett.* **426**: 111–116.

Petronilli, V., Miotto, G., Canton, M., Brini, M., Colonna, R., Bernardi, P. and DiLisa, F. (1999) Transient and long-lasting openings of the mitochondrial permeability transition pore can be monitored directly in intact cells by changes in mitochondrial calcein fluorescence *Biophys. J.* **76**: 725–734.

Qian, T., Nieminen, A.L., Herman, B. and Lemasters, J.J. (1997) Mitochondrial permeability transition in pH-dependent reperfusion injury to rat hepatocytes. *Am. J. Physiol.* **273**: C1783–C1792.

Reimer, K.A. and Jennings, R.B. (1992) Myocardial ischemia, hypoxia and infarction. In: *The Heart and Cardovascular System*, 2nd edn (eds H.A. Fozzard, R.B. Jennings, E. Huber, A.M. Katz, H.E. Morgan) Raven Press, New York, pp. 1875–1973.

Rück, A., Dolder, M., Wallimann, T. and Brdiczka, D. (1998) Reconstituted adenine nucleotide translocase forms a channel for small molecules comparable to the mitochondrial permeability transition pore. *FEBS Lett.* **426**: 97–101.

Schreiber, S.L. and Crabtree, G.R. (1992) The mechanism of action of cyclosporin-A and FK506. *Immunol. Today* **13**: 136–142.

Servin, F., Desmonts, J.M., Haberer, J.P., Cockshott, I.D., Plummer, G.F. and Farinotti, R. (1988) Pharmacokinetics and protein binding of propofol in patients with cirrhosis. *Anesthesiology* **69**: 887–891.

Shiga, Y., Onodera, H., Matsuo, Y. and Kogure, K. (1992) Cyclosporin-A protects against ischemia-reperfusion injury in the brain. *Brain Res.* **595**: 145–148.

Susin, S.A., Lorenzo, H.K., Zamzami, N., Marzo, I., Snow, B.E., Brothers, G.M., Mangion, J., Jacotot, E., Costantini, P., Loeffler, M. *et al.* (1999) Molecular characterization of mitochondrial apoptosis-inducing factor. *Nature* **397**: 441–446.

Sztark, F., Ichas, F., Ouhabi, R., Dabadie, P. and Mazat, J.P. (1995) Effects of the anaesthetic propofol on the calcium-induced permeability transition of rat heart mitochondria: Direct pore inhibition and shift of the gating potential. *FEBS Lett.* **368**: 101–104.

Thornberry, N.A. and Lazebnik, Y. (1998) Caspases: Enemies within. *Science* **281**: 1312–1316.

Uchino, H., Elmer, E., Uchino, K., Li, P.A., He, Q.P., Smith, M.L. and Siesjo, B.K. (1998) Amelioration by cyclosporin A of brain damage in transient forebrain ischemia in the rat. *Brain Res.* **812**: 216–226.

VanderHeiden, M.G., Chandel, N.S., Williamson, E.K., Schumacker, P.T. and Thompson, C.B. (1997) Bcl-x(L) regulates the membrane potential and volume homeostasis of mitochondria. *Cell* **91**: 627–637.

Wicker, U., Bucheler, K., Gellerich, F.N., Wagner, M., Kapischke, M. and Brdiczka, D. (1993) Effect of macromolecules on the structure of the mitochondrial inter-membrane space and the regulation of hexokinase. *Biochim. Biophys. Acta.* **1142**: 228–239.

Woodfield, K.Y., Price, N.T. and Halestrap, A.P. (1997) cDNA cloning of rat mitochondrial cyclophilin. *Biochim. Biophys. Acta.* **1351**: 27–30.

Woodfield, K.-Y., Rück, A., Brdiczka, D. and Halestrap, A.P. (1998) Direct demonstration of a specific interaction between cyclophilin-D and the adenine nucleotide translocase confirms their role in the mitochondrial permeability transition. *Biochem. J.* **336**: 287–290.

Yang, J.C. and Cortopassi, G.A. (1998) Induction of the mitochondrial permeability transition causes release of the apoptogenic factor cytochrome c. *Free Radical Biol. Med.* **24**: 624–631.

Yang, J., Liu, X.S., Bhalla, K., Kim, C.N., Ibrado, A.M., Cai, J.Y., Peng, T.I., Jones, D.P. and Wang, X.D. (1997) Prevention of apoptosis by Bcl-2: Release of cytochrome c from mitochondria blocked. *Science* **275**: 1129–1132.

Ysebaert, D.K., DeGreef, K.E., Nouwen, E.J., Verpooten, G.A., Eyskens, E.J. and DeBroe, M.E. (1997) Influence of cyclosporin A on the damage and regeneration of the kidney after severe ischemia/reperfusion injury. *Transplant Proc.* **29**: 2348–2351.

7

Regulation of apoptosis by cell metabolism, cytochrome c and the cytoskeleton

John M. Garland

1. Introduction

The characteristic morphological features of apoptosis such as membrane blebbing and DNA fragmentation suggest that the cell undergoes a dynamic re-direction of internal processing resulting in autodestruction. This raises questions concerning how cell dynamic architecture is maintained or regulated and its relationship to growth and/or maintenance signaling. Much of this remains to be established. This chapter deals with how metabolism and the cytoskeleton may impinge on apoptotic internal re-arrangement. The first sections deal with apoptosis, cytochrome c and the cytoskeleton. This is followed by a discussion of the application of these principles to a well-defined haematopoietic cell model.

2. Apoptosis: the general scene

2.1 *The two major effector pathways*

The effector enzymes (caspases) of apoptosis (see Chapter 2 for detailed discussion) have been broadly classified into three groups based on substrate preferences; effectors appear in group 2 and activators in group 3 (Thornberry *et al.*, 1997). However, it has become clear that at least two major apoptotic signalling pathways exist which are capable of considerable 'cross-talk'. These pathways are not mutually exclusive, but differ in one fundamental aspect: one is 'external' in that it is promoted by a series of specific external ligands operating through defined trans-membrane receptors ('Death Receptors': see Chapter 3); the other is an internal system where activation of the effector enzymes is induced by intracellular changes, particularly involving mitochondria. Both pathways may use the same caspases, but the mechanism of caspase activation involves different intermediates. There is a third mechanism, restricted however to cytotoxic T cells, which uses direct activation of caspases by a serpin (granzyme B) which is directly delivered into the target cell (Chapter 5).

Programmed Cell Death in Animals and Plants, edited by J.A. Bryant, S.G. Hughes and J.M. Garland.
© 2000 BIOS Scientific Publishers Ltd, Oxford.

2.2 *Death receptors*

Death receptors belong to a family of transmembrane proteins represented by the tumour necrosis factor receptor (TNF-R) and FAS. Signalling is through their cytoplasmic peptide domains and apoptosis is initially induced by the formation of a cytoplasmic death-inducing signaling complex (DISC; Ashkenazi and Dixit, 1998). DISC is formed when the external domain binds its cognate ligand and induces multimerization of the receptor. The DISC complex is formed from adaptor proteins and pro-caspase 8 which are recruited to the receptor cytoplasmic domain and interact with each other through specific sequence motifs (death domains and death effector domains) as shown in *Figure 1*. Formation of the DISC results in activation of caspase 8, which can then activate the major effector caspase, caspase 3, by cleaving its pro-enzyme form. However, two types of DISC formation have been recognized: a fast, type I in which caspase 8 is activated independently of mitochondria; and type II in which mitochondria are involved (Scaffidi *et al.*, 1998). Recently, a large DISC protein FLASH has been identified which appears to be the direct activator of caspase 8 (Imai *et al.*,1999), although the precise regulation of the complex is unclear. Apoptosis induced by DISC thus involves specific surface membrane receptors, ligands and distinct signalling pathways.

   The formation of DISC suggests that there is active movement of DISC components into the complex. At present, there is no information on the relationship between DISC and the cytoskeleton; however, it is almost certain that the cytoskeleton plays roles in both the movement of receptors and cytoplasmic proteins required during apoptosis.

2.3 *Cytochrome c*

The second pathway is much less specific in its mode of induction and involves mitochondria. The initial observations linking mitochondria with apoptosis were made using cell extracts and isolated nuclei (Li *et al.*, 1997; Liu *et al.*, 1996; Zou *et al.*, 1997). Nuclear fragmentation was induced by cell extracts supplemented with cytochrome c and dATP. The requirement for cytochrome c is highly specific: only certain isoforms of cytochrome c are effective and yeast cytochrome c is, for example, inactive; further, cytochrome c needs to be haem-complexed and removal of the haem group renders it inoperative (Kluck *et al.*, 1997). However, the redox function of cytochrome c is not essential since substitution of the Fe group by other non-redox metals does not cause loss of apoptogenic activity.

   The key event is the release of cytochrome c which can interact with a cytoplasmic protein, Apaf-1. Cytochrome c normally resides in the intermembrane space, but can be released if the outer membrane is permeabilized. Interaction of cytochrome c with Apaf-1, which also requires ATP, recruits pro-caspase 9 which is activated. Activated caspase 9 can efficiently cleave pro-caspase 3, leading to its activation and hence proteolysis of caspase 3 substrates (see *Figure 2*). Apaf-1 has several domains which regulate this activation cascade. The Trp–Asp repeats probably act to sterically inhibit the caspase recruitment domain (CARD) for caspase 9. Binding of ATP to Apaf-1 changes the conformation such that the Trp–Asp repeats no longer mask the CARD domain, and caspase 9 can be activated. The mechanism involves oligomerization of caspase 9 which in this conformation has a high autoprocessing capability (Srinivasula *et al.*, 1998; see also

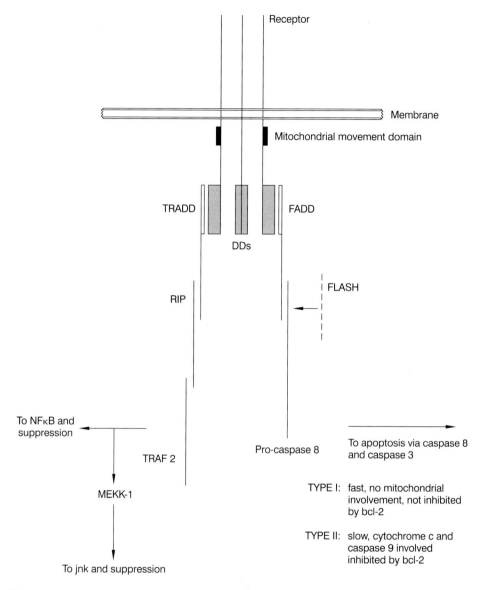

Figure 1. DISC complex formation. The receptors (e.g. FAS, TNF, Apo 3) are trimerized in the presence of ligand. The death domains (DD) of the receptor engage the adaptor protein FADD. This results in interaction with pro-caspase 8 and possibly FLASH, pro-caspase 8 is activated by cleavage and initiates the cascade. In certain receptors (TNF, Apo3), the signal can result in apoptosis suppression. This is mediated by recruitment of another adaptor TRADD which can bind RIP or FADD. Binding of RIP is associated with recruitment of TRAF which leads to apoptosis suppression via activation of the transcription factor NF-κB, or via activation of jun kinase (jnk) which activates the jun transcription factor. The detailed mechanism of apoptosis suppression is currently unclear. The death domain is represented by an open box. In type I DISC, activation of caspases is rapid and does not involve mitochondria. In type II, activation is slower and appears to require involvement of mitochondria.

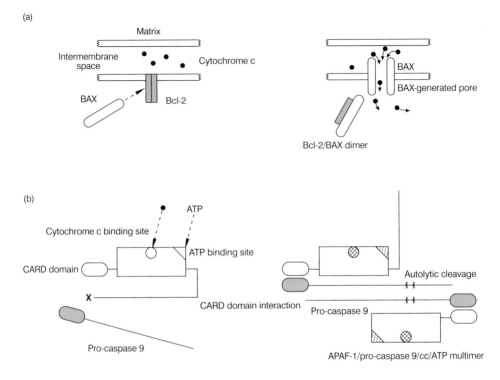

Figure 2. *Mitochondria and cytochrome c release. (a) Schematic model involving Bcl-2 and Bax. Cytochrome c is retained in the intermembrane space (L). Interaction of membrane-inserted Bcl-2 with Bax results in bcl-2/Bax dimerization and replacement of membrane-inserted Bcl-2 with Bax. Bax forms channels which allows cytochrome c to escape and activate caspase 3. Also, permeability transition pores and other factors which cause mitochondrial swelling result in discontinuities of the outer membrane and cytochrome c release. (b) Activation of pro-caspase 9 by Apaf-1 and cytochrome c/ATP. In the absence of cytochrome c and ATP the CARD domain is sterically hindered. Binding of cytochrome c and ATP results in unmasking of the CARD domain and multimerization of Apaf-1, which is able to bind pro-caspase 9. The proximity of several pro-caspase 9 molecules allows autolytic cleavage and activation of caspase 9 (see Chapter 1).*

the discussion of the apoptosome in Chapter 1). A very large multimer (1.3 megaDa) containing caspase 9, Apaf-1 and cytochrome c has been isolated, which probably represents the active complex (Zou *et al.*, 1999). Inhibition of this autoprocessing forms an important constitutive negative regulation. Subsequent studies showed that in certain cells cytochrome c was released into the cytosol during apoptosis, and that apoptosis could be induced in whole cells by introducing exogenous cytochrome c (see later).

In addition to cytochrome c, mitochondria may release other apoptosis-inducing factors, AIFs (Susin *et al.*, 1999); further there is evidence for mitochondrial caspases (Krajewska *et al.*, 1999). Thus mitochondria appear to have a dual role in orchestrating cell integrity, controlling both survival through metabolism and apoptosis through a metabolic co-factor that induces caspase activation.

Regulation of apoptosis by cytochrome c; checks and balances. Although the mechanism by which cytochrome c is released is still unknown, release of cytochrome c is inhibited by Bcl-2 and related anti-apoptotic proteins (Chapter 4), whereas it is stimulated by pro-apoptotic members of the Bcl-2 family such as Bax (see below). The current view, substantiated by experimentation and modelling, is that the Bcl-2 type proteins are distantly related to bacterial toxins which induce pores in membranes, and that members such as Bax create pores in the outer mitochondrial membrane. Pore formation by Bax has been verified in *in vitro* systems. Bax may be directed to the membrane by the protein Bid. The efficiency of Bid in directing Bax is regulated by its phosphorylation, thus linking Bid/Bax with signal transduction. The ability of Bcl-2 family members to heterodimerize with differing affinities has prompted the model of pro-apoptotic members being sequestered by anti-apoptotic members through this heterodimerization, thus removing pro-apoptotic members from the mitochondrial interactive pool. For example, another member, Bad, is thought to regulate Bcl-2 levels, again through alternative phosphorylation states. When phosphorylated, Bad heterodimerizes with a cytoplasmic serine kinase of 14–3–3 protein family. When unphosphorylated, Bad has higher affinity for Bcl-2 and thus heterodimerizes with Bcl-2. This sequesters Bcl-2 and thus removes its protective effect, allowing pro-apoptotic members such as Bax to homodimerize and form pores, leading to mitochondrial swelling and thus release of cytochrome c.

Mitochondria, membrane potential and free radicals. The involvement of mitochondria in apoptosis is a surprise since they have been considered to be organelles concerned solely with metabolism. Following the discovery that cytochrome c was the activity stimulating apoptosis in cell-free systems, it was reported that apoptosis in intact cells was accompanied by loss in the mitochondrial inner membrane potential, $\Delta\psi_m$ (Zamzami *et al.*, 1995a, 1995b). This potential is generated by electron flow through the cytochrome chain on the matrix side and partitioning of $[H]^+$ into the inter-membrane space on the other side during oxidative phosphorylation (see *Figure 3*). This generates a net negative charge of around 100 mV on the mitochondrial matrix side. ATP is generated by proton reflux through the F_oF_1 ATPase, a rotary molecular motor which uses the proton gradient to generate ATP from bound ADP and P_i (Junge, 1999; Walker, 1998). $\Delta\psi_m$ is dependent on inner membrane integrity, and can be discharged by physical damage to the insulating inner membrane, or by protonophores (uncoupling agents) such as dinitrophenol (DNP) which dissipate the proton motive force by allowing the protons to move through the inner mitochondrial membrane back into the matrix, thus bypassing the F_oF_1 ATPase. Further this allows the ATPase to function in the direction of hydrolysis, rather than as an ATP synthase. The consequence of this is that mitochondrial ATP synthesis falls and already-formed cellular ATP can be hydrolysed by re-entry into the mitochondrial matrix, thus lowering cellular ATP levels still further. Mitochondria are very sensitive to changes in milieu and are liable to swell, thus stressing the membranes and altering their permeability. Also, mitochondrial inner membranes can generate 'permeability transition pores' which appear to be formed from distinct proteins and create physical channels allowing passage of small molecules ($< 6000\ M_r$) (see Chapter 6). A number of pH- and charge-sensitive vital dyes whose fluorescence is determined by alternative redox states have been shown to concentrate in mitochondria, and thus act as indicators of $\Delta\psi_m$. Commonly used are rhodamine-123 and Di-O-(C6)3. Using such dyes, it was found that T cells lost $\Delta\psi_m$ prior to apoptosis, and that inhibition of the permeability transition pore by cyclosporin A significantly reduced apoptosis induced by several

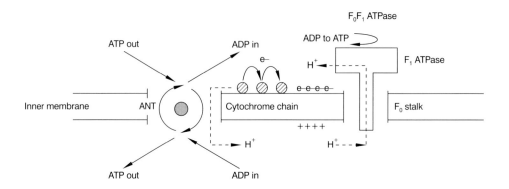

Figure 3. *Mitochondrial inner membrane potential. The membrane potential is normally generated through oxidative phosphorylation. During electron transport along the cytochrome chain, protons are transported out of the matrix, thus creating a net negative charge on the matrix side of the membrane. The F_oF_1 ATPase is a molecular stepping motor acting like a generator and is able to use this charge difference to channel protons back and convert ADP to ATP. Rotation is brought about by entry of protons into the F_o portion followed by their exit the other side. Protons captured by F_o and the charge difference across the membrane supply the energy needed for rotation and conversion of the ATPase catalytic sites into alternative states for synthesis of ATP. However, the generator can rotate in the reverse direction, when the ATPase works to hydrolyse ATP. In this condition (e.g. with oligomycin), cytoplasmic ATP can be decreased. Oligomycin blocks the proton re-entry mechanism and hence causes a hyperpolarization of the membrane. ANT, adenine nucleotide transporter.*

agents (Zamzami *et al.*, 1995a, 1995b). Further studies suggested that apoptosis induction was related to free radical production generated during oxidative phosphorylation and that Bcl-2 proteins operated on a free radical axis (Hockenbery *et al.*, 1993). However, this was before the involvement of cytochrome c was discovered, although this does not rule out a role for free radicals.

Bcl-2 family genes: positive and negative regulators of mitochondrially-induced apoptosis. The *Bcl-2* gene was originally defined as a proto-oncogene involved in the 14/18 chromosome translocation in B-cell lymphoma. It is now well established that it belongs to a family of genes encoding proteins which may be pro- (BAX, BAD, BAK) or anti- (Bcl-2 members, MCL) apoptotic (Adams and Cory, 1998). These proteins all contain at least one homologous BH domain (see Chapter 4) which determines interaction with each other; Bcl-2 contains BH- 1, 2 and 3 domains (*Table 1*). The BH3 domain is characteristic of pro-apoptotic members, BH1/2 of anti-apoptotic members. The Bcl-2 family members are capable of homo- or heterodimerization and are distributed in mitochondrial outer, endoplasmic and nuclear membranes, although some protein appears to be cytoplasmic. First, their overall effect on controlling apoptosis appears to be determined by their dimerization; each can abstract the other by heterodimerization and thus remove or reduce the amount of homodimerization. Second, the proteins are related to pore-forming proteins and have been shown to

Table 1. Members of the Bcl-2 family.

Bcl-2 type:								
	1 BH4	2 BH3	3	4 BH1	5	6 BH2	7	MA
		Regulation domains		Dimerization domain	Pore formation			
Bcl-2	+	+		+		+	+	
Bcl-XL		+		+		+	+	
Mcl-1		+		+		+	+	
E1B19K				+				
A1				+		+		
Bax-type								
BAX		+		+		+	+	
BAK		+		+		+	+	
BH3-only								
BAD		+						
BID		+						
BIK		+					+	

Numbers indicate α-helical structures. MA, membrane anchor.

inhibit cytochrome c release (Bcl-2/XL) or to increase outer membrane permeability and release cytochrome c (BAX); channel-forming ability has been demonstrated in lipid bilayers by BAX (Antonsson *et al.*, 1997; Muchmore *et al.*, 1996). Further tiers of regulation have been described: the targeting of BAX-type proteins to mitochondria by accessory proteins such as BID (Gross *et al.*,1999) and sequestration of pro-apoptosis members by phosphorylation (BAD); in the latter, phosphorylated BAD is sequestered to a cytoplasmic kinase of the 14–3–3 type and thus is removed from Bcl-2 heterodimerization where it would negate Bcl-2's anti-apoptotic activity (Zha *et al.*, 1996). However, there is also growing evidence that Bcl-2 has other ways of inhibiting apoptosis, for example by direct interaction with caspase activators (Hu *et al.*, 1998).

Recently, a new pro-apoptotic member of the family, BIM, has been described. Of particular interest is that it appears to be complexed to a microtubule motor protein, dynein (Puthalakath *et al.*, 1999). During apoptosis, BIM is displaced from the dynein light chain and then interacts with Bcl-2 where it negates Bcl-2's anti-apoptotic function. Of course, this immediately raises the question of the proximity of BIM and the dynein light chain to mitochondrial Bcl-2, see later. Similarly, pro-apoptotic BAK has been found significantly associated with the cytoskeleton (R. Brown, unpublished data and Chapter 8).

The major distinction of these mechanisms from DISC is that the stimulus to apoptosis via mitochondria is in response to intracellular perturbations, which can have many and varied origins, whereas DISC induction is limited to highly specific external ligands produced by other cells or by the same one (T cells display both the FAS receptor and its ligand). There is one further point. DISC formation is less likely to be influenced by signals arising from other receptors than cytochrome c release. This is because DISC activation is relatively direct, whereas changes in mitochondria are more likely to have highly indirect origins. Control in both systems may also involve a whole series of positive and negative regulators whose overall 'arbitration' over survival and death resides in relatively universal ways of modifying protein function such as phosphorylation and signalling involving protein kinase activation.

2.4 *Granzyme B*

In this mechanism, cytotoxic T cells expressing the FAS ligand and receptor mutually interact with target cells. This interaction mediates delivery of perforin from the T cell to the target cell followed by discharge of granzyme B from granules within the T cell through the pores made by perforin (Kagi *et al.*, 1994). Granzyme B activates the endogenous apoptosis programme in the target cell (Atkinson *et al.*, 1998) through its universal activation of caspases (Harris *et al.*, 1998). This is a highly specialized mechanism because expression of granzyme B is restricted to cytotoxic T cells (see Chapter 5) but it clearly has implications for the cytoskeleton because the granules are delivered by the cytoskeleton through physical recruitment of the granules to the target cell contact area.

2.5 *The XIAP-type inhibitors of apoptosis*

Recently, a series of protein inhibitors of apoptosis has been described, related by possession of a motif (BIR) found in a baculovirus anti-apoptotic protein. These proteins function by directly inhibiting caspases either in activated form or by preventing their cleavage and activation (LaCasse *et al.*, 1998). The mechanism of BIR inhibition is not yet known, but of interest here is that a potent single-BIR-domain inhibitor of apoptosis, survivin, normally restricted to tumour and embryonic cells, binds to the mitotic spindle (Li *et al.*, 1998) and thus must be intimately associated with the cytoskeleton. Association of such proteins with normal physiological activities suggests that their targets, the caspases, may have, or have had, a role in normal physiology as well.

3. Apoptosis and the cytoskeleton

The observation that disruption of cell-matrix adhesion induced apoptosis (Frisch and Francis, 1994) demonstrated that the cytoskeleton played more than a passive role in apoptosis. However, only recently has any specific information been obtained on cytoskeletal involvement in apoptosis, and it is still very incomplete. A summary of the key observations follows, preceded by a brief resumé of cytoskeletal organization.

The cytoskeleton provides a highly elastic framework coupling the outer membrane to all internal parts of the cell. Both its charged surface and its dynamic organization allow it to respond to mechanical changes but also it carries an array of associated proteins and molecular motors which continually traffic proteins and even RNA (Janmey, 1998; Nobutaka, 1998). The cytoskeleton communicates with the exterior through transmembrane receptors which activate numerous kinases (ERKs) concerned with cytoskeletal regulation. Mention has been made above of the movement of receptors for soluble ligands, and many receptors will affect the cytoskeleton through recruitment of kinases (mitogen-activated protein kinases, MAPKs) which eventually activate cytoskeletal regulators through kinase casades. More direct communication is

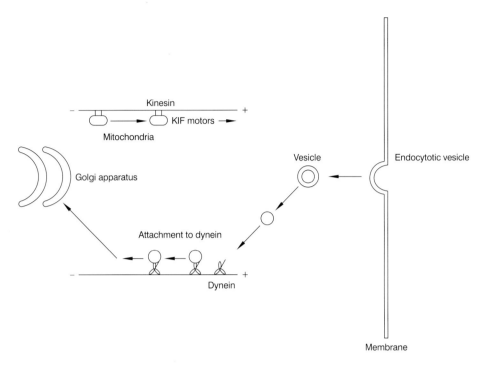

Figure 4. *Microtubules and cell trafficking. Microtubules are dynamically re-modelled from the + end by addition and removal from the – end, although addition can take place at the – end. Kinesin-type motors traffic organelles, particles and molecules towards the membrane (anterograde), dynein motors in the reverse direction (retrograde). Endocytotic vesicles from the membrane are complexed to dynein and move to the Golgi apparatus, while mitochondria attach to kinesin through specific kinesins. e.g. KIF1b and move outwards. During mitosis, the entire cytoskeleton is re-modelled around the spindle, but the microtubule system ensures that mitochondria are distributed to daughter cells. In yeasts, mutations to mitochondrial outer membranes and certain cytoplasmic proteins inhibits this movement resulting in abnormal distribution, or can affect the shape and division of mitochondria resulting in giant mitochondria.*

through the integrin receptor system which couples focal cell adhesion plaques (FAPs) to the actin cytoskeleton (Schoenwaelder and Burridge, 1999). The FAPs are generated through clustering of integrin receptors, and both these and similar clustering which occurs in membrane ruffles or tubular filipodia extensions are regulated by a family of small G-proteins represented by rho which have multiple effects on the cytoskeleton (Aspenstrom, 1999). Interaction of integrins with extracellular matrix proteins results in linkage of the integrins to the actin cytoskeleton and formation of stress fibres. The actin cytoskeleton provides a dynamic framework of fibres capable of ATP-dependent sliding movements and thus the actin cytoskeleton can physically propel the cell, alter its shape and extrude or internalize parts of it, as seen for example in membrane ruffles, filipodia, discharge of synaptic and lysosomal vesicles and internalization of coated pits and ligand–receptor complexes.

The microtubule system is a highly dynamic system of small tubules generated by polymerization/depolymerization of tubulin. The microtubule system permeates the cell and establishes contact with all the intracellular organelles. Microtubules have polarity in the sense that they polymerize from one end (+ or peripheral end) and are effectively depolymerized from the other (– or nuclear end). They are constantly being re-modelled, and they can be viewed as small rail tracks which passenger a lot of intracellular traffic. Their ability to do this is governed by small molecular motor proteins, dynein and kinesin that travel in opposite directions, dynein towards the nucleus, kinesin towards the cell periphery (Nobutaka, 1998). These motors can attach a large variety of proteins, other molecules and organelles such as secretory vesicles and mitochondria, and physically transport them across the cell. It should be noted that the nucleus also has its own internal scaffold and filamentous structural system which is vital to organizing DNA for replication, segregation and transcription (*Figure 4*).

These two cytoskeletal systems work in cooperation to ensure appropriate distribution of internal compartments or molecules in response to environmental cues or internal demands, and the cell is constantly undergoing 'self-testing'. During apoptosis drastic internal and external rearrangements occur, and it is clear that the cytoskeleton must be involved.

3.1 *Cytoskeletal aspects of apoptosis*

The physical manifestations of apoptosis, namely surface blebbing, nuclear fragmentation and extrusion of vesicles containing cytoplasmic and nuclear contents all confirm that both microtubules and the actin cytoskeleton are involved in the cell's autodestruction. Most current work focuses on actin, and this will be dealt with first. It will, however, be apparent that there are many gaps in our understanding, not least because different cells may have different cross-talk, and therefore activate different secondary cascades.

Actin and apoptosis. Both actin itself and some of its polymerization regulators, e.g. actin-severing gelsolin, are targets for activated caspases and have been associated with the apoptotic process (Geng *et al.*, 1998; Kothakota *et al.*, 1997; Mashima *et al.*, 1999; Meredith *et al.*, 1998). The caspase-generated gelsolin fragment cleaves actin in a calcium-dependent manner, and inhibition of this cleavage or reduction in the amount of fragment itself delays apoptosis. Similarly, caspases cleave some of the cell–cell adherence-associated proteins such as β-catenin (Brancolini *et al.*, 1997) which links cadherin to α-catenin and actin. Such cleavage is associated with apoptosis. Cadherins are calcium-dependent transmembrane receptors involved in the formation of intercellular tight junctions and thus a similar link between cell–cell adhesion exists as with cell–extracellular matrix protein attachment. More details are known concerning actin modulation and focal cell adhesion plaques and/or focal adhesion itself during apoptosis.

Activated rho-type GTPases may induce or inhibit apoptosis (Esteve *et al.*, 1998; Nishida *et al.*, 1999). In the case of inhibition, this is associated with activation of a kinase known as Akt, which is known to modulate apoptosis by phosphorylation of BAD and thus implicates the mitochondrial–cytochrome c axis. Induction on the other hand is dependent on ceramide production which is also known to operate, partly at least, through cytochrome c release. That this induction is inhibited by Bcl-2 would strengthen the conclusion that both mechanisms operate through regulation of cytochrome c release. In other studies, inactive rho, which might stabilize the actin cytoskeleton, induced apoptosis, a result consistent with enhancement of apoptosis by stabilizing the cytoskeleton with jasplakinolide (Moorman *et al.*, 1999; Posey and Bierer, 1999).

How this mechanism works is not clear, but some insight has been obtained into how some of the actin-orientated kinases may contribute to apoptosis, particularly membrane blebbing. Formation of focal plaques depends on the cooperative action of the kinases MAPKAP-2 and ERK (Huot *et al.*, 1998). MAPKAP-2 phosphorylates HSP27, a protein which appears to cap the ends of actin filaments. Phosphorylation of HSP27 allows actin polymerization by removing the cap. MAPKAP 2 is itself activated by a stress-activated kinase (SAPK), in turn activated by MAP kinase (MAPK). Whether actin filaments can organize properly depends on a second kinase cascade operating through MEKK1 and ERK. ERK can be activated by a number of membrane receptors, including the epidermal growth factor receptor (EGFr) and the vascular endothelial growth factor receptor (VEGFr). If ERK is inactivated, focal actin organization does not occur and the membrane blebs intensely. Further, MEKK1 is subject to cleavage by caspase 3, resulting in an activated but truncated product. This too results in increased apoptosis (Yujiri *et al.*, 1998), perhaps through interfering with normal MEKK1 function. This is illustrated in *Figure 5*. Other work shows that membrane blebbing in apoptosis is also linked to myosin light chain phosphorylation (Mills *et al.*, 1998). Phosphorylated myosin interacts with F-actin and provides the ATPase-based stepping motor of the actin–myosin complex. In this model, blebbing is caused by myosin light chain (MLC) phosphorylation mediated by activation of MLC kinase. G-proteins of the rho family also seem to exert a regulatory role in this phosphorylation step and in addition regulatory phosphatases may contribute to the overall status of free versus phosphorylated myosin light chains.

Thus, several enzyme systems intersect on actin organization and activity, and there is growing evidence that apoptotic blebbing is related to activity in the actin

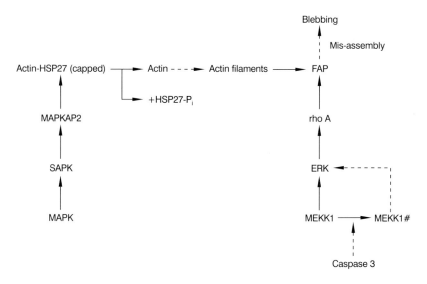

Figure 5. *Model for cytoskeletal regulation and blebbing during apoptosis. On the left, actin polymerization is prevented by capping through HSP27. Phosphorylation by MAPKAP2 releases HSP27 and actin polymerization can occur. Proper assembly into focal adhesion requires signalling through an ERK receptor or another stress induced kinase, and rho A. Failure to assemble results in blebbing. Caspase 3 can cleave MEKK1 and the fragment can inhibit the normal function of MEKK1 in assembling actin fibres, thus leading to blebbing.*

cytoskeleton and in general to the apoptotic process. However, it is not clear how primary alterations in the actin cytoskeleton link with signals for apoptosis, nor whether such changes are secondary as opposed to inductive.

Microtubules and apoptosis. Less information is available on the role of microtubules. The variety of microtubule-associated proteins is very high and most work has been conducted on neurons. Microtubules, or at least disruption thereof, appear to regulate expression of apoptosis-related genes such as *MCL-1* and *BAX* (Srivastava *et al.*, 1998; Townsend *et al.*, 1998). In both cases, activation of protein kinases (PK-C and -A) is clearly involved, and with MCL-1, MAPK is activated via ERK. In breast tumour cells, microtubule binding agents can differentially regulate MAP kinases (Shtil *et al.*, 1999), illustrating the cross-talk that can occur in the cytoskeleton. There is indeed an increasing number of observations linking apoptosis with cytoskeletal kinases, and undoubtedly there will be many more. DAPK is a microfilament-associated calcium-activated kinase which is pro-apoptotic and whose loss is correlated with both low apoptosis induction and high metastatic potential (Inbal *et al.*, 1997). ASK is a MAP kinase that activates stress-activated kinases (SAPKs) and is itself regulated by the redox protein thioredoxin (Saitoh *et al.*, 1998). This is of added interest because caspases are themselves subject to redox regulation through their active site cysteine, and a growing number of redox moities (NO, peroxides, glutathione) are being related to apoptotic signalling. Stress-activated kinases are involved in regulating HSP 27 (see above) hence potentially linking microtubules to actin-based blebbing. Vav is a protein accessory factor for GTP exchange by cytoskeletal rho proteins, and is involved in apoptotic thymocyte selection through activation of protein kinase c (Kong *et al.*, 1998). In part, association of enzymes with the cytoskeleton is governed by micro-tubule affinity-regulating kinases (MARKs), a family of kinases possessing a conserved catalytic region and tail (Drews *et al.*, 1998).

Perhaps one of the key areas in understanding cytoskeletal signalling revolves around linking mechanoreception (changes in tension and shape) to activation of critical enzymes associated with the mechanical structural framework (Chicurel *et al.*, 1998). Finally, although caspases are regarded as specifically directed to apoptosis, it is possible that at least some of them may be normal regulators, for example of the cytoskeleton (Watanabe and Akaike, 1999). In this context it is interesting that formation of a novel, non-tubulin/actin/intermediate filament filamentous structure by caspases during apoptosis (death effector filaments, DEF) has been described (Siegal *et al.*, 1998). These are interesting because an enzymically inactive fragment of caspase 10 containing only the death effector domain can still form death effector filaments (Ng *et al.*, 1999).

Cytoskeletal motors. This discussion would not be complete without some further reference to motors. Most work has been done in neurons where axonal transport features prominently in the peripheral distribution of components synthesized in the cell body (which may be several metres away). There are two super-families of motors, the kinesins which move organelles and molecules towards the periphery (antero-grade) and the dyneins which move in the opposite direction (retrograde) (Hirokawa, 1998; Walker and Sheetz, 1993). Kinesins are ATP-based motors with an ATP-binding head, middle section and bifurcated tail in which the motor part is either N-terminal, C-terminal, or in the middle. The tail links kinesin to the target organelle. Very many

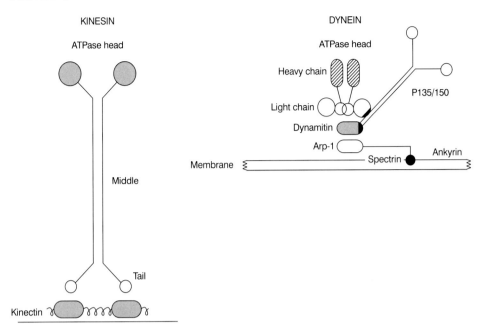

Figure 6. *Cytoskeletal motors. Kinesins contain a head, middle and tail portion. Interaction with the target organelle is through kinectin which binds to membranes possibly with other accessory proteins. Dyneins are multimers of heavy chains intermediate and light chains. The probable arrangement of the subunits is as follows. The dynein heavy chain interacts with intermediate chains which interact with light chains. The light chains complex dynamitin, a GTPase, which attaches to a short actin-like protein, Arp-1. Arp-1 interacts with spectrin, an elongated molecule on the cell surface to where it is bound by the ankyrin receptor. The cargo is attached to the motor through the p150/135 glued hetero-dimer.*

forms of kinesin exist, mostly as dimers, but the KIF family of kinesins are unusual in that they are monomeric. Dyneins contain several heavy chains together with variable numbers of intermediate and light chains and exist in two major classes, flagellar and cytoplasmic. Again, there are multiple forms of dynein. Both kinesin and dynein appear to link with organelles through attachment or targeting protein complexes such as kinectin and dynactin. In addition to their role in organelle movement, both appear to have roles in mitosis and the movement of chromosomes (see review by Vallee and Sheetz, 1996).

Despite such seeming importance, only recently has very limited evidence emerged that motors may be associated with the apoptotic process. The TNF receptor which operates through the DISC has been found to induce mitochondrial movement via its cytoplasmic membrane-proximal domain (De Vos *et al.*, 1998). Ability to translocate mitochondria synergized with the cell death programme. However, Fas, another DISC protein did not translocate mitochondria, therefore its significance is uncertain. Kinectin has also been identified as a substrate for caspases, probably caspase 7 (Machleidt *et al.*, 1998). Cytoplasmic light chain dynein mutations have been recorded as causing apoptosis in *Drosophila* (Dick *et al.*, 1996) whilst dynein heavy chain is down-regulated in brain glioma cells during OH-vitamin-D3-induced apoptosis (Baudet *et al.*, 1998). However, more intriguing is the association of a BH3 domain-only pro-apoptotic protein, BIM, with the dynein light chain 8 (Puthalakath *et al.*, 1999). During apoptotic induction, BIM dissociated from the light chain and complexed to Bcl-2, thus sequestering Bcl-2 and permitting apoptosis.

Although at present the role of motors in apoptosis has been barely examined, it should be remembered that intracellular movement of organelles and molecules is a fundamental cell activity; it will be surprising if these movements and the molecules controlling them do not have a very significant regulatory or even inductive role, as evidenced in the preceding paragraphs.

Mitochondria and the cytoskeleton. Discussion of apoptosis and the cytoskeleton would not be complete without mentioning an area that is receiving more attention. The establishment of mitochondria as regulators of apoptosis through cytochrome c release and other factors has meant a re-appraisal of their connection with other structures and their distribution. Reference has been made previously to the cytoskeletal motors kinesin and dynein. These are targeted to organelles by specific accessory proteins such as dynamin and kinectin (Vallee and Sheetz, 1996).

It has been well-established that mitochondria are associated with microtubules (Heggeness *et al.*, 1978), and recently a kinesin-family motor which specifically translocates mitochondria, KIF-1B, has been identified (Nangaku *et al.*, 1994). This is an unusual motor in that it is monomeric rather than possessing the usual dimer form of globular head/intertwined middle string/globular tail organization. Recent work has demonstrated that specific genes regulate mitochondrial inheritance and behaviour, including fusion and distribution (Yaffee, 1999). First, the microtubule-associated GTPase dynamin has been shown to alter mitochondrial morphology and distribution in yeast and human cells (Otsuga *et al.*, 1998; Smirnove *et al.*, 1998). This is interesting because dyamin is concerned with formation of vesicles where it functions as a 'pinchase' to bud off the nascent vesicles (McNiven, 1998). A series of mutations have been described in yeast which alter distribution and morphology (mitochondrial distribution and morphology, *mdm*), morphology and movement (*mmm*) and genome maintenance (*mgm*). The proteins encoded by *mdm10* and *mmm1* appear to attach mitochondria to actin filaments (Burgess *et al.*, 1994; Sogo and Yaffe, 1994). These and other related mitochondrial proteins are reviewed by Hermann *et al.* (1998). It is clear that a variety of mitochondrial outer membrane and cytoplasmic proteins regulate mitochondrial fusion, cell distribution and retention of the mitochondrial genome which codes for a small number of essential mitochondrial proteins.

Recently, attention has been focused on the growing number of known atypical cytoplasmic myosins (Mermall *et al.*, 1998). They are implicated in certain diseases but also appear to be involved in organelle movement, intracellular trafficking and signalling. Little is known about their function in mammalian cells, but undoubtedly this will change.

4. Apoptosis and metabolism

Despite the central role of mitochondria in both generation of ATP and apoptosis, linkage between their respiratory and cell death functions has received little attention. The two principle questions are whether or not apoptosis is ATP-dependent, and whether interference with the respiratory function causes apoptosis through secondary changes (mitochondrial swelling due to pH, calcium or nucleotide induction/suppression of the permeability transition pore). Furthermore, any change in ATP metabolism is likely to affect a large variety of processes. Several studies (Nieminen *et al.*, 1994; Simbla *et al.*, 1997) suggest that ATP depletion by inhibition of the respiratory chain, reversal of the mitochondrial ATPase by oligomycin or respiratory uncoupling induces apoptosis, which may be regulated through Ca^{2+}-dependent permeability transition pores. Other studies suggest the opposite (Stefanelli *et al.*, 1997). In some cases, initiation of/increase in glycolysis may prevent apoptosis induced by ATP depletion (Chalmers-Redman *et al.*, 1999). During apoptosis induced by TNF (tumour necrosis factor), cyclical changes in ATP occur which are linked to both respiratory uncoupling and generation of free radicals (Sanchez-Alcazar *et al.*, 1997). Direct interaction of free radicals with the ATPase has been implicated in cell differentiation, although not specifically in apoptosis (Comelli *et al.*, 1998). Interestingly, the two different types of FAS responses, I and II (see above), are also differentiated by caspase activation. Caspase activation in type 1 cells is insensitive to ATP reduction while in type II cells, which involve mitochondria during caspase activation, ATP depletion blocked caspase activation (Eguchi *et al.*, 1999). However, in both types cytochrome c was still released and ATP depletion inhibited apoptosis. This suggests that type I and II cells have different ATP depletion targets in the induction programme.

De-regulating metabolism also has other effects: ATP depletion in some cells can cause the blebbing characteristic of apoptosis (Gabai *et al.*, 1992) and respiratory blockade also appears to decrease the affinity of TNFr for ligand (Sanchez-Alcazar *et al.*, 1995). ATP concentration has also been implicated as a deciding factor between apoptosis or necrosis (Eguchi *et al.*, 1997; Leist *et al.*, 1997), low levels of ATP producing necrosis while normal levels produce apoptosis. Clearly, total loss of ATP may be expected to result in gross intracellular imbalances, and since apoptosis represents structural re-direction, some ATP may be necessary to preserve normal architecture during apoptosis. Interestingly, in neither of these papers was the residual glycolysis measured and therefore whether cells started from the same metabolic level by removing glucose cannot be determined (see below).

The metabolic status of cells is important in the study of apoptosis control. Cells with primary (or inducible) glycolysis in which mitochondria are essentially inactive whilst considerable amounts of lactic acid are produced in the cytoplasm are likely to behave differently with respect to the many factors that affect oxidative phosphorylation. For example, in primary glycolysis, free radical mechanisms and production locations will be different, the mitochondrial ATPase is likely to be operating in reverse, membrane transporters will be re-aligned for lactate transport and regulation of the permeability transition pore less likely to be influenced by mitochondrial adenine nucleotide transferase activity. Many tissue culture cell lines used in apoptosis studies in fact use primary glycolysis and rarely if ever are the fundamental parameters of ATP production fully quantitated. Consequently, interfering with mitochondrial function in isolation may be difficult to interpret.

4.1 *Nitric oxide*

The involvement of nitric oxide, NO, in many cell processes and responses is now well known. It is synthesized by a series of synthase (NOS) isoforms (Michel and Feron,

1997) which are found throughout the cytoplasm and in organelles such as mitochondria (Bates *et al.*, 1996) where it may regulate oxidative phosphorylation. Of relevance to this chapter is that a protein inhibitor of NO synthase, PIN, has been
identified as a dynein light chain (Jaffrey and Snyder, 1996). Whether PIN structurally
inactivates NOS or merely translocates NOS is uncertain (Rodriguez-Crespo *et al.*,
1998). However, mention has already been made of NO in apoptosis, and recently
S-nitrosylation of caspases has been invoked as a mechanism for regulation (Mannick
et al., 1999); *S*-nitrosylation of the catalytic site cysteine in pro-caspase 3 was found in
FAS-unstimulated cells, whereas after stimulation, active enzyme was de-nitrosylated.
Since *S*-nitrosylation inhibited caspase 3 activity, nitrosylation/de-nitrosylation could
be a tertiary mechanism for caspase activity control.

5. Measurement of apoptosis

Before going on to discuss apoptosis in a particular model system, it is necessary to
consider briefly the methods used in measuring apoptosis since these may be critical
in placing significance on particular investigations. There are numerous ways of
measuring apoptosis. Phosphatidyl serine exteriorization (annexin positivity) is a relatively early event. Cleavage of poly(ADP-ribose) polymerase (PARP) and DNA fragmentation are late indicators at an irreversible stage of the process. Caspase cleavage
and activation measured by either Western blotting or enzymic activity in cell extracts
using peptide substrates which release fluorogenic indicators on cleavage, occurs prior
to DNA fragmentation and PARP cleavage. Several methods are employed for
detecting DNA cleavage. The most reliable is formation of a DNA ladder exhibiting
the typical 180 bp internucleosome 'ladder' or a TUNEL-type assay where free DNA
ends are labelled by terminal nucleotide transferase using fluorescent nucleotides.
Measuring mitochondrial membrane potential is not a reliable indicator of apoptosis
because apoptosis may occur without it and changes in potential may occur without
apoptosis.

The most direct way to quantitate apoptosis is to count apoptotic cells, either by
morphology or after TUNEL staining. Vital staining is not a good method because
changes in membrane permeability are late events; this method may miss apoptotic
cells whose membrane is still intact. Much caution is needed in using flow cytometry.
First, gates on the instrument must be set appropriately. Second, cytometry done
either with propidium iodide or DAPI stains relies on establishing a 'pre-G1' peak
created by expulsion of DNA after nuclear fragmentation. This pre-G1 peak cannot be
used to quantitate apoptosis because loss of DNA shifts the whole DNA curve to the
left; DNA from fragmented G_2/M and S nuclei re-appears in the S or G_1 channels
respectively as well as in pre-G_1. If cytometric quantitation is needed, TUNEL assays
should be used.

6. Apoptosis in a defined haematopoietic cell system

Section 4 dealt briefly with major areas of key importance in apoptosis regulation. This
section takes a particular *in vitro* cell line model and describes how some of the
concepts can be applied to an intact apoptosis-inducing system. The cells are derived
from bone marrow and are considered to represent some kind of intermediate stem
cell. The principles governing their growth and survival are thought to be reflected in

the normal marrow environment and to be of importance in understanding growth dysregulation, for example in leukaemia. The study of apoptosis and its dysregulation in leukaemias and haematopoietic disorders has mushroomed over the last 3 years (over 1100 papers have been published between 1996–1999), illustrating the growing importance of understanding its mechanisms. As with other cell systems, marrow dysfunction often has severe consequences and the role of apoptosis is becoming increasingly clear (Wickremasinghe and Hoffbrand, 1999).

6.1 Supply and demand: keeping the balance in peripheral blood

Each day approximately 100 000 000 granulocytes are generated from bone marrow to replace those removed by the reticulo-endothelial system. Similarly, red cells have a life of about two months and the entire peripheral blood cellular compartment is replaced at least once a year. To this must be added acute destabilizing episodes, for example infection or blood loss, which need rapid but controlled responses, and the widely differing requirements for the widely different cell lineages. The system therefore requires an exquisite balance between cell removal and compensatory production; a miscalculation of less than 0.1% either side of what is needed would have catastrophic consequences, leading to either leukaemia or panleucopenia within a few months. How the peripheral blood communicates its demand for new cells to the marrow supply depot is still unknown. However, the hierarchical system of cell production in the marrow is sufficiently understood to begin to shed some light on control points and where apoptosis might exert a critical role.

6.2 Marrow kinetics: stem cells

Mature cells arise in the marrow from progressive division and differentiation of undifferentiated progenitor cells, creating an inverted pyramid with a small number of stem cells at the inverted apex and a vastly expanded mature population at the top (*Figure 7*). The most immature cell in the tier has the capacity to generate any kind of haematopoietic lineage, i.e., it is truly a multipotential progenitor or stem cell. Once engaged on the path to differentiation, however, it loses its capacity for multi-potentiality and becomes committed to a lineage: granulocyte, erythrocyte, lymphocyte, etc. Once committed, however, the progenitor can still clonally expand to a limited extent. At this stage, the multi-potential stem cell has differentiated into a committed progenitor/stem cell. Both multi-potential and committed progenitors can be functionally recognized in *in vitro* colony-type assays, and in *in vivo* assays using lineage re-population studies with chromosome markers or reconstitution of complete haematopoiesis in lethally-irradiated recipients. A diagrammatic arrangement of cell progenitors is given in *Figure 7*. Because expansion is exponential at each stage, one multipotential stem cell can produce very large numbers of differentiated progeny *via* committed stem cells, the size of the eventual clone being determined by the number of cell divisions: thus, 100 stem cells each dividing 10 times would supply the daily need for granulocytes. Conversely, removal of a single stem cell would have a major impact on the final clone size depending on where in the tier this happened. Although this model is highly simplistic and does not take into account the many variables that are known to influence marrow kinetics, it does offer an accurate scheme on which to model haemostasis.

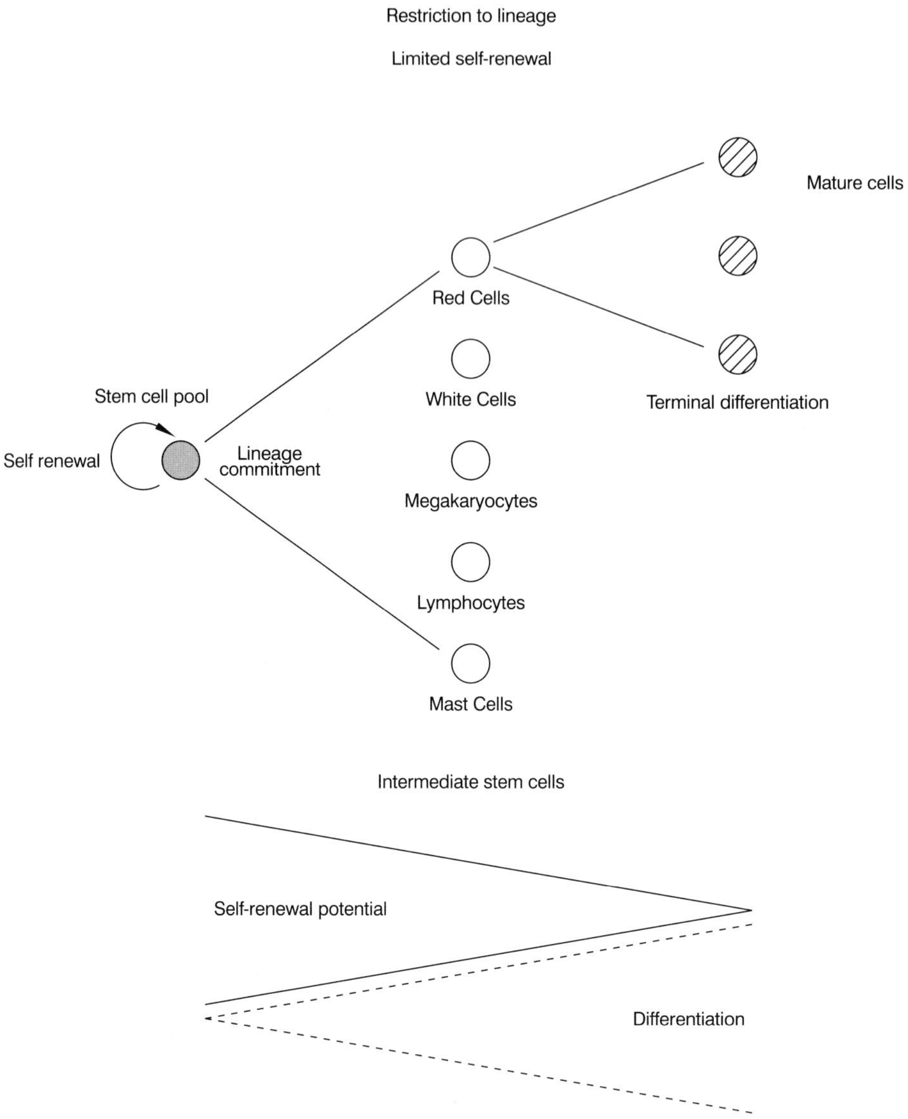

Figure 7. Stem cell arrangement in the marrow. The multipotential stem cell population (far left) can produce intermediate stem cells (middle) for all marrow lineages. The intermediate stem cells have strictly limited potential for expansion and once formed can only produce cells of that lineage, i.e. they are lineage-committed and referred to as colony forming cells or units-committed, CFU/C-C. The multipotential stem cell population has the potential to self-renew (SRP) or exit into committed stem cells. Thus the SRP progressively reduces towards the right (continuous line) while differentiation to full phenotype increases (dotted line). Mature cells, extreme right, have virtually no capacity to replicate, i.e. they are end cells. The exception is lymphoid cells which have the capacity to form 'memory' cells and clonally re-expand on exposure to antigen again. The mechanisms for control are unknown but include cytokines which promote growth and proliferation. Apoptosis is a means for deleting cells which will affect the final clone size depending on where in the hierarchy the deletion has occurred. Leukaemic cells represent nearly all stages of differentiation and their growth is a consequence of dysregulation in both apoptosis and proliferation signalling.

cloned and given the International Cytokine Classification of IL (interleukin) 3. These lines all have mixed phenotypes, sharing myeloid, lymphoid and mast-cell antigens, but are phenotypically blast-cell-like. Some can differentiate into more mature phenotypes, usually myeloid, but several lines have been described with characteristics of true stem cells in that they will reconstitute haematopoiesis in mice whose marrow has been ablated by irradiation. They all have in common an absolute requirement for IL3, and this generated intense research into both the factor and its mode of action. The eventual cloning of IL3 led to similar cloning of other haematopoietic growth factors, when it was shown that IL3 was a 33 kDa protein which signalled through specific heterodimeric receptors belonging to a family of similar ones. The cloning of receptors showed that the IL3 receptor shared one chain with the GM-CSF receptor, IL3 specificity being due to a second chain. Subsequently, the shared chain was found to be present in receptors for other growth factors, thus establishing that cells used conserved peptides in signalling complexes, generating specificity by expressing other peptides to form heterodimeric receptors.

The phenomenon of absolute factor dependency was the first clue that cell survival, and not only proliferation, was regulated in a specific way and offered a potential route to understanding how haematopoiesis could be controlled. Previously, it was generally thought that cell division required some kind of stimulus, but that without it cells either went into a quiescent state or simply died. In fact, it was uncertain whether cells needed cycling to survive, and that quiescence or extended G_1 (the G_0 state) was simply just a temporary halt in cycle progression. With the recognition that a growth factor acutely regulated survival within one cycle, a hint of haematopoietic control emerged: clonal expansion would be stimulated by the presence of specific growth factors, clonal deletion by their absence. Thus, marrow could function through various feedback loops regulating factor production by accessory cells in contact with stem cells. Whilst this model is certainly correct in some aspects, growth factor production in LTBMC is surprisingly modest, and paradoxically neither IL3 protein nor mRNA has been consistently described in them despite the fact that IL3-dependent cells originated from there. Nevertheless, IL3-dependent cells have become universal models for understanding survival and proliferation regulatory mechanisms in stem cells, and in fact represent the only intermediate stem-cell-like continuous cultures available. They are not exact replicas, nor are they true stem cells; therefore caution should be exercised in equating results to 'real' stem cells. Nevertheless, their properties are sufficiently similar to allow general principles to be arrived at, and results with them have proven their value in translating such principles to freshly-isolated bone marrow cells and indeed other cell types. The rest of this chapter therefore concentrates on IL3 dependency and apoptosis, referring only briefly to other haematopoietic cells.

7.1 *IL3 dependency* in vitro: *cell death by factor withdrawal*

Marrow cells can proliferate in semi-solid assays for several days, but require supplementation with either conditioned medium or purified growth factors such as granulocyte/macrophage colony-stimulating factors or macrophage colony-stimulating factors; isolated cells do not grow in liquid culture. However, the exceptions are IL3-dependent cells. These were first isolated from LTBMC by growth in WEHI-3b cell conditioned medium, from which cells IL3 was eventually cloned. Although considered myeloid, they express mixed antigens (Garland, 1984) and many seem related to the mast cell lineage. Characteristically, they grow rapidly (cycle time 8–10 hours) in many media supplemented with IL3. Withdrawal of IL3 results in cell death within a few hours and it was this rapid death which highlighted that cells could be deleted from marrow very quickly by simply removing a survival signal. Specificity was shown by inability of any other factor or of conditioned medium to replace IL3, although WEHI-3b conditioned medium could be replaced by T-cell condition medium. Subsequently it was shown that activated murine T cells produce relatively large amounts of IL3. The earliest studies (Garland, 1984) showed that cell death induced by IL3 withdrawal was associated with growth arrest throughout the cell cycle (consistent with the rapid death) and morphological changes to both mitochondria and nuclei, particularly nuclear DNA redistribution. Although cells changed from a very pleomorphic shape to spheroidal, there was no evidence for changes in membrane fluidity. These changes are now recognizable as characteristic of apoptosis.

7.2 IL3 withdrawal induces apoptosis

That IL3 withdrawal induced apoptosis was established by Vaux et al. (1988) who showed that over-expression of the Bcl-2 gene known to promote cell survival in B cells, also delayed (but did not inhibit) cell death induced by IL3 withdrawal in FDC-P1 cells. IL3 withdrawal induced the DNA laddering characteristic of apoptosis which was delayed by Bcl-2, hence it was established both that Bcl-2 was a potential survival gene for other cells than just B cells, and also that IL3 signalling was suppressive for whatever mechanism resulted in apoptosis in its absence. Despite the fact that the apoptotic nature of IL3-induced cell death has been known for approaching 10 years, the mechanism is still unknown.

7.3 IL3 and metabolic control

One of the key questions to answer is what cell processes does IL3 signalling regulate? Amongst the candidates are cell cycle progression, metabolism, cellular repair and housekeeping gene expression. First, the random cycle arrest excluded a specific linkage with cycle progression, although there was some preferential arrest in G2-M. Next, it was considered that IL3 regulated metabolism, particularly oxidative phosphorylation. The first studies showed, however, that IL3-dependent cells did not use oxidative phosphorylation to generate ATP (Palacios and Garland, 1984); ATP is generated entirely by glycolysis as shown by lack of oxygen consumption, production of lactic acid and resistance to inhibitors of oxidative phosphorylation such as cyanide and antimycin A (Garland and Halestrap, 1997). That IL3 was at least indirectly regulating metabolism was shown by reduction in lactic acid production within 30 minutes of IL3 withdrawal. The establishment of IL3-dependent cells stably transfected with Bcl-2 (Perkins et al., 1996) allowed extended studies on the full metabolic consequences of IL3 withdrawal, since cells were protected from early apoptosis. The studies were conducted on three cell lines: B0, a parental line, B15, transfected with Bcl-2 and A15, doubly transfected with Bcl-2 and an activated MAPKK which had been shown to delay apoptosis similarly to Bcl-2 (Perkins et al., 1996). The results were of some surprise. First, despite the rapid reduction in lactic acid production, intracellular ATP (and ADP) levels did not change, implying that reduction in glycolysis was due to loss in demand for ATP, not the other way round (Table 2). Second, in Bcl-2

Table 2. Intracellular ATP content of B0 and B15 cells after incubation in the presence and absence of IL-3 and dinitrophenol.

	ATP content of cell as a percentage of that at time zero				
	2 h	4 h	6 h	8 h	22 h
B0	174 ± 16	202 ± 13	197 ± 27	229 ± 10	56 ± 29
B0 + IL3	183 ± 12	213 ± 4	226 ± 36	243 ± 52	623 ± 183
B0 + DNP	125 ± 5	143 ± 10	102 ± 12	94 ± 35	ND
B0 + IL3 + DNP	141 ± 25	146 ± 6	155 ± 35	117 ± 36	ND
B15	184 ± 24	196 ± 12	226 ± 13	289 ± 36	211 ± 17
B15 + IL3	188 ± 31	220 ± 18	237 ± 11	323 ± 18	629 ± 129
B15 + DNP	122 ± 15	105 ± 21	237 ± 39	187 ± 3	61 ± 9
B15 + IL3 + DNP	133 ± 15	108 ± 6	239 ± 18	157 ± 7	104 ± 15

Conditions of incubation were the same as described in the legend to Fig. 8. Data for intracellular ATP content are presented as a percentage of the value found at time zero for each incubation. Mean values for ATP content of B0 and B15 cells were (±S.R.) 100 ± 18 and 121 ± 28 pmol/10^5 cells, respectively. Each data point is the mean ± S.E. from at least three separate experiments; the table summarizes seven different experiments. The increase in ATP at 2 h represents the cells recovering their metabolic status after washing in ice-cold buffer, the increase between 8 and 22 h reflects the doubling of cell number over this time period. ND, none detected.

transfected cells, identical reductions in lactate production occurred and progressed, such that cells rapidly established a state of metabolic arrest with no detectable lactate production and no loss of ATP (*Figure 8*). These cells recovered quickly on re-supplying IL3, establishing that withdrawal of IL3 was not by itself a lethal event, and that cells could remain perfectly intact and viable for many days providing the apoptotic signal was inhibited. Thus, apoptotic induction is a highly focused event suppressed by an IL3-signalling process which requires ATP.

Elsewhere it was found that apoptosis induced by IL3 withdrawal in BAF-3 cells, an established pro-B-type IL3-dependent line, could be reduced by antioxidants, suggesting that free radicals could be involved (Hockenbery *et al.*, 1993). Clearly, oxidative phosphorylation is a potential source of free radicals, which would be expected to induce PTP as well as other mitochondrial damage. First, all IL3-dependent cells tested so far have glycolytic metabolism, do not exhibit measurable oxidative phosphorylation and are resistant to inhibitors of oxidative phosphorylation. During the studies on metabolism it was considered that Bcl-2 might function by switching on oxidative phosphorylation. However, no evidence could be found that mitochondrial activity as measured by $\Delta\psi_m$ was altered by IL3 withdrawal (Garland and Halestrap, 1997; Garland *et al.*, 1997) except during early visible apoptosis when significant internal re-arrangements are already occurring. Further, antioxidants promoted apoptosis whilst mild pro-oxidants inhibited it. However, protonophores (uncoupling agents) such as FCCP and DNP both accelerated apoptosis and reduced intracellular ATP (*Figure 9*). This shows that the mitochondrial ATPase was present and had been acting as an ATP synthase until the addition of the uncouplers dissipated the proton motive force. Also, the reduction in ATP concentration induced by FCCP and DNP was accompanied by increases in lactate production some

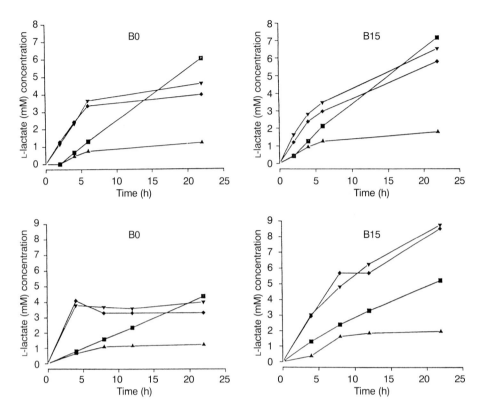

Figure 8. *Effects of IL3 withdrawal and uncoupling agents on lactic acid production in B0 and B15 cells. L-Lactate concentrations in the supernatant from incubations of B0 (left panels) or B15 (right panels) cells were measured at the times indicated. Two separate experiments are shown, with each data point representing the mean of duplicate determinations.* ■, *with IL3;* ▲, *without IL3;* ▼, *with IL3 and DNP;* ◆, *without IL3 with DNP. The cell concentration was 2 × 10⁵ cells ml⁻¹. Reproduced with permission from Mezger, J. (1997)* J. Biol. Chem. **272**, *p. 4680. © The American Society for Biochemistry and Molecular Biology.*

twofold above normal, demonstrating that glucose transport and glycolysis during IL3 signalling was under-saturated. That a reduction in ATP concentration generated a compensatory increase in glycolytic flux even in the absence of IL3 proved that the reduction in lactate concentration during IL3 withdrawal could not be due to reduced ATP concentration *per se* but must be related to ATP demand. How then is the mito-chondrial membrane potential generated in these cells? One explanation is that the ATPase can function in reverse, essentially transporting protons out of the matrix instead of channelling them back in, thus creating the normal potential across the membrane. Alternatively, there may be a small amount of residual oxidative phosphorylation which does not contribute to ATP production but maintains mitochondrial integrity. As for a role for the permeability transition pore, the lack of change in $\Delta\psi_m$ would virtually exclude this; this is supported by the finding that neither cyclosporin A nor mitochondria-specific inhibitors of the permeability transition pore had any effect on apoptosis.

Further study of the role of free radicals showed that more free radicals (as assessed by fluorescein diacetate) were generated by IL3 signalling than during IL3 withdrawal,

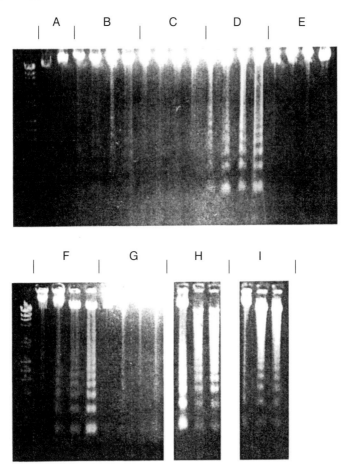

Figure 9. *Induction of apoptotic DNA ladders by the uncoupling agent, dinitrophenol. Lane A, control cells at commencement of experiment; B0, left; B15, right. Lanes in groups B through I read, from left to right, with IL3 without IL3, with IL3 with DNP, without IL3 with DNP. B, B0 after 2 h; C, B15 after 2 h; D, B0 after 4 h; E, B15 after 4 h; F, B0 after 8 h; G, B15 after 8 h; H, B0 cells after 20 h. From left to right, without IL3, with IL3 and DNP, without IL3 with DNP. I, as for H, B15 cells (controls with IL3 showed no apoptotic ladders and have been omitted from H and I). Molecular weight marker, λ Hind III digest. Reproduced with permission from Mezger, J. (1997) J. Biol. Chem.* **272,** *p. 4680. © The American Society for Biochemistry and Molecular Biology.*

which can be accounted for by the increase in production of reduced NAD/NADP during glycolysis. Thus free radicals *per se* are not inducers of apoptosis in these cells. The finding that antioxidants increased apoptosis even in the presence of IL3 now also has a likely explanation. Apoptosis is induced by activation of cysteine-dependent proteases (caspases). Increasing the reducing environment favours reduction of the critical SH group required in the caspase active site, and thus potentiates caspase activity. Conversely, pro-oxidant conditions would tend to oxidize the active site cysteine and favour apoptosis suppression. The redox potential within the cell can thus help to determine the efficacy of cleaved, potentially activated caspases and therefore control the outcome of caspase activation. It is therefore perhaps no surprise that resistance of tumour cells to cell death induction has been linked to levels of reduced glutathione or thioredoxin.

7.4 Induction of apoptosis by cytochrome c in IL3-dependent cells

Key questions to answer are whether IL3 withdrawal induces apoptosis through cytochrome c, and if so what is the route through which mitochondria are involved? Formation of DISC seems unlikely since IL3-dependent cells do not respond to TNF or to FAS ligand. It is also unlikely that oxidative phosphorylation or ATP concentration could be involved since oxidative phosphorylation is absent or very low and ATP concentration does not change. Previous studies had shown that IL3-dependent cells rapidly expressed reporter constructs introduced by electroporation (Garland *et al.*, 1992). This technique was therefore used to investigate apoptosis induced by cytochrome c in whole cells (Garland and Rudin, 1998).

   Cytochrome c efficiently induces apoptosis in growing IL3-dependent cells (*Figure 10*). The amounts needed are similar to those calculated as effective in cell-free systems on a per cell basis. There is a steep induction curve, and plateau levels of apoptosis are rapidly reached. Only certain isoforms of cytochrome c are effective, as in cell-free systems, and apoptosis is induced in cells growing in IL3. This observation suggests that cytochrome c induction is not repressible by IL3 signalling, and therefore IL3 signalling must at least prevent apoptosis by suppressing cytochrome c release. This does not necessarily mean, however, that IL3 withdrawal automatically induces cytochrome release. This conclusion was strengthened by showing that cytochrome c was still effective in cells protected from apoptosis by expression of Bcl-2 and MAPKK, both of which delay by several days apoptosis induced by IL3 withdrawal (*Figure 11*). Further, apoptosis was inhibited by peptide inhibitors of caspase 3, showing that activated caspases were necessary. This shows that cytochrome c acts downstream of both Bcl-2 and MAPKK and is consistent with the known role of Bcl-2 in preventing cytochrome release but not inhibiting activated caspases. The role of over-expressed MAPKK in these cells is not known but clearly cytochrome c acts downstream of this as well. Cytochrome c also induced apoptosis in fully metabolically quiescent cells (*Figure 12*), showing that cytochrome c-induced apoptosis is independent of metabolism. Thus, IL3-dependent cells are 'primed' for apoptosis induction by cytochrome c regardless of growth, metabolism and anti-apoptotic protein expression. The results suggested an overall model (*Figure 13*) in which the mitochondrial axis is operative but in which IL3 withdrawal has a major effect on cytoskeletal dynamics, resulting in reduced ATP demand, alterations in MAP kinases and levels of Bcl-X, and the previously-defined signalling through BAD.

7.5 Membrane potential during induction of apoptosis by cytochrome c

Reference has been made above to $\Delta\psi_m$ and its linkage to apoptosis. IL3-dependent cells maintain their $\Delta\psi_m$ despite their lack of oxidative phosphorylation. More detailed studies were therefore done to determine what, if any, relationship there might be between $\Delta\psi_m$ and IL3 withdrawal. Use was made of JC1, a carbocyanine dye which localizes to mitochondria and fluoresces orange in the presence of an intact potential, but green in its absence. These studies confirmed that mitochondria retained $\Delta\psi_m$ up until visible

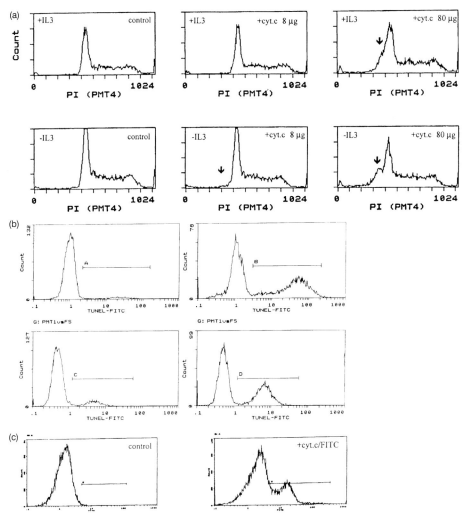

Figure 10. *Induction of apoptosis by cytochrome c in B0 cells. (a) DNA analysis by flow cyometry. Cells were electroporated with bovine cytochrome c (Sigma Ltd) at different concentrations or control protein (FCS) and examined by flow cytomery after 4 h of incubation. (Top) With IL3; (bottom) without IL3. Apoptosis is detectable at 8 μg ml⁻¹ cytochrome c (middle panels) and is significant at 80 μg ml⁻¹ (left panels, arrow). Apoptotic fractions (gate a, arrow) were as follows: with IL3 control, <1%; 8 μg ml⁻¹, 2%; 80 μg ml⁻¹, 11%; without IL3: control, <1%; 8 μg ml⁻¹, 5%; 80 μg ml⁻¹, 20%. (b) Apoptosis identified by TUNEL-labeling and flow cytometry. Intact cells were gated by forward- and side-scatter profiles. Fluorescence gates (A,B,C, and D) were then set for all positive events determined by Fluorospheres (Coulter Inc). The percentage of TUNEL-positive cells is in brackets. All cells were incubated with IL3 and were electroporated with 80 μg ml⁻¹ cytochrome c or equivalent amounts of FCS (control cells). (A and B) B0 cells examined by flow cytometry after 2 hours. (A) Controls (4%); (B) with cytochrome c (42%); (C and D) B15 cells after 4 h of incubation. (C) Controls (11%); (D) with cytochrome c (36%). (c) Uptake of cytochrome c by electroporated cells. B0 cells were electroporated as described above with 80 μg ml⁻¹ cytochrome mixed with 2 μg ml⁻¹ biotinylated cytochrome c labeled with FITC-streptavidin. Cells were washed and recultured for 1 h without IL3 before flow cytometric analysis. (Left) Control cells electroporated with unlabeled cytochrome c; (right) with labeled cytochrome c. 24% of cells were labeled; 27% of cells were apoptotic by direct microscopy at 2 h. In 3 experiments, 31 ± 5.5% were labeled by FITC-cytochrome c and 36 ± 6% were apoptotic after 2 h. Controls showed 2 ± 1.4% apoptotic cells. Reproduced from Garland and Rudin, Blood, vol. 92, pp. 1235–1246, 1998 with permission of W.B. Saunders and Co., Philadelphia.*

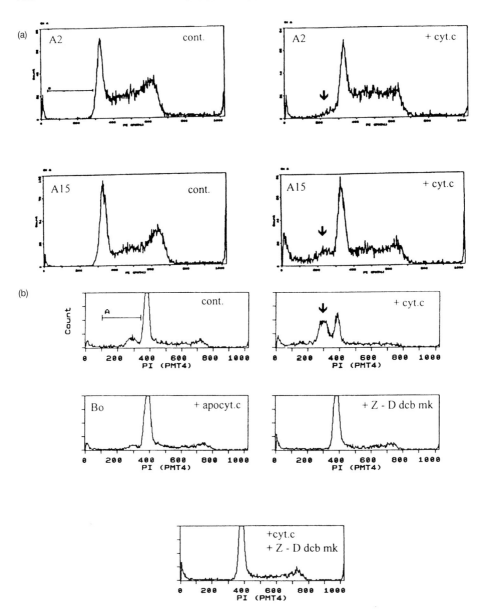

Figure 11. Cytochrome c overrides survival-signalling pathways to induce apoptosis. Cytochrome c-induced apoptosis is dominant to bcl-2 and activated MAP-KK survival signals and is mediated by caspases. (a) Cytochrome c overrides IL3, bcl-2, and activated MAP-KK. A2 cells (upper panels) expressing activated MAP-KK and A15 cells (lower panels) co-expressing activated MAP-KK and bcl-2 were electroporated with control protein (left) or bovine cytochrome c at 80 μg ml⁻¹ (right) and recultured with IL3. Pre-G1 apoptotic fractions were as follows: A2 control <1%; with cytochrome c 6%; A15: control <1%; with cytochrome c 16%. (b) Cytochrome c-induced apoptosis is dependent on caspases. B0 cells were electroporated with 80 μg ml⁻¹ apo- or holo-cytochrome c and recultured with IL3, with or without z-Ddcbmk at 50 μg ml⁻¹ (100 μg ml⁻¹). Cells were analyzed after 3 h. Pre-G1 apoptotic fractions were as follows: control cells, 16%; with holo-cytochrome c, 45%; with apo-cytochrome c, 9%; with z-Ddcbmk only, 7%; with holo-cytochrome c and z-Ddcbmk, 4%. z-Ddcbmk abolishes cytochrome c-induced DNA fragmentation. Reproduced from Garland and Rudin, Blood, vol. 92, pp. 1235–1246, 1998 with permission of W.B. Saunders and Co., Philadelphia.

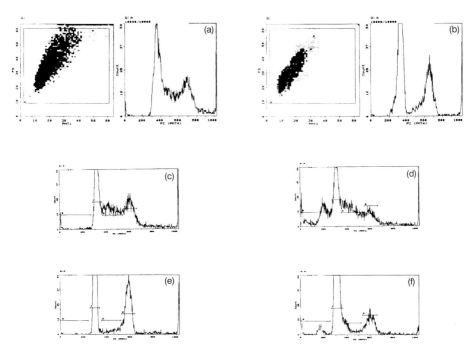

Figure 12. *Induction of apoptosis by cytochrome c in metabolically quiescent cells. Forward/side scatter plots (gate A) and DNA histograms of A15 cells quiesced for 24 h without IL3 then electroporated with bovine cytochrome c at 80 μg ml⁻¹. In growing cells, IL3 was present throughout all manipulations. DNA analysis was performed after 2 h. (a) Cells growing in IL3. (b) Cells quiesced for 24 h. (c) Growing cells electroporated with FCS. (d) Growing cells electroporated with 80 μg ml⁻¹ bovine cytochrome c. (e) Quiesced cells electroporated with FCS. (f) Quiesced cells electroporated with 80 μg ml⁻¹ cytochrome c. Gates B through E set for pre-G1, G1, S, and G2/M. The left shift in (b) is due to cell contraction. Note the pre-G1 apoptotic DNA, broadening of G1 with right shoulder in quiesced cells, and reduction in G2/M in cells electroporated with cytochrome c. Apoptosis induced by cytochrome c by direct microscopy was as follows: growing, 51%; quiesced, 43%. Controls showed less than 3% apoptosis. Reproduced from Garland and Rudin, Blood, vol. 92, pp. 1235–1246, 1998 with permission of W.B. Saunders and Co., Philadelphia.*

apoptosis occurred. To further establish the lack of involvement of membrane potential, cells were exposed to known modulators of mitochondrial function and metabolism. First, atractyloside, which inhibits the adenine nucleotide transporter and thus prevents electron transport-related ATP synthesis, had no effect on growth or apoptosis. This supported the concept that these cells do not use oxidative phosphorylation. Next, membrane potential was uncoupled with the protonophores DNP and FCCP. These resulted in rapid loss of $\Delta\psi_m$ and a decrease in ATP concentration (see above). Although apoptosis was increased with or without IL3, it was not greatly accelerated. Next, cells were exposed to oligomycin, an inhibitor of the F_oF_1 ATPase, in a medium with or without glucose. In cells with active oxidative phosphorylation, oligomycin prevents re-entry of protons into the mitochondrial matrix and thus hyperpolarizes the membrane. Oligomycin also allows the ATPase to work as an ATPase rather than synthase, and hence can reduce ATP concentration. Oligomycin induced severe morphological changes and accelerated apoptosis, but as expected did not reduce the membrane potential. In cells grown without glucose, $\Delta\psi_m$ was lost within 2 hours but this did not

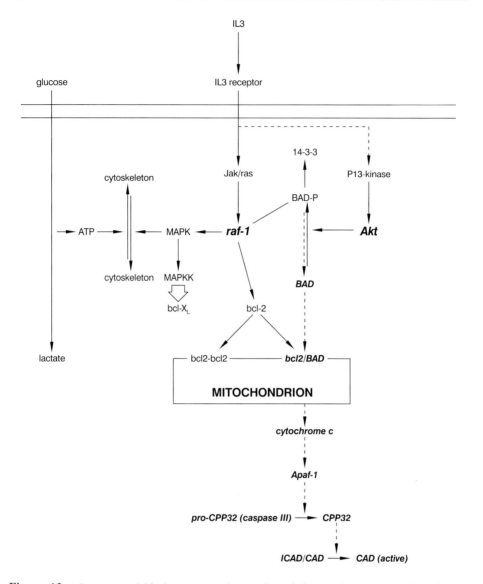

Figure 13. Schematic model linking IL3 regulation of metabolism with apoptosis and cytochrome c release. IL3 signalling activates multiple pathways, but includes activation of JAK-type kinases, STAT transcription factors, Ras, Raf, MAP-K, and Akt. MAP-K activation and increase in cytoskeletal assembly/remodelling during growth creates an ATP demand and therefore an increase in glyolysis (left side, solid lines). Simultaneously, Raf targeted to the mitochondria by bcl-2, and/or Akt, phosphorylates BAD that sequesters it to cytoplasmic 14-3-3 protein (right side, solid lines). Withdrawal of IL3 leads to reduced Raf, Akt, and MAP-K activity, loss in cytoskeletal assembly, decline in ATP demand, and downregulated glycolysis. Coincidentally, unphosphorylated BAD is redirected to heterodimerise with mitochondrial bcl-2/bcl-X_L which releases cytochrome c irrespective of $\Delta\psi_m$, activates caspase (CPP32), and CAD. Exogenous cytochrome c by-passes IL3 signalling and directly activates caspase III. Although activation of MAP kinases in the MAP-KK A12 and A15 overexpression mutants has been difficult to demonstrate, overexpressed activated MAP-KK upregulates Bcl-X_L. This would explain why A2 and A15 cells have increased survival on IL3 withdrawal but are still induced by exogenous cytochrome c. Reproduced from Garland and Rudin, Blood, vol. 92, pp. 1235–1246, 1998 with permission of W.B. Saunders and Co., Philadelphia.

result in the accelerated apoptosis seen with cytochrome c. Thus, apoptosis was dissociated from changes in membrane potential. The loss in $\Delta\psi_m$ shows that it is dependent on ATP production and suggests that in these cells it may be generated by reversal of the ATPase. However, it cannot be excluded that a low level of oxidative phosphorylation is present, sufficient to maintain potential but not contributing to metabolism, nor, if disturbed, inducing cytochrome c release. These experiments overall showed that mitochondrial potential was dependent on metabolism but not directly related to apoptosis.

Do these results have any bearing on bone marrow and normal haematopoiesis? In that growth factor signalling by IL3 appears to correspond to similar signalling by other growth factors in normal marrow it is probable that they do. Further, it has been shown that stem cells from normal marrow appear to have a component population with a primary glycolytic pathway for ATP synthesis (Garland and Katz, 1987). Whether leukaemic cells also exhibit primary glycolysis and how they respond in detail to cytokine withdrawal has not been significantly investigated. However, the study with electroporated cytochrome c above also showed that one leukaemic line (CEM) was resistant to cytochrome c induction of apoptosis, so malignant haematopoietic cells may, for example, have similar defects in caspases as in breast tumour cells (Li *et al.*, 1997). Interestingly, however, in CEM cells, resistance to cancer drugs is linked to changes in mitochondrial respiratory activity (Jia *et al.*, 1996). This may relate of course to ATP concentration but could also be more subtle and involve relationships with the F_oF_1 ATPase at present not understood.

7.6 *The IL3-dependent cell model is incomplete and has added complexity*

Recent work (J.M. Garland, unpublished data) has revealed some unexpected complexities in this system. First, these cells express granzyme B which activates all known caspases. Second, there are two forms of both pro- and activated caspase 3, only one of which is recognized by currently available antibodies. Third, the levels of pro- and activated caspase 3 are nearly ten times that found in other cells. Fourth, cell extracts are very poorly inducible by cytochrome or ATP and finally apoptosis induced by other mechanisms is not inhibited by caspase inhibitors. The expression of granzyme B considerably affects interpretation regarding the role of any caspase in this system, since, as mentioned already, granzyme B will activate all known caspases. Thus, it is not possible to distinguish mitochondrial from granzyme B activation of caspases during IL3 withdrawal. Further, the presence of non-recognizable forms of caspase 3 raises questions about biological significance, cell distribution and function. The interpretation of apoptosis mechanisms in these cells according to current concepts is therefore difficult and further work is in progress to clarify the relative contributions of these factors to apoptosis induced by growth factor withdrawal or any other agent. This illustrates an important point, that conclusions regarding apoptosis in whole cells must take account of the whole cell, not isolated portions of it, and assumptions made on current concepts may not be universally applicable.

7.7 *Conclusions from study*

These studies suggest that removing a growth factor leads to reduction in ATP demand, and subsequently to apoptosis which may be related to the cytoskeleton and/or mitochondria, but that other factors may play an important role. Certainly,

ATP is not involved directly and metabolism can be abrogated without automatic cell destruction, providing Bcl-2 is present. This is an important point because it might be expected that removing signalling and arresting metabolism might initiate necrosis. The model here behaves somewhat like tumour cells in that they can survive in adverse environments (low pO_2, low pH, poor nutrient supply); metabolic arrest in the absence of a growth stimulus could be a mechanism by which they have a survival advantage.

7.8 *Relevance of the model to disease*

The studies above illustrate the complexity of apoptosis in cells as opposed to extracts or homogenates. Further, a cell line represents one cell derived from many millions in a heterogeneous population. Apoptosis is a clear target for intervention and is being intensely investigated. The requirement for growth factors is a general principle for survival of haematopoietic cells, including leukaemias (Lotem and Sachs, 1996). Expression of anti-apoptotic genes such as Bcl-2, p53, has been widely studied (McKenna and Cotter, 1997) in leukaemias. Molecular pathology is, however, usually very diverse (Meinhardt *et al.*, 1999). This is probably because disease reflects end-stages of progressive and multiple defects, affecting cells at different stages of maturation/differentiation/'stem cell-ness'. Apoptosis and survival promotion, for example by cytokines play key roles in determining the overall balance of differentiated cells (Clarkson *et al.*, 1997; Greenberg, 1998) and it is likely that the first step is failure of apoptosis which leads to genomic instability and further disruption of normal regulation.

It is important to remember, however, that the apoptosis failure may relate only to a particular pathway and usually many tumour cells are susceptible to apoptosis; indeed in many tumours the apoptotic rate can be measured and can be high. The critical parameter is the overall balance, and in this respect tumour cells reflect the control needed in bone marrow to accurately regulate self-renewal. From this viewpoint, tumours are a problem in population dynamics in which apoptosis plays a central role. Expression of one apoptosis-related gene such as Bcl-2, may therefore have little importance if this is compensated for by another pro-apoptotic gene; or the self-renewal programme is enhanced and accelerated while the apoptosis programme is unchanged. The goal therefore is to manipulate the balance rather than attempt to activate suicide in every cell. The increased understanding also allows examination of drug treatments: could they actually *increase* anti-apoptotic gene expression in tumour cells as they can do in normal cells? Evidence is accruing that this can be the case (see also Chapters 19 and 20).

8. Final comments

This chapter has only touched on certain aspects of understanding apoptosis, which has undoubtedly been one of the most significant biological discoveries in recent years. Much progress has been made in elucidating the molecules and some of the pathways involved, and these have direct consequences for biomedical research and even to some extent for plant science (see Chapter 1). There is much left to be resolved, however, and there are likely to be many surprises ahead.

References

Adams, J.M. and Cory, S. (1998) The Bcl-2 protein family: arbiters of cell survival. *Science* **281**: 1322–1326.

Antonsson, B., Conti, F., Ciavatta, A., Montessuit, S., Lewis, S., Martinou, I., Bernasconi, L., Bernard, A., Mermod, J., Mazzei, G. *et al.* (1997) Inhibition of Bax channel forming ability by Bcl-2. *Science* **277**: 370–372.

Ashkenazi, A. and Dixit, V. (1998) Death receptors: signaling and modulation. *Science* **281**: 1305–1308.

Aspenstrom, P. (1999) The Rho GTPases have multple effects on the actin cytoskeleton. *Exp. Cell Res.* **246**: 20–25.

Atkinson, E.A., Barry, M., Darmon, A.J., Shostak, I., Turner, P., Moyer, R. and Bleakley, R. (1998) Cytotoxic T lymphocyte-assisted suicide. *J. Biol. Chem.* **273**: 21261–21266.

Bates, T.E., Loesch, A., Burnstock, G. and Clark, J.B. (1996) Mitochondrial nitric oxide synthase: a ubiquitous regulator of oxidative phosphorylation? *Biochem. Biophys. Res. Comm.* **218**: 40–44.

Baudet, C., Perret, E., Delpech, B., Kaghad, M., Brachet, P., Wion, D. and Caput, D. (1998) Differentially expressed genes in C6.9 glioma cells during vitamin D-induced cell death program. *Cell Death Diff.* **5**: 116–125.

Bjornson, C.R., Rietze, R.L., Reynolds, B.A., Magli, M. and Vescovi, A.L. (1999) Turning brain into blood: a hematopoietic fate adopted by adult neural stem cells in vivo. *Science* **283**: 534–537.

Brancolini, C., Dean, L., Rodriguez, J. and Schneider, S. (1997) Dismantling cell–cell contacts during apoptosis is coupled to a caspase-dependent proteolytic cleavage of beta catenin. *J. Cell Biol.* **139**: 759–771.

Burgess, S., Delannoy, M. and Jensen, R. (1994) MMM-1 encodes a mitochondrial outer membrane protein essential for establishing and maintaining the structure of yeast mitochondria. *J. Cell Biol.* **126**: 1375–1391.

Chalmers-Redman, R.M., Fraser, A., Carlile, G., Pong, A. and Taton, W.G. (1999) Glucose protection from MPP+-induced apoptosis depends on mitochondrial membrane potential and ATP synthase. *Biochem. Biophys. Res. Comm.* **257**: 440–447.

Chicurel, M., Chen, C.S. and Ingber, D.E. (1998) Cellular control lies in the balance of forces. *Curr. Opin. Cell Biol.* **10**: 232–239.

Clarkson, B.D., Strife, A., Wisniewski, D., Lambek, C. and Carpino, N. (1997) New understanding of the pathogenesis of CML: a prototype of early neoplasia. *Leukemia* **11**: 1404–1428.

Comelli, M., Londero, D. and Mavelli, I. (1998) Severe energy impairment consequent to inactivation of the mitochondrial ATP synthase as an early event in cell death: a mechanism for the selective sensitivity to H202 of differentiating erythroleukaemia cells. *Free Rad. Biol. Med.* **24**: 924–932.

De Vos K., Goosens, V., Boone, E., Vercammen, D., Vancompernolle K., Vandenabele, P., Haegeman, G., Fiers, W. and Grooten, J. (1998) The 55 kDa tumour necrosis factor induces clustering of mitochondria via its membrane proximal region. *J. Biol. Chem.* **273**: 9673–9690.

Dexter, T.M., Garland, J.M., Scott, D., Skolnick, E. and Metcalf, D. (1980) Growth of factor-dependent hematopoietic cell lines. *J. Exp. Med.* **149**: 1036–1043.

Dexter, T.M., Testa, N.G. and Garland, J.M. (1991) *Colony Stimulating Factors: Molecular and Cell Biology*. Marcel Dekker, New York.

Dick, T., Ray, K., Salz, H. and Chia, W. (1996) Cytoplasmic dynein (ddlcl) mutations cause morphogenetic defects and apoptotic cell death in *Drosophila melanogaster*. *Mol. Cell Biol.* **16**: 1966–1977.

Drews, G., Ebheth, A. and Mandelkow, E.-V. (1998) MAPs, MARKs and microtubule dynamics. *Trends Biochem. Sci.* **23**: 307–309.

Eguchi, Y., Shimizu, S. and Tsujimoto, Y. (1997) Intracellular ATP levels determine cell death fate by apoptosis or necrosis. *Cancer Res.* **57**: 1835–1840.

Eguchi, Y., Srinivasan, A., Tomaselli, K., Shimizu, S. and Tsujimoto, Y. (1999) ATP-dependent steps in apoptotic signal transduction. *Cancer Res.* **59**: 2174–2181.

Enver, T., Heyworth, C. and Dexter, T.M. (1998) Do stem cells play dice? *Blood* **92**: 348–349.

Esteve, P., Mbade, N. and Perona, R. (1998) Rho-regulated signals induce apoptosis in vitro and in vivo by a p53-independent but bcl-2 dependent pathway. *Oncogene* **17**: 1855–1869.

Frisch, S. and Francis, H. (1994) Disruption of epithelial cell-matrix interactions induces apoptosis. *J. Cell Biol.* **124**: 619–625.

Gabai, V.L., Kabakov, A.E and Mosin, A. (1992) Association of blebbing with assembly of cytoskeletal proteins in ATP-depleted EL-4 ascites tumour cells. *Tiss. Cell.* **24**: 171–178.

Garland, J.M. (1984) *Lymphokines 9.* Academic Press, London and New York.

Garland, J.M. and Halestrap, A. (1997) Energy metabolism during apoptosis: Bcl-2 promotes survival in haematopoietic cells induced to apoptose by growth factor withdrawal by stabilising a form of metabolic arrest. *J. Biol. Chem.* **272**: 4680–4691.

Garland, J.M. and Katz, F. (1987) Relationship of factor-induced proliferation to respiratory status in marrow progenitor cells. *Leukaemia* **1**: 558–563.

Garland, J.M. and Rudin, C. (1998) Cytochrome c induces caspase-dependent apoptosis in intact haematopoietic cells and over-rides apoptosis suppression mediated by Bcl-2, MAP kinase kinase and malignant change. *Blood* **92**: 1235–1246.

Garland, J.M., Robin, P. and Harel-Bellan, A. (1992) Haematopoietic stem cell lines activate novel enhancer-dependent expression of reporter DNA immediately after transfection by mechanisms involving interleukin 3 and protein kinase c. *Leukaemia* **6**: 729–734.

Garland, J.M., Sondergaard, K. and Jolly, J. (1997) Redox regulation of apoptosis in interleukin 3 dependent haematopoietic cells. Absence of alteration in both mitochondrial membrane potential (delta psi-m) and free radical production during apoptosis induced by IL3 withdrawal. *Brit. J. Haematol.* **99**: 756–763.

Geng, Y., Azuma, T., Tang, J., Hartwig, J., Muszynski, M., Kwiatkowski, D. and Libby, P. (1998) Caspase-3-induced gelsolin fragmentation contributes to actin cytoskeletal collapse, nucleolysis, and apoptosis of vascular smooth muscle cells exposed to proinflammatory cytokines. *Eur. J. Cell Biol.* **77**: 294–302.

Greenberg, P.L. (1998) Apoptosis and its role in the myelodysplastic syndromes: implications for disease natural history and treatment. *Leuk. Res.* **22**: 1123–1136.

Gross, A., Yin, X-M., Wang, K., Wang, K., Wei, M., Jockel, J., Milliman C., Bromage, H., Tempst, P. and Korsmeyer, S. (1999) Caspase cleaved BID targets mitochondria and is required for cytochrome c release while BCL-XL prevents this release but TNFR1/Fas death. *J. Biol. Chem.* **274**: 1156–1163.

Harris, J.L., Peterson, E., Hudig, D. and Thornberry, N. (1998) Definition and redesign of the extended substrate specificity of granzyme B. *J. Biol. Chem.* **273**: 27364–27373.

Heggeness, M., Simon, M. and Singer, J. (1978) Association of mitochondria with microtubules in cultured cells. *Proc. Natl Acad. Sci. USA* **75**: 3863–3868.

Hermann, G.J. and Shaw, J.M. (1998) Mitochondrial dynamics in yeast. *Annu. Rev. Cell Dev. Biol.* **14**: 265–303.

Hirokawa, N. (1998) Kinesin and dynein superfamily proteins and the mechanism of organelle transport. *Science* **279**: 519–526.

Hockenbery, D.M., Oltvai, Z., Yin, X-M., Milliman C. and Korsmeyer S. (1993) Bcl-2 functions in an antioxidant pathway to prevent apoptosis. *Cell* **75**: 241–252.

Hu, Y., Benedict, M.A., Wu, D., Inohara, N. and Nunez, G. (1998) Bcl-XL interacts with Apaf-1 and inhibits Apaf-1-dependent caspase 9 activation. *Proc. Natl Acad. Sci. USA* **95**: 4386–4391.

Huot, J., Houle, F., Rousseau, S., Deschesnes, R., Shah, G., and Landry, J. (1998) SAPK/p38-dependent F-actin reorganization regulates early membrane blebbing during stress-induced apoptosis. *J. Cell Biol.* **143**: 1361–1373.

Imai, Y., Kimura, T., Murakami, A., Yajima, N., Sakamaki, K. and Yonehara, S. (1999) The CED-4-homologous protein FLASH is involved in FAS-mediated activation of caspase 8 during apoptosis. *Nature* **398**: 777–785.

Inbal, B., Cohen, O., Polak-Charcon, S., Kopolovic, J., Vadai, E., Eisenbach, L. and Kimchi, K. (1997) DAP kinase links control of apoptosis to metastasis. *Nature* **390**: 180–184.

Jaffrey, S.R. and Snyder, S. (1996) PIN: an associated protein inhibitor of neuronal nitric oxide synthase. *Science* **274**: 774–778.

Janmey, P.A. (1998) The cytoskeleton and cell signaling: component localisation and mechanical coupling. *Phys. Rev.* **78**: 763–781.

Jia, L., Kelsey, S.M., Grahn, M., Jiang, X-R. and Newland, A. (1996) Increased activity and sensitivity of mitochondrial respiratory enzymes to tumour necrosis factor α-mediated inhibition is associated with increased cytotoxicity in drug-resistant leukaemic cell lines. *Blood* **87**: 2401–2410.

Junge, W. (1999) ATPase and other motor proteins. *Proc. Natl Acad. Sci. USA* **96**: 4735–4737.

Kagi, D., Vignaux, F., Ledermann, B., Burki, K., Depraetere, V., Nagata, S., Hengartner, H. and Golstein, P. (1994) Fas and perforin pathways as major mechanisms of T-cell mediated cytotoxicity. *Science* **265**: 528–530.

Kluck, R., Martin, S., Hoffman, B.M., Zhou, J., Green, D. and Newmeyer, D. (1997) Cytochrome c activation of Cpp-32 proteolysis plays a critical role in a Xenopus cell-free apoptosis system. *EMBO J.* **16**: 4639–4649.

Kong, Y-Y, Fischer, K-D., Bachmann, M., Mariathasan, S., Kozieradski, I., Nghiem, M., Bouchard, D., Bernstein, A., Ohashi, P. and Penninger, J. (1998) Vav regulates peptide-specific apoptosis in thymocytes. *J. Exp. Med.* **188**: 2099–2111.

Kothakota, S., Azuma, T., Reinhard, C., Klippel, A., Tang, J., Chu, K., McGarry, T., Kirschner, M., Koths, K., Kwiatkowski D. and Williams, L. (1997) Caspase-3-generated fragment of gelsolin: effector of morphological change in apoptosis. *Science* **278**: 294–298.

Krajewska, S., Krajewska, M., Ellerby, L., Welsh K., Zie, Z.H., Deveraux, Q.L., Salvesen, G.S., Bredesen, D.E., Rosenthal, R.E., Fiskum, G. and Reed, J.C. (1999) Release of caspase-9 from mitochondria during neuronal apoptosis and cerebral ischaemia. *Proc. Natl Acad. Sci. USA* **96**: 5752–5757.

LaCasse, E., Baird, S., Korneluk, R.G. and MacKenzie, A. (1998) The inhibitors of apoptosis (IAPs) and their emerging roles in cancer. *Oncogene* **17**: 3247–3259.

Leist, M., Single, B.S., Castoldi, A.F., Kuhnle, S. and Nicotera, P. (1997) Intracellular adenosine triphosphate (ATP) concentration: a switch in the decision between apoptosis and necrosis. *J. Exp. Med.* **185**: 1481–1486.

Li, F.L., Srinivasan, A., Wang, Y., Armstrong, R., Tomaselli, K. and Fritz, L. (1997) Cell-specific induction of apoptosis by micro-injection of cytochrome c. *J. Biol. Chem.* **272**: 30299–30305.

Li, F.L., Ambrosini, G., Chu, E., Plescia, J., Tognin, S., Marchisio, P. and Altieri, D. (1998) Control of apoptosis and mitotic spindle checkpoint by survivin. *Nature* **396**: 580–584.

Li, P., Nijhawan, D., Budihardjo, I., Srinivasula, S., Ahmed, M., Alnemri, E. and Wang, X. (1997) Cytochrome c and ATP-dependent formation of Apaf-1/caspase 9 complex initiates an apoptotic cascade. *Cell* **91**: 479–489.

Liu, X., Kim, C.N., Yang, J., Jemmerson, R. and Wang, X. (1996) Induction of apoptotic programme in cell-free extracts. Requirement for dATP and cytochrome c. *Cell* **86**: 147–157.

Lotem, J. and Sachs, L. (1996) Control of apoptosis in haematopoiesis and leukaemia by cytokines, tumour suppressor and oncogenes. *Leukemia* **10**: 925–931.

Lowenberg, B. and Touw, I.P. (1993) Haematopoietic growth factors and their receptors in acute leukaemia. *Blood* **81**: 281–285.

Machleidt, T., Geller, P., Schwandner, R., Scherer, G. and Kronke, M. (1998) Caspase 7-induced cleavage of kinectin in apoptotic cells. *FEBS Lett.* **436**: 51–54.

McKenna, S. and Cotter, T.G. (1997) Functional aspects of apoptosis in haematopoiesis and consequences of failure. *Adv. Cancer Res.* **71**: 121–164.

McNiven, M. (1998) Dynamin: a molecular motor with pinchase action. *Cell* **94**: 151–154.

Mannick, J.B., Hausladen, A., Liu, L., Hess, D., Zeng, M., Maio, Q., Kane, L., Gow, A. and Stamler, J. (1999) Fas-induced caspase denitrosylation. *Science* **284**: 651–654.

Mashima, T., Naito, M. and Tsuruo, T. (1999) Caspase-mediated cleavage of cytoskeletal actin plays a positive role in the process of morphological apoptosis. *Oncogene* **18**: 2423–2430.

Meinhardt, G., Wendtner, C.M. and Hallek, M. (1999) Molecular pathogenesis of chronic leukaemia: factors and signaling pathways regulating growth and survival. *J. Mol. Med.* **77**: 282–293.

Meredith, J., Mu, Z., Saido, T. and Du, X. (1998) Cleavage of the cytoplasmic domain of the integrin beta3 subunit during endothelial cell apoptosis. *J. Biol. Chem.* **273**: 19525–19531.

Mermall, V., Post, P. and Mooseker, M. (1998) Unconventional myosins in cell movement, membrane traffic and signal transduction. *Science* **279**: 527–533.

Metcalf, D. (1993) Haematopoietic regulators: Redundancy or subtlety? *Blood* **82**: 3515.

Metcalf, D. (1998) Lineage commitment and maturation in haematopoietic cells: the case for extrinsic regulation. *Blood* **92**: 345–348.

Michel, T. and Feron, O. (1997) Nitric oxide synthases: Which, where, how and why? *J. Clin. Invest.* **100**: 2146–2152.

Mills, J.C., Stone, N., Erhardt, J. and Pittman, R.N. (1998) Apoptotic blebbing is regulated by myosin light chain phosphorylation. *J. Cell Biol.* **140**: 627–636.

Moorman, J.P., Luu, D., Wickham, J., Bobak, D. and Hahn, C. (1999) A balance of signaling by Rho family small GTPases Rho, Rac-1 and Cdc-42 coordinates cytoskeletal morphology but not cell survival. *Oncogene* **18**: 47–57.

Muchmore, S.W., Sanier, M., Liang, H., Meadows, R.P., Harlan, J.E., Yoon H., Nettesheim, D., Chang, B., Thompson C., Wong, S. *et al.* (1996) X-ray and NMR structure of human Bcl-XL, an inhibitor of programmed cell death. *Nature* **381**: 335–341.

Nangaku, M., Sato-Yoshitaka, R. and Okada, Y. (1994) KIF-1B, a novel microtubule plus end-directed monomeric motor for transport of mitochondria. *Cell* **79**: 1209–1220.

Ng, P.W., Porter, A.G. and Janicke, R.U. (1999) Molecular cloning and characterisation of two novel pro-apoptotic isoforms of caspase 10. *J. Biol. Chem.* **274**: 10301–10308.

Nieminen, A.L., Saylor, A., Herman, B. and Lematers, J.J. (1994) ATP depletion rather than mitochondrial depolarisation mediates hepatocyte killing after metabolic inhibition. *Am. J. Physiol.* **267**: C67.

Nishida, K., Kaziro, Y. and Satoh, T. (1999) Anti-apoptotic function of Rac in hematopoietic cells. *Oncogene* **18**: 407–415.

Nobutaka, H. (1998) Kinesin and dynein superfamily proteins and the mechansim of organelle transport. *Science* **279**: 519–526.

Otsuga, D., Keegan, B., Brisch, E., Thatcher, J., Hermann, G., Bleazard, W. and Shaw, J. (1998) The dynamin-related GTPase Dnm1p control mitochondrial morphology in yeast. *J. Cell Biol.* **143**: 333–349.

Palacios, R. and Garland, J.M. (1984) Distinct mechanisms may account for the growth promoting activities of interleukin 3 on cells of lymphoid and myeloid origin. *Proc. Natl Acad. Sci. USA* **81**: 1208–1211.

Perkins, G.R., Marshall, C.J. and Collins, M. (1996) The role of MAP kinase kinase in interleukin 3 stimulation of proliferation. *Blood* **87**: 3669–3675.

Posey, S.C. and Bierer, B.E. (1999) Actin stabilization by jasplakinolide enhances apoptosis induced by cytokine deprivation. *J. Biol. Chem.* **274**: 4259–4265.

Puthalakath, H., Huang, D., O'Reilly, L.A., King, S.M. and Strasser, A. (1999) The proapoptotic activity of the bcl-2 family member Bim is regulated by interaction with the dynein motor complex. *Molec. Cell* **3**: 287–295.

Rodriguez-Crespo, I., Straub, W., Gavilanes, F. and Ortiz de Montelano, P.R. (1998) Binding of dynein light chain (PIN) to neuronal nitric oxide synthase in the absence of inhibition. *Arch. Biochem. Biophys.* **359**: 297–304.

Saitoh, M., Nishitoh, M., Fujii, K., Tobiume, Y., Sawada, M., Kawabata, M., Miyazono, K. and Ichijo, H. (1998) Mammalian thioredoxin is a direct inhibitor of apoptosis-signal-regulating kinase (ASK). *EMBO J.* **17**: 2596–2606.

Sanchez-Alcazar, J.A., Hernandez, I. and De la Torre, M.P. (1995) Down-regulation of tumour necrosis factor receptors by blockade of mitochondrial respiration. *J. Biol. Chem.* **270**: 23944–23950.

Sanchez-Alcazar, J.A., Ruiz-Cabello, J. and Hernandez-Munoz, I. (1997) Tumor necrosis factor-alpha increases ATP content in metabolically inhibited L929 cells preceding cell death. *J. Biol. Chem.* **272**: 3016–3021.

Scaffidi C., Fulda S., Srinivasan A., Friesen C., Li, F., Tomaselli, K., Debatin, K., Krammer P. and Peter, M. (1998) Two CD95 (APO-1/FAS) signaling pathways. *EMBO J.* **17**: 1675–1687.

Schoenwaelder, S. and Burridge, K. (1999) Bidirectional signaling between the cytoskeleton and integrins. *Curr. Opin. Cell Biol.* **11**: 274–286.

Shtil, A.A., Mandlekar, S., Yu, R., Walter, R.J., Hagen, K., Tan, T., Roninson, I. and Kong, A. (1999) Differential regulation of mitogen-activated protein kinases by microtubule binding agents in breast cancer cells. *Oncogene* **18**: 377–384.

Siegel, R.M., Martin, D.A., Zheng, L., Ng, S., Bertin, J. and Cohen, J. (1998) Death-effector filaments: novel cytoplasmic structures that recruit caspases and trigger apoptosis. *J. Cell Biol.* **141**: 1243–1253.

Simbla, G., Glascott, P.A., Akita, S., Hoek, J.B. and Farber, J. (1997) Two mechanisms by which ATP depletion potentiates induction of the mitochondrial permeability transition. *Am. J. Physiol.* **273**: C479.

Smirnove, S., Shurland, D-L., Ryazantsev, S. and van der Bliek, A. (1998) A human dyamin-related protein controls the distribution of mitochondria. *J. Cell Biol.* **143**: 351.

Sogo, L.F. and Yaffe, M. (1994) Regulation of mitochondrial morphology and inheritance by Mdm10p, a protein of the mitochondrial outer membrane. *J. Cell Biol.* **6**: 1361–1373.

Srinivasula, S.M., Ahmed, M., Fernandez-Alnemri, T. and Alnemri, E.S. (1998) Autoactivation of procaspase 9 by Apaf-1 mediated oligomerisation. *Molec. Cell* **1**: 949–957.

Srivastava, R.K., Srivastava, A., Korsmeyer, S., Nesterova, M., Cho-Chung, Y. and Longo, D. (1998) Involvement of microtubules in the regulation of Bcl-2 phosphorylation and apoptosis through cyclic AMP-dependent protein kinase. *Mol. Cell Biol.* **18**: 3509–3517.

Stefanelli, C., Bonavita, F. and Stanic, I. (1997) ATP deletion inhibits glucocorticoid-induced thymocyte apoptosis. *Biochem. J.* **322**: 909–917.

Susin, S.A., Lorenzo, H., Zamzami, N., Marzo, I., Snow, B., Brothers, G., Mangion, J., Jacotot, E., Costantini, P., Loeffler, M. *et al.* (1999) Molecular characterisation of mitochondrial apoptosis-inducing factor. *Nature* **397**: 441–446.

Thornberry, N., Rano, T., Peterson, E.P. and Rasper, D. (1997) A combinatorial approach defines specificities of members of the caspase family and granzyme B. *J. Biol. Chem.* **272**: 17907–17912.

Townsend, K., Trusty, J., Traupman, M., Eastman, A. and Craig, R. (1998) Expression of the anti-apoptotic MCL-1 gene product is regulated by mitogen activated protein kinase-mediated pathway triggered through microtubule disruption and protein kinase C. *Oncogene* **17**: 1223–1234.

Vallee, R.B. and Sheetz, M.P. (1996) Targeting of motor proteins. *Science* **271**: 1539–1544.

Vaux, D., Cory, S. and Adams, J.M. (1988) Bcl-2 gene promotes haematopoietic cell survival and co-operates with c-myc to immortalise pre-B cells. *Nature* **335**: 440–443.

Walker, J. (1998) ATP synthesis by rotary catalysis (Nobel Lecture). Angewandte Chemie-International Edition **37**: 2309–2319.

Walker, R. and Sheetz, M.P. (1993) Cytoplasmic microtubule-associated motors. *Annu. Rev. Biochem.* **62**: 429–451.

Watanabe, Y. and Akaike, T. (1999) Possible involvement of caspase-like family in maintenance of cytoskeletal integrity. *J. Cell Physiol.* **179**: 45–51.

Whitlock, C.A. and Witte, O. (1987) Long term culture of murine bone marrow precursors of B lymphocytes. *Meth. Enzymol.* **150**: 275–286.

Wickremasinghe, R.G. and Hoffbrand, V. (1999) Biochemical and genetic control of apoptosis: relevance to normal haematopoiesis and hematological malignancies. *Blood* **93**: 3587–3600.

Yaffee, M.P. (1999)The machinery of mitochondrial inheritance and behaviour. *Science* **283**: 1493–1496.

Yujiri, T., Sather, S., Fanger, G. and Johnson, G. (1998) Role of MEKK1 in cell survival and activation of JNK and ERK pathways defined by targeted gene expression. *Science* **282**: 1911–1914.

Zamzami, N., Marchetti, P., Castedo, M., DeCaudin, D., Macho, A., Hirsch, T., Susin, S., Petit, P.X., Mignotte, B. and Kroemer, G. (1995a) Sequential reduction of mitochondrial transmembrane potential and generation of reactive oxygen species in early programmed cell death. *J. Exp. Med.* **182**: 367–377.

Zamzami, N., Marchetti, P., Castedo, M., Zanin, C., Vayassier, J-L., Petit, P. and Kroemer, G. (1995b) Reduction in mitochondrial potential constitutes an early irreversible step of programmed cell death. *J. Exp. Med.* **181**: 1661–1672.

Zha, J., Harada, H., Yang, E., Jockel, J. and Korsmeyer, S.J. (1996) Serine phosphorylation of the death antagonist BAD in response to survival factor results in binding to 14–3–3 not Bcl-XL. *Cell* **87**: 619–628.

Zou, H., Henzel, W.J., Liu, X., Litschg, A. and Wang, X. (1997) Apaf-1, a human protein homologous to *C. elegans* CED 4 participates in cytochrome c activation of caspase 3. *Cell* **90**: 405–413.

Zou, H., Li, Y., Liu, X. and Wang, X. (1999) An APAF-1-cytochrome c multimeric complex is a functional apoptosome that activates procaspase 9. *J. Biol. Chem.* **274**: 11549–11556.

8

Organelle-specific death pathways

K. T. Pun and R. J. Brown

1. Introduction

Apoptosis, or programmed cell death (PCD), is a mode of cell death in multicellular organisms which allows the efficient removal of unwanted cells during development and tissue homeostasis. All eukaryotic cells that undergo apoptosis are characterized by morphological changes that include chromatin condensation, DNA fragmentation, membrane blebbing, and cell shrinkage. These features imply a uniform execution of the death machinery and underline the strict evolutionary conservation of apoptosis (Kerr *et al.*, 1972; see also Chapter 1). We now know that these changes are brought about by a conserved family of enzymes called caspases, which proteolytically cleave selected intracellular proteins to allow the cell to be physically 'dismantled' (Chapter 2; reviewed in Cohen, 1997; Cryns and Yuan, 1998; Nicholson and Thornberry, 1997).

While the execution phase of apoptosis, which gives rise to the pathology outlined above is well understood, less is known about how pro-apoptotic signals are propagated from the initiating event. Apoptosis induced by DNA-damage, dysfunction of endoplasmic reticulum (ER), microtubule disruption or kinase inhibition may not necessarily share upstream components, although the mechanisms will converge at the level of caspase activation. In order to understand these upstream events it will be necessary to start from the compartment in the cell that first receives the signal, and identify the earliest critical responses that determine whether the cell will undergo apoptosis.

While searching for these upstream regulators of apoptotic commitment, it is important to remember that higher eukaryotic cells are greatly compartmentalized. This adaptation allows important biochemical reactions to take place separately and free from cross-talk. It also increases the efficiency of each process by allowing the cell to group together the key enzymes on specialist membranes. This division of function within the cell brings with it the requirement to have an ability to monitor each compartment and ensure that no mistakes detrimental to the cell as a whole are made. Thus the cell requires compartment-specific sensors. An example of such a sensor, resident in the nucleus, is the protein p53, which has the dual ability to either induce

Programmed Cell Death in Animals and Plants, edited by J.A. Bryant, S.G. Hughes and J.M. Garland.
© 2000 BIOS Scientific Publishers Ltd, Oxford.

cell cycle arrest to facilitate repair, or commit the cell to apoptosis under conditions in which the genome has undergone irreparable damage (see Chapter 16). If p53 is one of the nuclear sensors with access to the apoptotic pathway, are there proteins responsible for sensing damage to the ER or mitochondria, and how do these proteins gain access to the apoptotic machinery?

This review will consider the evidence so far to suggest that multiple locations within the cell play key roles in generating signals that commit cells to apoptosis, and will focus attention on the endoplasmic reticulum (ER) and nuclear envelope.

2. Bcl-2 family proteins are present in multiple subcellular compartments

The Bcl-2 family of proteins were some of the earliest proteins identified to take part in the regulation of apoptosis (see Chapter 4). They include the founder anti-apoptotic regulators Bcl-2 and Bcl-X_L, as well as pro-apoptotic members typified by Bax, Bak, and the BH3-only proteins Nbk/Bik, Bad, Bid and Bim. These proteins are characterized by possession of one or more signature domains called bcl-2 homology domains 1 to 4 (BH1–4). Pro-apoptotic members bind to anti-apoptotic members via their BH3 domains in an antagonistic fashion to promote cell death (Chapter 4; reviewed in Adams and Cory, 1998; Brown, 1997; Kelekar and Thompson, 1998). A great deal of work has been aimed at understanding the mechanism of action of these proteins, which are able to modulate apoptosis initiated by a wide range of insults affecting different subcellular compartments. At least some of this ability is attributed to their intracellular distributions.

The anti-apoptotic protein Bcl-2 is localized to the cytoplasmic face of the ER, nuclear envelope, and the outer mitochondrial membrane (Akao et al., 1994; Jacobson et al., 1993; Krajewski et al., 1993; Lithgow et al., 1994). Each of these subcellular locations plays an important part in inhibiting apoptosis, depending on the inducing signal. Clues to the role of these organelles in mediating the apoptotic signal came from organelle targeting experiments, in which Bcl-2 expression was restricted to either the mitochondria or ER, via exchanging its C-terminal membrane insertion sequence with those from cytochrome b5 (cb5) or ActA respectively. When localized to the mitochondria, Bcl-2 protected the cell from apoptotic signals in some but not all cases. For example, MDCK cells were protected from apoptosis induced by serum deprivation, by both mitochondria targeted and wild-type Bcl-2, but not by the ER targeted mutant; whereas rat-1/myc cells were protected from apoptosis more effectively by ER-targeted Bcl-2 (Zhu et al., 1996). In contrast a cytosolic Bcl-2 mutant deleted in its transmembrane domain failed to protect at all.

One of the means by which Bcl-2 is thought to protect against apoptosis is by binding to and inhibiting the activity of Bax at the mitochondria (Oltvai et al., 1993). Although this explains the protection offered by wild-type Bcl-2 and the Bcl2-cb5 mutant, it does not explain why Bcl-2-cb5 fails to protect in some circumstances, or how Bcl-2-actA protects. Both Bcl-2 mutants were able to immunoprecipitate Bax as effectively as wild-type, showing that Bax binding alone was insufficient to explain survival promoted by Bcl-2 (Zhu et al., 1996). Furthermore, genetic approaches using gain-of-function and loss-of-function models of Bcl-2 and Bax in transgenic mice indicate that Bcl-2 is able to regulate apoptosis in the absence of Bax (Knudson and Korsmeyer, 1997). In the light of these experiments, we cannot conclude that Bcl-2 suppression of apoptosis in cells depends solely on its mitochondrial distribution.

At the ER membrane, the major targets for Bcl-2 and Bcl-X$_L$ are the pro-apoptotic proteins Bak and Nbk/Bik. By immunofluorescence, both Bak and Nbk/Bik are localized almost entirely to the ER and the nuclear envelope, and presumably exerts its pro-apoptotic effects on these membranes (Boyd et al., 1995; Han et al., 1996b; R. Brown, unpublished data). Two-hybrid analysis and co-immunolocalization data show that Bak interacts specifically with proteins that are normally localized to the ER; these include E1B-19K (Farrow et al., 1995), the S. pombe calnexin homologue Cnx-1 (Torgler et al., 1997), and a novel protein that we have designated Clone 1/24 (R. Brown, unpublished data). Cnx-1 was identified in a yeast two-hybrid screen against an S. pombe cDNA library following the observation that Bak over-expression resulted in lethality. Mammalian calnexin is involved in the normal folding of proteins and quality control in the ER lumen, and the significance of this interaction will be discussed further in the next section. No new interactors have been reported for the BH3-only protein Nbk/Bik additional to E1B-19K, Bcl-2 and Bcl-X$_L$, and it is now thought that Nbk/Bik's ability to regulate apoptosis purely depends on its dominant negative inhibition of the protective Bcl-2 family members. Since Bcl-2 family proteins act upstream of caspase activation (Boulakia et al., 1996; Chinnaiyan et al., 1996), these observations suggest that signals that impinge on the caspases must originate from both the ER and mitochondrial compartments.

In addition to endogenous cellular proteins, studying the targeting of viral proteins has proved instructive. Prevention of apoptosis following viral infection is critical for replication in the host cell. The adenoviral protein E1B-19K, a viral homologue of Bcl-2, is a potent inhibitor of apoptosis in cells (see Chapter 18). It is able to block signals that initiate from different regions of the cell, including apoptosis induced by E1A and p53-dependent mechanisms, growth factor deprivation, and by over-expression of Bax, Bak, Nbk/Bik and Fadd (reviewed in White, 1994, 1996). E1B-19K is located predominantly in the ER and perinuclear regions (White and Cipriani, 1989), even though, unlike Bcl-2, it does not possess a transmembrane domain. Its ability to block apoptosis depends on its interactions with the pro-apoptotic members of the Bcl-2 family, and on proper localization (Rao et al., 1997; see also Chapter 18). Since E1B-19K was originally described as a lamin-binding protein and nuclear/ER-localized, the majority of its activity must be centred on this organelle. So how does the ER signal for apoptosis?

3. ER dysfunction induces apoptosis

3.1 Defects in N-linked glycosylation and protein processing

The ER plays a major role in the detection of cellular stress. Many agents which induce apoptosis in the cell have been shown to adversely affect the essential functions of the ER, namely protein processing and translocation, vesicle trafficking and calcium homeostasis. These include the drugs tunicamycin, an inhibitor of N-linked glycosylation, dithiothreitol (DTT), a reducing agent which reverses the normal oxidizing environment of the ER lumen, thapsigargin which inhibits the ATP-dependent Ca^{2+} pump, and calcium ionophore.

The main mechanism engaged by cells to overcome stress in the ER, often manifested by the accumulation of unfolded or misfolded proteins, is the unfolded protein response (UPR). This involves the rapid upregulation of a group of ER-resident chaperones including the glucose-regulated proteins (Grps) Grp78/BIP and Grp94 (reviewed in Lee, 1992; Sidrauski *et al.*, 1998). This transcriptional response is mediated by the bifunctional protein kinase/endonuclease Ire1, that has a sensor function in the ER lumen (Tirasophon *et al.*, 1998; Wang *et al.*, 1998). The over-expression of Grp78/Bip in cells is able to attenuate ER stress and apoptosis in response to tunicamycin and calcium ionophore (Dorner *et al.*, 1992; Morris *et al.*, 1997). However, it is now clear that in addition to the protective response of Grp upregulation, Ire1 also controls the upregulation of another protein Chop (c/ebp homologous protein), which predisposes the cell to apoptosis. The balance between these two branches of the UPR, depending on the extent of insult, is thought to determine the survival or death outcome of the cell (Brewer *et al.*, 1997; Wang *et al.*, 1998; *Figure 1*).

Chop (also called GADD153) is a basic region leucine zipper (bZIP) transcription factor, which exists normally at very low levels but is upregulated by ER stress to initiate pro-apoptotic signals (Fornace *et al.*, 1988; Ron and Habener, 1992; Wang *et al.*, 1996). Direct evidence for the involvement of Chop in apoptosis comes from a *chop* knockout mouse. Both *in vitro* data using mouse embryonic fibroblasts and *in vivo* data showed that *chop–/–* cells were much more resistant to apoptosis compared to the heterozygote or wild-type when challenged with the *N*-glycosylation inhibitor tuni-camycin. These data demonstrated that Chop plays a crucial role in the ER stress pathway to apoptosis (Zinszner *et al.*, 1998).

The most compelling evidence for apoptosis caused by defects in ER function comes from studies on the protein Dad-1 (defender against apoptotic death). Dad-1 was identified as a temperature-sensitive mutation in BHK21 cells. At the restrictive

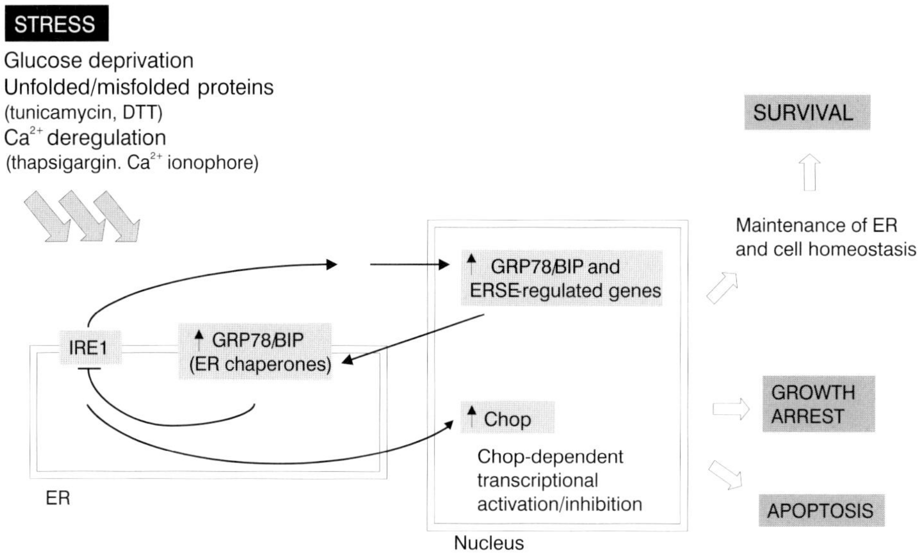

Figure 1. ER dysfunction induces the unfolded protein response (UPR). The protein Ire-1 induces up-regulation of Grp78/Bip and other ER-stress response element (ERSE) regulated genes, and the transcription factor, Chop. The balance between these protective and pro-apoptotic branches of the response determines whether the cell survives or dies.

temperature, these cells rapidly undergo apoptosis, due to instability and subsequent degradation of the mutant protein, implying that the presence of Dad-1 was essential for cell survival (Nakashima *et al.*, 1993). Sequence homology and subsequent biochemical analysis showed that Dad-1 is the mammalian homologue of the yeast oligosaccharyltransferase Ost2, an essential component of the N-glycosylation apparatus (Kelleher and Gilmore, 1997; Sanjay *et al.*, 1998; Silberstein *et al.*, 1995). The instability of Dad-1 in these mutant BHK21 cells results in dysregulation of *N*-linked glycosylation, increase in the level of unfolded or misfolded proteins in the ER, and apoptosis. The time course of disappearance of the Dad-1 protein directly correlated with upregulation of both Bip and Chop proteins (Wang *et al.*, 1996), suggesting that the signal to apoptosis is again through Ire1 and Chop.

A third link in to ER dysfunction and apoptosis is provided by studies on the proapoptotic protein Bak in the yeast *S. pombe*. Bak was expressed in *S. pombe* in an attempt to define the apoptotic pathway in a simple system, where the cell background was not hampered by the presence of other members of the Bcl-2 family. Expression of Bak in *S. pombe* resulted in complete loss of viability (Ink *et al.*, 1997; Torgler *et al.*, 1997). The molecular basis of this defect was traced to the interaction of the Bak protein with the calnexin homologue Cnx-1, as mentioned above (Torgler *et al.*, 1997).

Mammalian calnexin is an ER-resident chaperone which facilitates the folding of mono-glucosylated protein intermediates during *N*-linked glycosylation. Cnx-1 is essential for survival in *S. pombe* (Jannatipour and Rokeach, 1995; Parlati *et al.*, 1995). The lethality caused by Bak expression was dependent on Cnx-1 interaction, and was rescued by co-expression of its dimerizing partners Bcl-2, Bcl-X_L and E1B-19K, which titrated Bak away from its intracellular interaction with calnexin. In contrast, overexpression of either wild-type Cnx-1 or a C-terminally truncated mutant unable to bind Bak, failed to rescue Bak lethality, although they were able to rescue *cnx-1* knockouts. This suggested that rather than just blocking Cnx-1 function, the interaction between Bak and Cnx-1 results in a dominant lethal effect, perhaps by the propagation of a lethal signal or the recruitment of additional interactors to the Bak/Cnx-1 complex (Torgler *et al.*, 1997). Electron micrographs of Bak-expressing *S. pombe* showed accumulation of large vesicular structures, reminiscent of those demonstrated in apoptotic mammalian cells (Vander Heiden *et al.*, 1997).

Similarly, in a functional screen for suppressors of the Bax-dependent phenotype in the yeast *S. cerevisiae*, the protein BI-1 was identified whose subcellular location was predominantly ER (Xu and Reed, 1998). Expression of BI-1 suppressed apoptosis in mammalian cells induced by Bax and other agents. This suggested to us that although the overwhelming evidence for Bax function in mammalian systems is that it induces apoptosis by acting on mitochondria to release cytochrome *c* and activate the caspase cascade (Wolter *et al.*, 1997; see also Chapters 6 and 7), there is a component of Bax activity which is mitochondria-independent, and resides in the ER. In yeasts, which do not possess caspases and so do not die by 'classical' apoptosis, death from expression of Bax may be mediated by other means, perhaps through manifestation of stress in the ER which can be overcome by BI-1. We do not know if BI-1 is stress-inducible. However, it is interesting that as with other stress induced proteins, close homologues of BI-1 exist in *C. elegans* and *Arabidopsis*, as well as mammals.

These findings show that the cell detects dysfunction in different subcellular organelles by using proteins with essential functions as sensors to initiate apoptotic events. The established example in the mitochondria is cytochrome *c*, which is essential in oxidative phosphorylation, but also has a second role in directly activating caspase processing. In comparison, the ER employs at least Grp78/Bip, involved in protein processing, and the proteins Dad-1 and Cnx-1, both essential components of the *N*-linked glycosylation pathway, to convey its 'status of health' and apoptotic messages.

3.2. *Defects in calcium homeostasis*

The ability of ER-localized Bcl-2 to protect cells from apoptosis has prompted many investigators to look at the role of calcium in apoptosis. The rationale for this is that the ER represents the major intracellular Ca^{2+} store, and maintenance of Ca^{2+} homeostasis is critical to a vast array of signalling and metabolic activities within the cell. The incredible finding was that Bcl-2 participates in maintaining Ca^{2+} homeostasis in both the ER and mitochondria to prevent apoptosis.

Sequestration of high concentrations of Ca^{2+} ions to the ER is an active process that depends on the ER-associated Ca^{2+}-ATPase. Specific inhibition of this ATPase by compounds such as thapsigargin results in sustained ER Ca^{2+} pool depletion, and as a consequence elevation of cytosolic Ca^{2+} concentration $[Ca^{2+}]$. Apoptosis results by at least two routes. First, depletion of the lumenal Ca^{2+} pool causes inhibition of glyco-protein processing and maturation, resulting in propagation of an ER stress signal (McCormick *et al.*, 1997). Second, the increase in cytosolic $[Ca^{2+}]$ above a threshold induces mitochondria to rapidly accumulate Ca^{2+}, and this overloading impairs electron transport and oxidative phosphorylation, permeability transition pore opening, and leads to cell death (Kroemer *et al.*, 1997; Murphy *et al.*, 1996). Endogenous mechanisms which exist to regulate Ca^{2+} homeostasis include activation of the UPR, as Grp78/Bip and Grp94 also function as Ca^{2+}-binding proteins (Lievremont *et al.*, 1997). Ca^{2+} is also bound by the lumenal proteins calreticulin and calnexin, and artificially increasing the levels of these proteins renders cells more resistant to Ca^{2+} depletion and death by helping to sequester ER Ca^{2+} (Liu *et al.*, 1997).

Over-expression of Bcl-2 protects the cell by preventing Ca^{2+} redistribution from the ER to mitochondria following growth factor withdrawal, and inhibiting apoptosis associated Ca^{2+} waves and nuclear Ca^{2+} uptake (Baffy *et al.*, 1993; Lam *et al.*, 1994; Marin *et al.*, 1996). The presence of Bcl-2 maintains a low but detectable level of Ca^{2+} accumulation in the ER which is sufficient to restore normal functions of Ca^{2+}-dependent protein processing and transport, even in the presence of the ATPase inhibitor thapsigargin, allowing cell growth to occur (Bian *et al.*, 1997; He *et al.*, 1997). At the same time, Bcl-2 expression also enhances the uptake capacity of mitochondria, and the increased resistance of mitochondria to Ca^{2+}-induced respiratory damage (Murphy *et al.*, 1996).

Since Bcl-2 itself is not a Ca^{2+}-selective channel, nor does it bind Ca^{2+}, it must be exerting a protective effect by modulating a Ca^{2+}-dependent pore or transporter on these membranes, either directly or indirectly. One study reports the upregulation of ER Ca^{2+}-ATPase levels by Bcl-2 to enhance uptake, although the contribution of this in the presence of thapsgargin is questionable. However the ability of Bcl-2 also to physically interact with the Ca^{2+}-ATPase raises the possibility that Bcl-2 may be able to maintain its activity (Kuo *et al.*, 1998). A second possible target for modulation by Bcl-2 may be the type I inositol 1,4,5 trisphosphate receptor (IP3R1) Ca^{2+}-release channel in the ER, whose inhibition confers resistance to diverse apoptotic stimuli (Marks, 1997).

Finally, resolution of the crystal structure of Bcl-X$_L$, and by analogy Bcl-2, showed it to be similar to the pore-forming domain of diphtheria toxin (Muchmore et al., 1996), and subsequent studies showed that Bcl-2-related proteins can function as cation- or anion-selective ion channels in synthetic membranes (Antonsson et al., 1997; Minn et al., 1997; Schendel et al., 1997). It is possible to speculate that by acting as an ion channel, Bcl-2 can alter the ionic balance and fluxes of these intracellular organelles to maintain survival.

4. Apoptosis initiated at other intracellular locations

In addition to the mitochondria and ER, apoptotic signals are generated at other compartments or structures within the cell, for example, the microtubules (see Chapter 7). As the main component of the mitotic spindle, microtubules are critical to chromosomal segregation during normal cell growth and division. Cells with abnormal microtubules, and thus defects in mitosis, need to be eliminated. This in fact forms part of the cell cycle control. Agents which disrupt the functional integrity of microtubules, for example the anti-cancer drug taxol which binds to and kinetically stabilizes microtubules, induce apoptosis in proliferating cells.

Bcl-2 family proteins are once again involved. One of the mechanisms by which taxol and similar microtubule targeted compounds mediate their effects involves the phosphorylation of Bcl-2 (Haldar et al., 1995, 1997). The phosphorylation of Bcl-2 is correlated with a reduced ability to bind Bax, and abrogates its normal anti-apoptotic function, allowing the apoptotic programme to proceed. Additionally, recent reports suggest that the potent pro-apoptotic inducer Bim, a BH3-only protein, acts as a sensor at the microtubules to detect abnormalities at this site. Bim is sequestered on microtubules in an inactive form by binding to the LC8 dynein light chain, a component of the microtubule-associated dynein motor complex, and is released in its active form on induction of apoptosis (O'Connor et al., 1998; Puthalakath et al., 1999). In support of the sensor hypothesis, Bim release occurs at a time when greater than 95% of cells are still viable and the microtubules are intact, and microtubule disrupting drugs such as taxol potently induce Bim translocation. Translocation of Bim results in binding to and neutralization of Bcl-2's anti-apoptotic effect.

The inhibitor of apoptosis protein (IAP), survivin, also localized to the microtubules, links the cell cycle with cell death (Li et al., 1998). It is a cell cycle-dependent protein expressed at high levels in the G$_2$ phase of the cell cycle and regulates whether the cell enters mitosis or undergoes apoptosis, by assessing the integrity of the mitotic spindle. Survivin may actually function to preserve the integrity of the mitotic apparatus, since disruption of survivin–microtubule interactions results in loss of survivin's anti-apoptotic function and progressive increase of caspase 3 activity leading to death.

The integrity of the cell nucleus is of paramount importance, both for preservation of the cell's DNA, and for maintaining proper nuclear functions such as transcription and splicing. The role of p53 as a sensor of DNA damage has been extensively studied (reviewed in Ko and Prives, 1996). The response of cells to irreparable DNA damage is growth arrest and/or apoptosis (see Chapter 16). The latter is in part mediated by the direct transcriptional up-regulation of Bax by p53 (Han et al., 1996a; Miyashita and Reed, 1995). This is confirmed by the observation that mutant p53 containing a disabled transactivation domain is unable to induce apoptosis. However, there are clearly Bax-independent events which lead to apoptosis, as p53-mediated apoptosis can be observed in the absence of Bax up-regulation. One alternative is the down-regulation of Bcl-2 (Haldar *et al.*, 1994; Miyashita *et al.*, 1994). Another mechanism may be mediated via the induction and trafficking of Fas-FasL to the cell surface. Apoptosis by irradiation and many chemotherapeutic agents, which act via p53, has been reported to show a requirement for a functional Fas pathway (Bennett *et al.*, 1998; Owen-Schaub *et al.*, 1995).

The requirement for maintaining proper nuclear function is demonstrated in a number of examples involving apoptosis that occur in neurodegenerative diseases. The syndrome spinal cerebellar ataxia (SCA-1) is characterized by loss of Purkinje cells in the cerebellum, and has been associated with accumulation of protein complexes within the nucleus as a result of polyglutamine expansion of the neuronal specific protein Ataxin-1 (Skinner *et al.*, 1997). Other polyglutamine-expansion conditions such as Huntington's disease are also characterized by the appearance of protein aggregates within the nucleus resulting in neuronal death. Experiments in which expression of these proteins were targeted to the *Drosophila* eye showed conclusively that it is the over-expression of these mutant proteins which forms the basis of cell loss (Jackson *et al.*, 1998; Warrick *et al.*, 1998). Moreover, expression of just the isolated polyglutamine repeats is sufficient for the formation of these nuclear inclusions and induction of apoptosis (Sanchez *et al.*, 1999). The presence of these aggregates must either alter the subnuclear environment in such a way as to disrupt essential nuclear processes, for example, DNA condensation and de-condensation, or cause sequestration of certain nuclear proteins required for normal nuclear functions such as DNA repair or pre-mRNA splicing. TUNEL staining and the co-localization of caspases to these structures confirmed that death was by apoptosis. These findings illustrate the importance of understanding the role of subcellular components of apoptosis in the context of disease.

5. The role of subcellular apoptotic complexes

A fundamental question which needs to be addressed is, how do apoptotic signals generated from different subcellular compartments gain access to the apoptotic machinery?

The recruitment of large protein complexes via specific protein–protein interactions has been reported to play a crucial role in co-ordinating the initiating signals that lead to activation of apoptosis (*Figure 2*). The most extensively characterized has been the formation of the DISC (death-inducing signalling complex) at the plasma membrane as a direct result of activation of death receptors such as fas (Kischkel *et al.*, 1995). Binding of ligand to the Fas receptor (Apo-1/CD95) causes receptor trimerization, and the sequential recruitment of the adapter molecule Fadd/Mort1, and pro-caspase 8 (Flice/Mach1) via specific interactions of death domains (DD) and death effector domains (DED) on these proteins (Boldin *et al.*, 1996; Muzio *et al.*, 1996). Components of the DISC complex were elegantly identified by 2D-gel electrophoresis of fas immunoprecipitated proteins, followed by micro protein sequencing by nano-electrospray tandem mass-spectrometry (nano-ES MS/MS) (Muzio *et al.*, 1996). Formation of this complex resulted in the induced proximity of pro-caspase 8 which catalysed their activation by auto- or trans-processing (Medema *et al.*, 1997). The ensuing activation of downstream effector caspases ends ultimately in the proteolytic processing of intracellular proteins and manifestation of apoptotic morphology. A

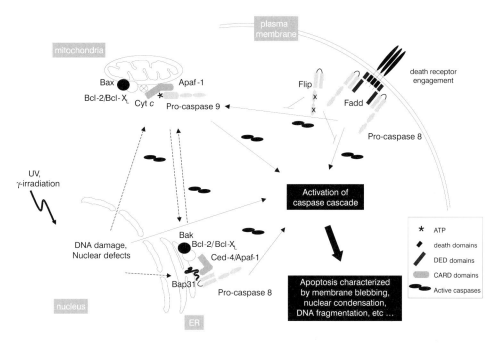

Figure 2. *Multiple subcellular complexes allow apoptotic signals initiated from different locations within the cell to gain access to the common apoptotic machinery, i.e. the effector caspases. Modulation of these complexes by adapter molecules and bcl-2 family proteins determine activation of the caspase cascade. Cross-talk between the complexes allow the cell to coordinate its apoptotic operation.*

similar complex exists downstream of the TNFR1 receptor, except that an additional adapter molecule TRADD is required for the initial recruitment of Fadd. The composition of the complex can be altered by the presence of additional proteins. These include FLIP, an inactive homologue and competitor of pro-caspase 8 which acts as a dominant negative to prevent the proteolytic cascade (Thome *et al.*, 1997; see also Chapter 2); and FLASH, a Ced-4/Apaf-1-like molecule that binds both FADD and pro-caspase 8, which takes part in complex activation and is believed to be the target for intervention by the protector E1B-19K (Imai *et al.*, 1999).

Increasing evidence now shows that in addition to the plasma membrane, protein complexes are assembled at the membrane surface of organelles to initiate the apoptotic cascade. Protein complexes have been reported on the mitochondria. The apoptosome complex contains at least the components cytochrome *c*, Bcl-X$_L$, Apaf-1, pro-caspase 9 (Li *et al.*, 1997; Zou *et al.*, 1997). On the cytoplasmic surface of the mitochondria, the Ced-4 homologue Apaf-1 serves as a docking protein for pro-caspase 9 and cytochrome c. The binding of released mitochondrial cytochrome *c* concomitant with dATP hydrolysis induces a conformation change in Apaf-1 to allow self-association, and exposes its caspase recruitment domain (CARD) to bind pro-caspase 9. As with caspase 8 in the DISC, the outcome is that caspase 9 is activated to cleave downstream effector caspases, including the direct cleavage of caspase 3 (Srinivasula *et al.*, 1998; Zou *et al.*, 1999). Composition of proteins in the complex determines whether the pathway is activated or not. It is negatively regulated by Bcl-X$_L$ whose binding to

Apaf-1 makes it unavailable to cytochrome *c* and dATP (Hu *et al.*, 1998; Huang *et al.*, 1998; Pan *et al.*, 1998). In contrast, pro-apoptotic Bax and Bak promote the active complex by disrupting the Bcl-X$_L$/Apaf-1 interaction.

On the ER membrane, the apoptotic complex is assembled by the ER-resident integral membrane molecule Bap31. Although its normal function is the trafficking of specific newly synthesized membrane proteins, including immunoglobulin and cellu-brevin, from the ER to the golgi apparatus(Annaert *et al.*, 1997), it is instrumental in the transmission of an ER apoptotic signal to downstream effectors. Bcl-X$_L$/Bcl-2, Ced-4 and pro-caspase 8 have been immunoprecipitated with Bap31. Cleavage of Bap31 by caspase 8 gives rise to an N-terminal p20 fragment that is a potent activator of cell death. As with the apoptosome, these events can be modulated by Bcl-2 family proteins. Bap31 binding by the anti-apoptotic protein Bcl-X$_L$ prevents its proteolysis by caspase 8, whereas Bax and presumably Bak prevents association of Bcl-X$_L$/Bcl-2 with the complex, promoting Bap31 cleavage at the caspase recognition site, leading to apoptosis (Ng *et al.*, 1997; Ng and Shore, 1998).

Thus it can be demonstrated that specific apoptotic complexes can be assembled at different sites in the cell in response to local pro-apoptotic signals. The complexes contain both proteins which are organelle-specific such as Bap31, as well as common molecules such as the adapter Apaf-1 and initiator procaspases, which mediate direct access to the common apoptotic machinery in the cytoplasm, i.e. the effector caspases. Through binding these common adapters, the Bcl-2 family proteins can modulate the activity at each complex to regulate survival of the cell.

6. Concluding remarks

We have shown above that cells may respond to specific apoptotic stimuli by initiating apoptosis from specific organelles. Two questions remain, (1) at which stage is the cell committed to die? and (2) at what point do the apoptotic pathways converge?

It is clear that all apoptotic signals channel down to the caspases which execute the elimination of the cell. However, commitment is clearly upstream of the caspases. Except in the worm *C. elegans*, where inactivation of its sole caspase Ced-3 leads to rescue of all 131 cells that are normally destined to die (see Chapter 1), in general, caspase inhibitors can only delay death or prevent the manifestation of apoptotic morphology, but not fully rescue vertebrate cells or retain their clonogenic potential. While *C. elegans* has been instructive in helping to define many genes involved in the apoptotic programme, it is important to remember that this is developmental cell death, which may differ from other forms of cell death in adult tissues. In the adult worm the programme is switched off, except in germ cells, and it becomes completely resistant to irradiation and chemotherapeutic inducers of apoptosis. In contrast, somatic cells of nearly all vertebrates have continuous access to the apoptotic machinery, and are able to respond to a variety of stimuli. The complexity of being a differentiated cell in a complex tissue of a multicellular organism brings with it higher demands for 'social' control (Raff, 1992). The cell needs to be good at responding to signals from the external environment such as availability of nutrients, survival factors, and death ligands, as well as to signals generated from within, i.e. from sensors that police functions within the cell. In vertebrate cells, therefore, organelle dysfunction may be the commitment, and it is the prevention or neutralization of this damage which is crucial for rescue to occur.

So where do the apoptotic pathways converge? It is significant that the vast majority of experimental data and studies of knockout mice segregate apoptotic signals into two groups: death receptor mediated signals, and 'other' apoptotic inducers that encompass stimuli as diverse as UV irradiation, growth factor withdrawal and tunicamycin. Implicit in this is that the latter group share a common denominator. For the death receptor pathway, it is clear that signals from the cell surface are able to directly activate the caspase cascade in a linear pathway leading to death. However, for signals that initiate from the mitochondria, ER and microtubules, the route is less apparent, as they rarely occur in isolation and spatially there is actually close apposition between these structures (Rizzuto *et al.*, 1998). Experiments in simple organisms such as yeast have helped to define the role of particular organelles in cell death. The view held by many is that apoptotic pathways converge at the mitochondria. Although we cannot rule out this model, in view of the complexity of the mammalian system, it seems likely that the cell may activate multiple sites by network signalling, regardless of the location of the initiating signal, to allow a more efficient apoptotic operation to take place. Mechanistically, cross-talk between the various initiating complexes can be achieved through common adapter molecules such as Apaf-1, and cytosolic caspases, which can be recruited onto different organelles to cleave Bid, Bap31, and other targets. Loss of ER and mitochondrial functional integrity may alter the intracellular ionic environment to facilitate activation of the caspases.

There is clearly much cross-talk between the various apoptotic signalling pathways within the mammalian cell. Future studies involving the critical timing of events in endogenous systems should yield further insights into how this intricate network is connected and the mechanisms involved.

References

Adams, J.M. and Cory, S. (1998) The Bcl-2 protein family: arbiters of cell survival. *Science* **281**: 1322–1326.

Akao, Y., Otsuki, Y., Kataoka, S., Ito, Y. and Tsujimoto, Y. (1994) Multiple subcellular localization of bcl-2: detection in nuclear outer membrane, endoplasmic reticulum membrane, and mitochondrial membranes. *Cancer Res.* **54**: 2468–2471.

Annaert, W.G., Becker, B., Kistner, U., Reth, M. and Jahn, R. (1997) Export of cellubrevin from the endoplasmic reticulum is controlled by BAP31. *J. Cell Biol.* **139**: 1397–1410.

Antonsson, B., Conti, F., Ciavatta, A., Montessuit, S., Lewis, S., Martinou, I., Bernasconi, L., Bernard, A., Mermod, J.J., Mazzei, G. *et al.* (1997) Inhibition of Bax channel-forming activity by Bcl-2. *Science* **277**: 370–372.

Baffy, G., Miyashita, T., Williamson, J.R. and Reed, J.C. (1993) Apoptosis induced by withdrawal of interleukin-3 (IL-3) from an IL-3-dependent hematopoietic cell line is associated with repartitioning of intracellular calcium and is blocked by enforced Bcl-2 oncoprotein production. *J. Biol. Chem.* **268**: 6511–6519.

Bennett, M., Macdonald, K., Chan, S.W., Luzio, J.P., Simari, R. and Weissberg, P. (1998) Cell surface trafficking of Fas: a rapid mechanism of p53-mediated apoptosis. *Science* **282**: 290–293.

Bian, X., Hughes, F.M., Jr., Huang, Y., Cidlowski, J.A. and Putney, J.W., Jr. (1997) Roles of cytoplasmic Ca^{2+} and intracellular Ca^{2+} stores in induction and suppression of apoptosis in S49 cells. *Am. J. Physiol.* **272**: C1241–C1249.

Boldin, M.P., Goncharov, T.M., Goltsev, Y.V. and Wallach, D. (1996) Involvement of MACH, a novel MORT1/FADD-interacting protease, in Fas/APO-1- and TNF receptor-induced cell death. *Cell* **85**: 803–815.

Boulakia, C.A., Chen, G., Ng, F.W., Teodoro, J.G., Branton, P.E., Nicholson, D.W., Poirier, G.G. and Shore, G.C. (1996) Bcl-2 and adenovirus E1B 19 kDa protein prevent E1A-induced processing of CPP32 and cleavage of poly(ADP-ribose) polymerase. *Oncogene* **12**: 529–535.

Boyd, J.M., Gallo, G.J., Elangovan, B., Houghton, A.B., Malstrom, S., Avery, B.J., Ebb, R.G., Subramanian, T., Chittenden, T., Lutz, R.J. *et al.* (1995) Bik, a novel death-inducing protein shares a distinct sequence motif with Bcl-2 family proteins and interacts with viral and cellular survival-promoting proteins. *Oncogene* **11**: 1921–1928.

Brewer, J.W., Cleveland, J.L. and Hendershot, L.M. (1997) A pathway distinct from the mammalian unfolded protein response regulates expression of endoplasmic reticulum chaperones in non-stressed cells. *EMBO J.* **16**: 7207–7216.

Brown, R. (1997) The bcl-2 family of proteins. *Br. Med. Bull.* **53**: 466–477.

Chinnaiyan, A.M., Orth, K., O'Rourke, K., Duan, H., Poirier, G.G. and Dixit, V.M. (1996) Molecular ordering of the cell death pathway. Bcl-2 and Bcl-xL function upstream of the CED-3-like apoptotic proteases. *J. Biol. Chem.* **271**: 4573–4576.

Cohen, G.M. (1997) Caspases: the executioners of apoptosis. *Biochem. J.* **326**: 1–16.

Cryns, V. and Yuan, J. (1998) Proteases to die for. *Genes Dev.* **12**: 1551–1570. [Published erratum appears in *Genes Dev* 1999 Feb 1;13(3): 371].

Dorner, A.J., Wasley, L.C. and Kaufman, R.J. (1992) Overexpression of GRP78 mitigates stress induction of glucose regulated proteins and blocks secretion of selective proteins in Chinese hamster ovary cells. *EMBO J.* **11**: 1563–1571.

Farrow, S.N., White, J.H., Martinou, I., Raven, T., Pun, K.T., Grinham, C.J., Martinou, J.C. and Brown, R. (1995) Cloning of a bcl-2 homologue by interaction with adenovirus E1B 19K. *Nature* **374**: 731–733. [Published erratum appears in *Nature* 1995 Jun 1;375(6530): 431].

Fornace, A.J., Jr., Alamo, I., Jr. and Hollander, M.C. (1988) DNA damage-inducible transcripts in mammalian cells. *Proc. Natl Acad. Sci. USA* **85**: 8800–8804.

Haldar, S., Negrini, M., Monne, M., Sabbioni, S. and Croce, C.M. (1994) Down-regulation of bcl-2 by p53 in breast cancer cells. *Cancer Res.* **54**: 2095–2097.

Haldar, S., Jena, N. and Croce, C.M. (1995) Inactivation of Bcl-2 by phosphorylation. *Proc. Natl Acad. Sci. USA* **92**: 4507–4511.

Haldar, S., Basu, A. and Croce, C.M. (1997) Bcl2 is the guardian of microtubule integrity. *Cancer Res.* **57**: 229–233.

Han, J., Sabbatini, P., Perez, D., Rao, L., Modha, D. and White, E. (1996a) The E1B 19K protein blocks apoptosis by interacting with and inhibiting the p53-inducible and death-promoting Bax protein. *Genes Dev.* **10**: 461–477.

Han, J., Sabbatini, P. and White, E. (1996b) Induction of apoptosis by human Nbk/Bik, a BH3-containing protein that interacts with E1B 19K. *Mol. Cell. Biol.* **16**: 5857–5864.

He, H., Lam, M., McCormick, T.S. and Distelhorst, C.W. (1997) Maintenance of calcium homeostasis in the endoplasmic reticulum by Bcl-2. *J. Cell Biol.* **138**: 1219–1228.

Hu, Y., Benedict, M.A., Wu, D., Inohara, N. and Nunez, G. (1998) Bcl-XL interacts with Apaf-1 and inhibits Apaf-1-dependent caspase-9 activation. *Proc. Natl Acad. Sci. USA* **95**: 4386–4391.

Huang, D.C., Adams, J.M. and Cory, S. (1998) The conserved N-terminal BH4 domain of Bcl-2 homologues is essential for inhibition of apoptosis and interaction with CED-4. *EMBO J.* **17**: 1029–1039.

Imai, Y., Kimura, T., Murakami, A., Yajima, N., Sakamaki, K. and Yonehara, S. (1999) The CED-4-homologous protein FLASH is involved in Fas-mediated activation of caspase-8 during apoptosis. *Nature* **398**: 777–785.

Ink, B., Zornig, M., Baum, B., Hajibagheri, N., James, C., Chittenden, T. and Evan, G. (1997) Human Bak induces cell death in *Schizosaccharomyces pombe* with morphological changes similar to those with apoptosis in mammalian cells. *Mol. Cell. Biol.* **17**: 2468–2474.

Jackson, G.R., Salecker, I., Dong, X., Yao, X., Arnheim, N., Faber, P.W., MacDonald, M.E. and Zipursky, S.L. (1998) Polyglutamine-expanded human huntingtin transgenes induce degeneration of *Drosophila* photoreceptor neurons. *Neuron* **21**: 633–642.

Jacobson, M.D., Burne, J.F., King, M.P., Miyashita, T., Reed, J.C. and Raff, M.C. (1993) Bcl-2 blocks apoptosis in cells lacking mitochondrial DNA. *Nature* **361**: 365–369.

Jannatipour, M. and Rokeach, L.A. (1995) The *Schizosaccharomyces pombe* homologue of the chaperone calnexin is essential for viability. *J. Biol. Chem.* **270**: 4845–4853.

Kelekar, A. and Thompson, C.B. (1998) Bcl-2-family proteins: the role of the BH3 domain in apoptosis. *Trends Cell Biol.* **8**: 324–330.

Kelleher, D.J. and Gilmore, R. (1997) DAD1, the defender against apoptotic cell death, is a subunit of the mammalian oligosaccharyltransferase. *Proc. Natl Acad. Sci. USA* **94**: 4994–4999.

Kerr, J.F., Wyllie, A.H. and Currie, A.R. (1972) Apoptosis: a basic biological phenomenon with wide-ranging implications in tissue kinetics. *Br. J. Cancer* **26**: 239–257.

Kischkel, F.C., Hellbardt, S., Behrmann, I., Germer, M., Pawlita, M., Krammer, P.H. and Peter, M.E. (1995) Cytotoxicity-dependent APO-1 (Fas/CD95)-associated proteins form a death-inducing signaling complex (DISC) with the receptor. *EMBO J.* **14**: 5579–5588.

Knudson, C.M. and Korsmeyer, S.J. (1997) Bcl-2 and Bax function independently to regulate cell death. *Nat. Genet.* **16**: 358–363.

Ko, L.J. and Prives, C. (1996) p53: puzzle and paradigm. *Genes Dev.* **10**: 1054–1072.

Krajewski, S., Tanaka, S., Takayama, S., Schibler, M.J., Fenton, W. and Reed, J.C. (1993) Investigation of the subcellular distribution of the bcl-2 oncoprotein: residence in the nuclear envelope, endoplasmic reticulum, and outer mitochondrial membranes. *Cancer Res.* **53**: 4701–4714.

Kroemer, G., Zamzami, N. and Susin, S.A. (1997) Mitochondrial control of apoptosis. *Immunol. Today* **18**: 44–51.

Kuo, T.H., Kim, H.R., Zhu, L., Yu, Y., Lin, H.M. and Tsang, W. (1998) Modulation of endoplasmic reticulum calcium pump by Bcl-2. *Oncogene* **17**: 1903–1910.

Lam, M., Dubyak, G., Chen, L., Nunez, G., Miesfeld, R.L. and Distelhorst, C.W. (1994) Evidence that BCL-2 represses apoptosis by regulating endoplasmic reticulum-associated Ca^{2+} fluxes. *Proc. Natl Acad. Sci. USA* **91**: 6569–6573.

Lee, A.S. (1992) Mammalian stress response: induction of the glucose-regulated protein family. *Curr. Opin. Cell Biol.* **4**: 267–273.

Li, F., Ambrosini, G., Chu, E.Y., Plescia, J., Tognin, S., Marchisio, P.C. and Altieri, D.C. (1998) Control of apoptosis and mitotic spindle checkpoint by survivin. *Nature* **396**: 580–584.

Li, P., Nijhawan, D., Budihardjo, I., Srinivasula, S.M., Ahmad, M., Alnemri, E.S. and Wang, X. (1997) Cytochrome c and dATP-dependent formation of Apaf-1/caspase-9 complex initiates an apoptotic protease cascade. *Cell* **91**: 479–489.

Lievremont, J.P., Rizzuto, R., Hendershot, L. and Meldolesi, J. (1997) BiP, a major chaperone protein of the endoplasmic reticulum lumen, plays a direct and important role in the storage of the rapidly exchanging pool of Ca^{2+}. *J. Biol. Chem.* **272**: 30873–30879.

Lithgow, T., van Driel, R., Bertram, J.F. and Strasser, A. (1994) The protein product of the oncogene bcl-2 is a component of the nuclear envelope, the endoplasmic reticulum, and the outer mitochondrial membrane. *Cell Growth Differ.* **5**: 411–417.

Liu, H., Bowes, R.C., 3rd, van de Water, B., Sillence, C., Nagelkerke, J.F. and Stevens, J.L. (1997) Endoplasmic reticulum chaperones GRP78 and calreticulin prevent oxidative stress, Ca^{2+} disturbances, and cell death in renal epithelial cells. *J. Biol. Chem.* **272**: 21751–21759.

McCormick, T.S., McColl, K.S. and Distelhorst, C.W. (1997) Mouse lymphoma cells destined to undergo apoptosis in response to thapsigargin treatment fail to generate a calcium-mediated grp78/grp94 stress response. *J. Biol. Chem.* **272**: 6087–6092.

Marin, M.C., Fernandez, A., Bick, R.J., Brisbay, S., Buja, L.M., Snuggs, M., McConkey, D.J., von Eschenbach, A.C., Keating, M.J. and McDonnell, T.J. (1996) Apoptosis suppression by bcl-2 is correlated with the regulation of nuclear and cytosolic Ca^{2+}. *Oncogene* **12**: 2259–2266.

Marks, A.R. (1997) Intracellular calcium-release channels: regulators of cell life and death. *Am. J. Physiol.* **272**: H597–H605.

Medema, J.P., Scaffidi, C., Kischkel, F.C., Shevchenko, A., Mann, M., Krammer, P.H. and Peter, M.E. (1997) FLICE is activated by association with the CD95 death-inducing signaling complex (DISC). *EMBO J.* **16**: 2794–2804.

Minn, A.J., Velez, P., Schendel, S.L., Liang, H., Muchmore, S.W., Fesik, S.W., Fill, M. and Thompson, C.B. (1997) Bcl-x(L) forms an ion channel in synthetic lipid membranes. *Nature* **385**: 353–357.

Miyashita, T. and Reed, J.C. (1995) Tumor suppressor p53 is a direct transcriptional activator of the human bax gene. *Cell* **80**: 293–299.

Miyashita, T., Harigai, M., Hanada, M. and Reed, J.C. (1994) Identification of a p53-dependent negative response element in the bcl-2 gene. *Cancer Res.* **54**: 3131–3135.

Morris, J.A., Dorner, A.J., Edwards, C.A., Hendershot, L.M. and Kaufman, R.J. (1997) Immunoglobulin binding protein (BiP) function is required to protect cells from endoplasmic reticulum stress but is not required for the secretion of selective proteins. *J. Biol. Chem.* **272**: 4327–4334.

Muchmore, S.W., Sattler, M., Liang, H., Meadows, R.P., Harlan, J.E., Yoon, H.S., Nettesheim, D., Chang, B.S., Thompson, C.B., Wong, S.L. *et al.* (1996) X-ray and NMR structure of human Bcl-xL, an inhibitor of programmed cell death. *Nature* **381**: 335–341.

Murphy, A.N., Bredesen, D.E., Cortopassi, G., Wang, E. and Fiskum, G. (1996) Bcl-2 potentiates the maximal calcium uptake capacity of neural cell mitochondria. *Proc. Natl Acad. Sci. USA* **93**: 9893–9898.

Muzio, M., Chinnaiyan, A.M., Kischkel, F.C., O'Rourke, K., Shevchenko, A., Ni, J., Scaffidi, C., Bretz, J.D., Zhang, M., Gentz, R. *et al.* (1996) FLICE, a novel FADD-homologous ICE/CED-3-like protease, is recruited to the CD95 (Fas/APO-1) death-inducing signaling complex. *Cell* **85**: 817–827.

Nakashima, T., Sekiguchi, T., Kuraoka, A., Fukushima, K., Shibata, Y., Komiyama, S. and Nishimoto, T. (1993) Molecular cloning of a human cDNA encoding a novel protein, DAD1, whose defect causes apoptotic cell death in hamster BHK21 cells. *Mol. Cell. Biol.* **13**: 6367–6374.

Ng, F.W. and Shore, G.C. (1998) Bcl-XL cooperatively associates with the Bap31 complex in the endoplasmic reticulum, dependent on procaspase-8 and Ced-4 adaptor. *J. Biol. Chem.* **273**: 3140–3143.

Ng, F.W., Nguyen, M., Kwan, T., Branton, P.E., Nicholson, D.W., Cromlish, J.A. and Shore, G.C. (1997) p28 Bap31, a Bcl-2/Bcl-XL- and procaspase-8-associated protein in the endoplasmic reticulum. *J. Cell Biol.* **139**: 327–338.

Nicholson, D.W. and Thornberry, N.A. (1997) Caspases: killer proteases. *Trends Biochem. Sci.* **22**: 299–306.

O'Connor, L., Strasser, A., O'Reilly, L.A., Hausmann, G., Adams, J.M., Cory, S. and Huang, D.C. (1998) Bim: a novel member of the Bcl-2 family that promotes apoptosis. *EMBO J.* **17**: 384–395.

Oltvai, Z.N., Milliman, C.L. and Korsmeyer, S.J. (1993) Bcl-2 heterodimerizes in vivo with a conserved homolog, Bax, that accelerates programmed cell death. *Cell* **74**: 609–619.

Owen-Schaub, L.B., Zhang, W., Cusack, J.C., Angelo, L.S., Santee, S.M., Fujiwara, T., Roth, J.A., Deisseroth, A.B., Zhang, W.W., Kruzel, E. *et al.* (1995) Wild-type human p53 and a temperature-sensitive mutant induce Fas/APO-1 expression. *Mol. Cell. Biol.* **15**: 3032–3040.

Pan, G., O'Rourke, K. and Dixit, V.M. (1998) Caspase-9, Bcl-XL, and Apaf-1 form a ternary complex. *J. Biol. Chem.* **273**: 5841–5845.

Parlati, F., Dignard, D., Bergeron, J.J. and Thomas, D.Y. (1995) The calnexin homologue cnx1+ in *Schizosaccharomyces pombe*, is an essential gene which can be complemented by its soluble ER domain. *EMBO J.* **14**: 3064–3072.

Puthalakath, H., Huang, D.C., O'Reilly, L.A., King, S.M. and Strasser, A. (1999) The proapoptotic activity of the Bcl-2 family member Bim is regulated by interaction with the dynein motor complex. *Mol. Cell* **3**: 287–296.

Raff, M.C. (1992) Social controls on cell survival and cell death. *Nature* **356**: 397–400.

Rao, L., Modha, D. and White, E. (1997) The E1B 19K protein associates with lamins in vivo and its proper localization is required for inhibition of apoptosis. *Oncogene* **15**: 1587–1597.

Rizzuto, R., Pinton, P., Carrington, W., Fay, F.S., Fogarty, K.E., Lifshitz, L.M., Tuft, R.A. and Pozzan, T. (1998) Close contacts with the endoplasmic reticulum as determinants of mitochondrial Ca^{2+} responses. *Science* **280**: 1763–1766.

Ron, D. and Habener, J.F. (1992) CHOP, a novel developmentally regulated nuclear protein that dimerizes with transcription factors C/EBP and LAP and functions as a dominant-negative inhibitor of gene transcription. *Genes Dev.* **6**: 439–453.

Sanchez, I., Xu, C.J., Juo, P., Kakizaka, A., Blenis, J. and Yuan, J. (1999) Caspase-8 is required for cell death induced by expanded polyglutamine repeats. *Neuron* **22**: 623–633.

Sanjay, A., Fu, J. and Kreibich, G. (1998) DAD1 is required for the function and the structural integrity of the oligosaccharyltransferase complex. *J. Biol. Chem.* **273**: 26094–26099.

Schendel, S.L., Xie, Z., Montal, M.O., Matsuyama, S., Montal, M. and Reed, J.C. (1997) Channel formation by antiapoptotic protein Bcl-2. *Proc. Natl Acad. Sci. USA* **94**: 5113–5118.

Sidrauski, C., Chapman, R. and Walter, P. (1998) The unfolded protein response: an intracellular signalling pathway with many surprising features. *Trends Cell Biol.* **8**: 245–249.

Silberstein, S., Collins, P.G., Kelleher, D.J. and Gilmore, R. (1995) The essential OST2 gene encodes the 16-kD subunit of the yeast oligosaccharyltransferase, a highly conserved protein expressed in diverse eukaryotic organisms. *J. Cell Biol.* **131**: 371–383.

Skinner, P.J., Koshy, B.T., Cummings, C.J., Klement, I.A., Helin, K., Servadio, A., Zoghbi, H.Y. and Orr, H.T. (1997) Ataxin-1 with an expanded glutamine tract alters nuclear matrix-associated structures. *Nature* **389**: 971–974. [Published erratum appears in *Nature* 1998 Jan 15;391(6664): 307].

Srinivasula, S.M., Ahmad, M., Fernandes-Alnemri, T. and Alnemri, E.S. (1998) Autoactivation of procaspase-9 by Apaf-1-mediated oligomerization. *Mol. Cell* **1**: 949–957.

Thome, M., Schneider, P., Hofmann, K., Fickenscher, H., Meinl, E., Neipel, F., Mattmann, C., Burns, K., Bodmer, J.L., Schroter, M. *et al.* (1997) Viral FLICE-inhibitory proteins (FLIPs) prevent apoptosis induced by death receptors. *Nature* **386**: 517–521.

Tirasophon, W., Welihinda, A.A. and Kaufman, R.J. (1998) A stress response pathway from the endoplasmic reticulum to the nucleus requires a novel bifunctional protein kinase/endoribonuclease (Ire1p) in mammalian cells. *Genes Dev.* **12**: 1812–1824.

Torgler, C.N., Detiani, M., Raven, T., Aubry, J.P., Brown, R. and Meldrum, E. (1997) Expression of Bak in *S. pombe* results in a lethality mediated through interaction with the calnexin homologue Cnx1. *Cell Death Differ.* **4**: 263–271.

Vander Heiden, M.G., Chandel, N.S., Williamson, E.K., Schumacker, P.T. and Thompson, C.B. (1997) Bcl-xL regulates the membrane potential and volume homeostasis of mitochondria. *Cell* **91**: 627–637.

Wang, X.Z., Lawson, B., Brewer, J.W., Zinszner, H., Sanjay, A., Mi, L.J., Boorstein, R., Kreibich, G., Hendershot, L.M. and Ron, D. (1996) Signals from the stressed endoplasmic reticulum induce C/EBP-homologous protein (CHOP/GADD153). *Mol. Cell. Biol.* **16**: 4273–4280.

Wang, X.Z., Harding, H.P., Zhang, Y., Jolicoeur, E.M., Kuroda, M. and Ron, D. (1998) Cloning of mammalian Ire1 reveals diversity in the ER stress responses. *EMBO J.* **17**: 5708–5717.

Warrick, J.M., Paulson, H.L., Gray-Board, G.L., Bui, Q.T., Fischbeck, K.H., Pittman, R.N. and Bonini, N.M. (1998) Expanded polyglutamine protein forms nuclear inclusions and causes neural degeneration in *Drosophila*. *Cell* **93**: 939–949.

White, E. (1994) Function of the adenovirus E1B oncogene in infected and transformed cells. *Semin. Virol.* **5**: 341–348.

White, E. (1996) Life, death, and the pursuit of apoptosis. *Genes Dev.* **10**: 1–15.

White, E. and Cipriani, R. (1989) Specific disruption of intermediate filaments and the nuclear lamina by the 19-kDa product of the adenovirus E1B oncogene. *Proc. Natl Acad. Sci. USA* **86**: 9886–9890.

Wolter, K.G., Hsu, Y.T., Smith, C.L., Nechushtan, A., Xi, X.G. and Youle, R.J. (1997) Movement of Bax from the cytosol to mitochondria during apoptosis. *J. Cell Biol.* **139**: 1281–1292.

Xu, Q. and Reed, J.C. (1998) Bax inhibitor-1, a mammalian apoptosis suppressor identified by functional screening in yeast. *Mol. Cell* **1**: 337–346.

Zhu, W., Cowie, A., Wasfy, G.W., Penn, L.Z., Leber, B. and Andrews, D.W. (1996) Bcl-2 mutants with restricted subcellular location reveal spatially distinct pathways for apoptosis in different cell types. *EMBO J.* **15**: 4130–4141.

Zinszner, H., Kuroda, M., Wang, X., Batchvarova, N., Lightfoot, R.T., Remotti, H., Stevens, J.L. and Ron, D. (1998) CHOP is implicated in programmed cell death in response to impaired function of the endoplasmic reticulum. *Genes Dev.* **12**: 982–995.

Zou, H., Henzel, W.J., Liu, X., Lutschg, A. and Wang, X. (1997) Apaf-1, a human protein homologous to *C. elegans* CED-4, participates in cytochrome c-dependent activation of caspase-3. *Cell* **90**: 405–413.

Zou, H., Li, Y., Liu, X. and Wang, X. (1999) An APAF-1.cytochrome c multimeric complex is a functional apoptosome that activates procaspase-9. *J. Biol. Chem.* **274**: 11549–11556.

9

Telomeres, telomerase and cellular immortalization

Christopher J. Jones

1. Introduction

It is a well documented observation that normal human cells have a limited ability to proliferate *in vitro* (Hayflick, 1965). They then enter a state of viable growth arrest known as cellular or replicative senescence (Campisi, 1997; Faragher and Kipling, 1998). Replicative senescence arises as a consequence of chromosomal telomere shortening. This review discusses the structure of telomeres and the effects of abrogating the function of the proteins that bind them. Cancer cells overcome a series of barriers that limit proliferation and the significance of the widespread detection of telomerase in tumours is discussed. Telomerase is a ribonucleoprotein complex that synthesizes new telomeric DNA at chromosome ends. The key components of telomerase (a catalytic subunit and RNA template) and the phenotype of a telomerase knockout mouse are described. Finally, evidence is presented to show that prevention of telomere shortening, by forced expression of telomerase in some normal somatic cell types, is sufficient to allow avoidance of senescence.

2. Cellular or replicative senescence

Cells are often considered to have the choice of dividing, differentiating or dying. Another possible fate, known as cellular or replicative senescence, is revealed when cells are grown in culture. It should be noted that this is an unrelated process to the mechanisms of plant senescence discussed elsewhere in this book (Chapters 10 and 11). Senescence becomes apparent when a culture fails to grow in conditions that previously allowed its proliferation. Eventually after a finite number of population doublings (pdl) the culture consists of almost completely permanently growth-arrested cells. The phenotype of senescent cells differs from their 'younger' counterparts in terms of gene activation or repression (Faragher and Kipling, 1998) and morphologically they appear larger and flatter with a pronounced cytoskeleton (see *Figure 2f*). Senescence should not be confused with quiescence, terminal differentiation or cell death. Indeed senescent cells show no change in the rate of apoptosis

Programmed Cell Death in Animals and Plants, edited by J.A. Bryant, S.G. Hughes and J.M. Garland.
© 2000 BIOS Scientific Publishers Ltd, Oxford.

compared to proliferating cells (Norsgaard *et al.*, 1996). Senescent cells can be detected by the presence of a senescence-associated β-galactosidase (SA β-gal) activity that is detectable at low pH (pH 6.0) (Dimri *et al.*, 1995). This assay has shown that senescence is not just an *in vitro* artefact as SA β-gal stained cells have been detected *in vivo*. Interestingly, the *in vivo* incidence of SA β-gal staining cells increases with age and tentatively suggests a link between replicative senescence and human ageing.

Cells that have entered a state of replicative senescence usually have a G_1 DNA content and fail to enter S phase after the addition of growth factors (Campisi, 1997). Senescence is a dominant state as somatic cell hybrids of old and young cells retain the senescent phenotype. Expression of the cyclin-dependent kinase inhibitor p21[WAF1] in senescent cells suggests that growth arrest is induced by a p53-mediated damage response (see Chapter 16). Abrogation of p53 function by SV40 large T antigen prior to senescence confers a limited extension of lifespan in fibroblasts of approximately 20 pdl (Shay *et al.*, 1991). Micro-injection of anti-p53 antibodies into senescent fibroblasts allows cells to re-enter the cell cycle and resume proliferation (Gire and Wynford-Thomas, 1998).

3. Telomeres: form and function

Telomeres are protein-bound DNA structures that define the ends of linear chromosomes. Telomeres are often considered as a cellular clock counting the number of cell divisions, serving as an indicator of the relative age of the culture and as a predictive marker for the remaining lifespan of the population (Allsopp *et al.*, 1992). In human cells telomeres are characterized by a DNA repeat sequence $(TTAGGG)_n$ varying in length between 2 and 30 kilobases (kb). Chromosomes maintain long overhangs (approximately 150 nucleotides) in the G-rich strand at either one (Wright *et al.*, 1997) or both ends (Makarov *et al.*, 1997). The unique nature of the telomeric repeat sequence promotes binding of two proteins TRF1 (see Section 4.1) and TRF2 (see Section 4.2). The amount of $(TTAGGG)_n$ present at chromosome termini greatly exceeds that required to simply supply an end-protective function. Telomeric DNA also offsets a detrimental effect of DNA replication (see *Figure 1a*). Considered in its most basic form, semi-conservative DNA replication requires the activity of an RNA primase followed by the action of DNA polymerases (Waga and Stillman, 1994). Lagging strand synthesis is a discontinuous process involving the synthesis of RNA primers that are in turn removed by a ribonuclease (RNase) and replaced by DNA, the polymerase requiring a 3'-OH end to initiate synthesis. However, following primer removal, synthesis at the chromosome terminus is not possible and this results in the production of an overhang of 50–200 bp. These overhangs are modified with the result that the chromosome is shortened (see *Figure 1a*). Telomere shortening, due to the 'end replication' problem, was predicted by Olovnikov (1973) and is detectable by Southern blotting (Harley *et al.*, 1990). Telomeric DNA, in the form of a terminal restriction fragment (TRF) resistant to enzymatic digestion, is detected as a smear representing a distribution of telomeric sequences of all the chromosomes from a population of cells (see *Figure 1b*). These DNA smears also contain DNA from sub-telomeric regions. It is clear from analyses of this nature that TRFs shorten in somatic cells such as fibroblasts as the number of pdl increases (Harley *et al.*, 1990) decreasing to a threshold of about 5 kb (including sub-telomeric regions) in senescent cells (Allsopp and Harley, 1995).

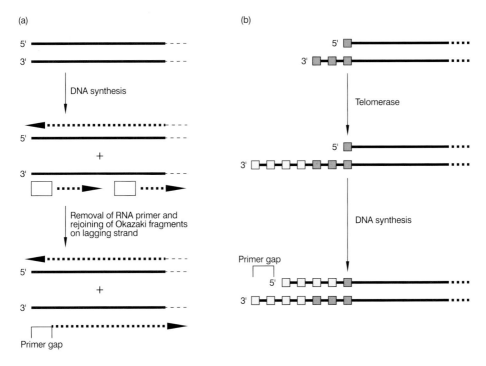

Figure 1. *The 'end replication' problem is countered by the action of telomerase. (a) During lagging-strand synthesis RNA primers (boxes) are extended by DNA polymerase to create Okazaki fragments. When the primers are finally removed DNA polymerase fails to synthesize lagging strand sequences at chromosome termini and leaves a gap. With continuing cell division DNA sequence is lost from linear chromosomes. (b) Telomerase extends the 3′ overhang at the end of chromosomes allowing it to be replicated by DNA polymerase and compensate for its inability to fill the primer gap.*

In a large number of cell types telomeres normally show progressive shortening, but cancer cells, with generally short telomeres, are able to proliferate indefinitely, avoiding replicative senescence. Germ, stem and most cancer cells have reactivated an enzyme, telomerase, that modulates telomere length. It is also apparent that cancer cells with shorter tracts of telomeric DNA have further deregulated their response to shortened telomeres, which is consistent with established models of multistep carcinogenesis (Lengauer *et al.*, 1998). The hypothesis that maintaining telomeres is required to avoid replicative senescence is supported by the observation that the germline and stem cells have much longer telomeres than somatic cells. When analysed by quantitative fluorescence *in situ* hybridization (Q-FISH) it is apparent that there is considerable variation of telomere length in the pre-senescent cell and even within opposite arms of the same chromosome (Lansdorp *et al.*, 1996). Short telomeres are observed on chromosomes 9p, 9q, 12p, 14p, 17p and 21p (Martens *et al.*, 1998). Thus it is not clear if the trigger for replicative senescence arises from one critically degraded telomere or through an integrated signal representing total telomeric DNA status. Lifespan extension induced by SV40 T is accompanied by further telomere shortening until the cells enter crisis due in part to chromosome fusion events. Extension of lifespan by abrogation of p53 function highlights a characteristic of senescent cells in that their telomeres are of sufficient length to preserve an end protective function and provide further potential for cell division. Because abrogation of p53 function encourages further cell division, removing the signal that invokes G_1 arrest via p53 (i.e. short telomeres) should have the same effect (see Section 8).

4. Telomere-binding proteins

Telomeres are DNA sequences and it is the proteins associated with them that define their function. In human cells two proteins, TRF1 and TRF2, have been isolated that bind to telomeric repeats directly. In addition, tankyrase, a protein that is associated with telomeres via its interaction with TRF1 has been identified.

4.1 *TRF1*

TRF1 (telomeric repeat binding factor), the first telomere-binding protein identified, was detected as an activity in HeLa cell extracts on the basis of its ability to retard the gel mobility of a double stranded DNA fragment $[TTAGGG]_{12}$ (Chong *et al.*, 1995). Using this assay, fractions eluted from a series of DNA affinity columns were screened and a protein with an electrophoretic mobility equivalent to a molecular weight of 60 kDa was purified. Amino acid sequencing of tryptic peptides uncovered a sequence corresponding to an uncharacterized partial cDNA in a nucleotide database. A full length cDNA was isolated containing an open reading frame (ORF) of 439 amino acids and a predicted M_r of 50.3 kDa. *In vitro* translated TRF1 binds $[TTAGGG]_{12}$ in the mobility shift assay, confirming that this cDNA encoded a telomere binding factor, and shows the same anomalous migration on sodium dodecyl sulphate polyacrylamide gel electrophoresis (SDS-PAGE) gels as does purified HeLa protein.

TRF1 possesses a number of functional motifs including an acidic N-terminal domain rich in Asp and Glu residues, a region containing two nuclear localization signals, and in the C-terminus a motif homologous to DNA-binding repeats similar to those present in Myb proto-oncogenes. Immunodetection of mouse TRF1 in interphase nuclei indicates a punctate distribution of the protein that is co-localized with $(TTAGGG)_n$ sequences. TRF1 is present at the telomeres of metaphase chromosomes (Chong *et al.*, 1995). TRF1 carries only a single copy of the Myb-type DNA domain, unlike most proteins in this family, and binds to telomeric DNA as a dimer (Bianchi *et al.*, 1997). A dimerization domain, specific to TRF1 is located in the N-terminal portion of the protein downstream of the acidic region and both Myb type domains of the dimer are required for stable DNA binding. TRF1 binding causes DNA to bend by an angle of 120° (Bianchi *et al.*, 1997).

Further investigation of TRF1 function reveals that the protein regulates telomere length. Overexpression of TRF1 in telomerase-positive, human HT1080 fibrosarcoma cells causes telomere shortening (van Steensel and de Lange, 1997). Deletion of the Myb type DNA binding domain (TRF1$^{\Delta M}$) produces a protein that will dimerize with wild type TRF1 but the resulting heterodimer is unable to complex with DNA. Overexpression of dominant negative TRF1$^{\Delta M}$ induces telomere extension. It should be noted that the levels of telomerase activity are unaffected in these cells suggesting that TRF1 may be affecting the accessibility of the enzyme to telomere ends. Therefore TRF1 may provide a homeostatic mechanism for regulating telomere length. Long telomeres with more TRF1 bound may downregulate telomerase while conversely short telomeres with little associated TRF1 many facilitate the action of the enzyme. TRF1 involvement in a putative homeostatic mechanism is unsurprising since telomere-binding proteins have similar functions in the yeasts *Saccharomyces cerevisiae* and *Schizosaccharomyces pombe*. In *S. cerevisiae* the Rap1p protein negatively regulates telomere elongation (Marcand *et al.*, 1997) as does Taz1p in *S. pombe* (Cooper *et al.*, 1997). Both proteins contain Myb type DNA binding domains.

An essential function of telomeres is to prevent the fusion of chromosome ends. Expression of the TRF1 Myb domain deletion mutation fails to induce the formation of dicentric chromosomes (Karlseder *et al.*, 1999) and it appears that the end protective function of a telomere resides with another protein, namely TRF2.

4.2 *TRF2*

TRF2 (TTAGGG repeat-binding factor) was isolated by 2 groups following the report of a TRF1-related Myb motif present in the database (Bilaud *et al.*, 1997; Broccoli *et al.*, 1997). The TRF2 ORF encodes a protein of 500 amino acids and a predicted M_r of 55.5 kDa. Like TRF1 the protein migrates with an anomalously high molecular weight on SDS-PAGE (65 kDa) and contains distinct domains. The Myb type domain of TRF2 shares 56% identity with that of TRF1. TRF2 also forms homodimers via a dimerization domain, although this shares only 27% identity with the similar motif in TRF1. Indeed the dimerization domains of TRF1 and TRF2 fail to interact evidenced by their failure to elicit a response in the yeast 2-hybrid system (Broccoli *et al.*, 1997). Another significant difference between TRF1 and TRF2 lies in their N-termini. TRF2 has a basic N-terminus whereas that of TRF1 is predominantly acidic. TRF2 localizes to telomeres *in vivo* but does not detectably bind $[TTAGGG]_{12}$ in an *in vitro* band shift assay. However, deletion of the basic N-terminus restores *in vitro* binding activity.

Over-expression of TRF2 in HT1080 cells has no detectable effect on telomere length although there is a minor change in the localization of TRF1 (van Steensel *et al.*, 1998). Expression of TRF2 lacking the N terminal basic domain (TRF2$^{\Delta B}$) exerts a mild dominant negative effect as it is able to displace a fraction of the wild type protein from telomeres. Expression of TRF2 lacking both the N-terminal basic and Myb type domains (TRF2$^{\Delta B \Delta M}$) exerts a strong dominant negative effect with little telomere-bound wild type TRF2 detectable. Both deletion mutants induce growth arrest in HT1080 cells with a phenotype similar to senescence and accompanied by expression of SA β-gal. Abrogation of TRF2 function by the dominant negative action of TRF2$^{\Delta B \Delta M}$ has a striking effect on chromosome dynamics as anaphase bridges and lagging chromosomes are detectable at a high frequency. This strongly suggests that the role of TRF2 in cells is to provide the chromosome end protection function of telomeres. Fused chromosome ends caused by the expression of TRF2$^{\Delta B \Delta M}$ contain telomeric DNA detectable by FISH. Analysis of genomic DNA from these cells indicates that there are telomeric fragments of double the mass of the non-fused population and that the appearance of fusions correlates with the loss of G-strand overhangs (van Steensel *et al.*, 1998). Therefore it can be concluded that the presence of $(TTAGGG)_n$ repeats alone is not sufficient to provide an end-protective function but requires the activity of telomere-binding proteins.

Another consequence of expression of TRF2$^{\Delta B \Delta M}$ in HT1080 derivatives was not uncovered due to a deficiency in p53 function. Expression of TRF2$^{\Delta B \Delta M}$ in cells that retain p53-mediated transcriptional activity undergo apoptosis characterized by detection of strand breaks in TUNEL assays, annexin-V staining and a sub G_1-DNA content in FACS analysis (Karlseder et al., 1999). As expected p53-dependent apoptosis is associated with increased expression of Bax. Expression of human TRF2$^{\Delta B\Delta M}$ also inactivates the corresponding wild type mouse TRF2 protein. Expression of TRF2$^{\Delta B\Delta M}$ in mouse embryo fibroblasts (MEFs) derived from knockout mice targeted for the cell cycle regulators pRb, INK4a/ARF and p21^{WAF1} induces apoptosis. The mutated protein fails to induce apoptosis in p53–/– fibroblasts although an increased rate of chromosome fusions is evident. Cells possess a panoply of repair and signalling mechanisms that respond to DNA damage; p53 is activated in the presence of DNA damage induced by ionizing radiation (e.g. strand breaks). The ATM protein (ataxia telangiectasia mutated) is involved in this signalling pathway and there is evidence to suggest that ATM directly phosphorylates p53 (Canman et al., 1998). B cells derived from AT patients fail to enter apoptosis when expressing TRF2$^{\Delta B\Delta M}$ whereas cultures derived from normal individuals undergo cell death. The exact nature of the signal that activates ATM is probably the presence of 'uncapped' telomeres, rather than DNA breaks formed when dicentric chromosomes undergo breakage during mitosis (Karlseder et al., 1999).

4.3 Tankyrase

Another protein localized to human telomeres was identified via its interaction with TRF1 in a yeast two-hybrid screen (Smith et al., 1998). The full length cDNA encodes a polypeptide of 1327 amino acids with a predicted M_r of 142 kDa. The protein contains a central domain of 24 ankyrin repeats and its C-terminal region has homology to the catalytic subunit of poly ADP-ribose polymerase (PARP). The protein has been named tankyrase on the basis of its properties (TRF1-interacting, ankyrin-related ADP-ribose polymerase). Tankyrase defines a function for the previously uncharacterized acidic N-terminal domain of TRF1. This domain is sufficient for interaction with tankyrase. The equivalent basic N-terminal domain in TRF2 does not interact with tankyrase. The function of the ankyrin domain in tankyrase is unknown although ankyrins are structural proteins that link the cytoskeleton to membrane bound proteins (Bennett, 1992). Recombinant tankyrase protein has PARP activity and can act as its own substrate. The activity is inhibited by the PARP inhibitor 3-amino benzamide (3-AB). TRF1 is an in vitro substrate for the PARP activity of tankyrase. The modification of TRF1 decreases its affinity for [TTAGGG]$_{12}$ in in vitro band shift assays suggesting a mechanism for allowing proteins access to telomeres.

It has been reported that PARP may associate with p53 to enhance expression of p21^{WAF1} and Mdm2 following DNA damage and its inhibition by 3-AB extends replicative lifespan (Vaziri et al., 1997). It is possible that the PARP activity inhibited in this context resides in tankyrase. It is tempting to postulate that a consequence of shorter telomeres in ageing cells is that there is less TRF1 available to localize tankyrase to the chromosome ends. Free tankyrase may provide the signal that activates p53.

5. Telomerase

The end replication problem can be overcome by reactivation of the enzyme telomerase (see Figure 1b). Mammalian telomerase synthesizes (TTAGGG)$_n$ repeats de novo onto chromosome ends. Reconstitution of telomerase activity requires an RNA template and a catalytic subunit. In human cells there is an additional telomerase-associated protein TLP1, the function of which is unknown but which is not required for *in vitro* telomerase activity (Weinrich *et al.*, 1997). Telomerase is detected in cell extracts by a telomeric repeat amplification protocol (TRAP assay; Kim *et al.*, 1994). This *in vitro* assay is based on the ability of telomerase to extend an oligonucleotide primer. Extension products are amplified by the polymerase chain reaction (PCR). Extracts with telomerase activity generate a DNA ladder on polyacrylamide gels (*Figure 2a*). Pre-treatment of extracts with RNase A destroys the template and consequently inactivates telomerase.

5.1 *RNA component of telomerase*

The RNA component of telomerase, known as hTR or hTERC (telomerase RNA component) was identified in 1995 through a combination of partial purification of the enzyme, subtractive hybridization and PCR cyclic selection to enrich for RNAs carrying the complement of the telomeric repeat sequence (Feng *et al.*, 1995). hTERC RNA matures to a length of 450 bases and the template region comprises 11 nucleotides (5'-CUAACCCUAAC). Mutation of this template region produces an equivalent mutation in the telomeric repeat sequence. Expression of antisense hTERC in HeLa cell cultures abolishes telomerase activity and subsequently causes cell death (Feng *et al.*, 1995).

5.2 *Catalytic subunit of telomerase*

An expressed sequence tag (EST) generated by the IMAGE consortium shows considerable homology to the telomerase catalytic subunit of the ciliate *Euplotes aediculatus*. On the basis of this a full length cDNA encoding the catalytic subunit of human telomerase was identified independently by a number of groups (Kilian *et al.*, 1997; Meyerson *et al.*, 1997; Nakamura *et al.*, 1997). The hTERT (telomerase reverse transcriptase but also known as TCS1, TP2, hTRT, hEST2) gene encodes a protein of calculated M_r 127 kDa which contains reverse transcriptase domains and a 'T' motif unique to members of the telomerase family (for review see Kipling, 1997a). Introduction of hTERT into normal diploid fibroblasts has a profound effect on replicative lifespan (see Section 8).

6. Telomerase and cancer

The limited replicative potential of cells, dictated by telomere shortening, acts as a tumour suppressor mechanism. Regardless of the cellular changes induced by oncogene activation or loss of tumour suppressor gene function the barrier posed by telomere shortening has to be overcome by cancer cells. It is of little surprise that telomerase activity has been detected in over 85% of tumour samples tested (Shay and Bacchetti, 1997). Even allowing for the possibility of false negatives in the TRAP assay (e.g. caused by infiltrating inflammatory cells) telomerase reactivation is an overwhelmingly important step in carcinogenesis. At present the use of telomerase as a diagnostic or prognostic marker is in its infancy (Kim, 1997). Despite the extreme sensitivity of the TRAP assay there is poor specificity because tissue morphology is lost in the preparation of cell extracts.

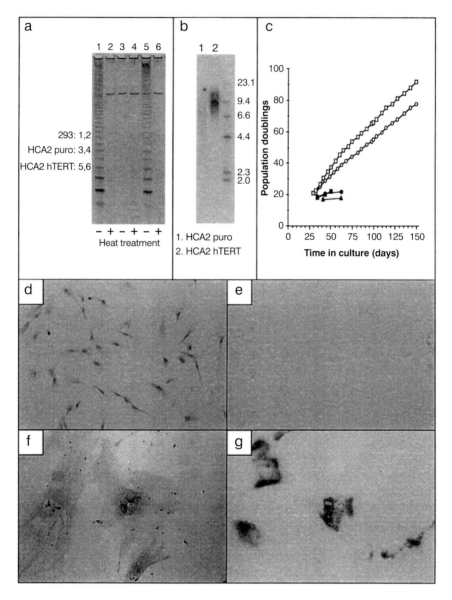

Figure 2. Immortalization of human diploid fibroblasts by the catalytic subunit of telomerase. A human fibroblast cell line HCA2 was infected with a retrovirus carrying an empty vector (puro) or the catalytic subunit of telomerase (hTERT). (a) Introduction of hTERT, but not the empty vector, restores telomerase activity to HCA2 cells. Cell extracts prepared from infected fibroblasts and a telomerase positive control cell line 293 were tested in a TRAP assay (see Section 5). In polyacrylamide gels stained with Sybr Gold a DNA ladder signifies telomerase activity. An internal standard of 150 bp serves as a control for PCR amplification. (b) Terminal restriction fragment analysis indicates that telomerase expressing cells have extended telomeres (high molecular weight smear) compared to vector controls (low molecular weight smear). (c) Telomerase expressing cells (open symbols) have an extended lifespan compared to vector controls (closed symbols). (d) Telomerase expressing HCA2 fibroblasts have a spindly 'youthful' morphology and fail to express senescence associated β-galactosidase (e). Vector controls have the larger and flattened appearance of senescent cells (f) and express SA β-gal (g).

An obvious marker is the RNA component of telomerase (hTERC), and is detectable by *in situ* hybridization but expressed in tissues with no detectable telomerase activity (Avilion *et al.*, 1996). However, the basal expression of hTERC has proved useful for *in vitro* immortalization studies (see Section 8). The identification of hTERT (the catalytic subunit) has provided new opportunities for telomerase detection based on probing for expression of the mRNA (Kolquist *et al.*, 1998). Expression of hTERT appears in early pre-malignant lesions increasing in strength as tumours progress. It is interesting to note that expression is also detected in normal non-malignant cells such as those of the breast lobular epithelium. Although mitotically quiescent, these cells can undergo rapid and widespread proliferation (Kolquist *et al.*, 1998). The next obvious step is the production of anti-telomerase antibodies that will provide an invaluable tool for analysis of clinical samples by immunocytochemistry.

Telomerase may have an important role in cancer treatment as a therapeutic target. Cancer cells generally possess much shorter telomeric tracts than germ and stem cells that also express telomerase. It is likely that the effect of telomerase inhibitors will be manifested first in tumour cells as critically shortened telomeres are produced and the proliferative capacity of cells exhausted. The fact that it is not until the third generations of telomerase-knockout mice that deficiency is observable in these cell types (see below) supports the potential 'safe' use of telomerase-directed therapies. In tumours with normal p53 function, disruption of the function of the telomere-binding protein TRF2 may induce apoptosis (see Section 4.2), although therapies based on this aspect of telomere function will require careful targeting.

7. Telomerase-knockout mouse

The gene encoding the RNA component of mouse telomerase mTERC (also known as mTR) has been targeted in the germline to generate a knockout mouse (mTR–/–). As this deletes the template for $(TTAGGG)_n$ synthesis this effectively produces a telomerase-deficient mouse (Blasco *et al.*, 1997). Although telomerase activity is present in stem cells and the germline it was argued that production of a viable mouse might be feasible due to the highly extended nature of mouse telomeres (Kipling, 1997b) providing enough of a reserve to generate an adult animal. Cells from initial cohorts of mTR–/– mice express no detectable telomerase activity. Normal MEFs senesce rapidly in culture in a non-telomere-dependent fashion. The rate of senescence in mTR–/– MEFs is no faster than wild type cells. Six generations (G6) of telomerase-negative mice have been produced, although from the fourth generation onwards there is an increased rate of aneuploidy associated with the loss of detectable telomere sequences at some chromosome ends. The loss of telomere sequence is approximately 4.8 kb per generation, consistent with the expected loss of $(TTAGGG)_n$ during the number of cell divisions required to generate an adult mouse (Blasco *et al.*, 1997).

Certain cellular processes reliant on the presence of active proliferative cell compartments become compromised in later generations of mTR–/– mice. No offspring are produced in matings of G6 mice and there is a low rate of successful reproduction when mated with wild type animals (Lee *et al.*, 1998). Male sterility is associated with decreased proliferation and increased apoptosis of spermatogenic cells. The testes of G6 mice are strikingly reduced in size compared to the wild type. Changes to the female reproductive system are less obvious although the weight of the ovaries is decreased in G6 compared to wild type. G6 uterine horns show decreased muscularity that may compromise function. G6 embryos fail to progress and the mice themselves are less likely to carry implanted wild type embryos to term (Lee *et al.*, 1998).

Self-renewing populations of stem cells replenish the haematopoietic system. Analyses of blood cell count and relative percentages of lymphocytes, neutrophils, eosinophils, basophils and monocytes show no difference between wild type and G6 mice. However bone marrow cells taken from G6 mice show decreased proliferative capacity in colony formation assays. G6 splenocytes are less likely to proliferate after stimulation and an increased rate of apoptosis is observed (Lee *et al.*, 1998).

Since the initial description of mTR–/– mice the animals have aged and lack of telomerase activity has induced some ageing-related phenotypes (Rudolph *et al.*, 1999). Mice from the third generation onwards show impaired wound healing, premature greying and hair loss. The lifespan of G6 animals is considerably shorter than wild type and G3 mice. Weight loss in G6 animals is associated with changes in the lining of the gastrointestinal tract. Although telomere shortening does not produce a complete ageing phenotype it uncovers cell systems where resistance to the effects of ageing requires an ability to replace lost cells (Rudolph *et al.*, 1999).

8. Immortalization of normal cells by telomerase

As hTERC, the RNA component, is present in most cell types, ectopic expression of hTERT is sufficient to restore telomerase activity in cells (Weinrich *et al.*, 1997). *Figure 2* demonstrates the effect of introduction of hTERT by retroviral gene transfer into a mortal human diploid fibroblast cell line. TRAP assays confirm that telomerase activity (*Figure 2a*) is present in transduced cells but it does not necessarily follow that this has an effect on telomere maintenance. Addition of a C-terminal HA epitope tag to hTERT produces active telomerase *in vitro* but is unable to extend telomeres *in vivo* (Counter *et al.*, 1998; Ouellette *et al.*, 1999). Introduction of wild type hTERT into fibroblasts causes telomere lengthening (*Figure 2b*). Fibroblasts expressing telomerase avoid senescence and continue to proliferate (*Figure 2c*; Bodnar *et al.*, 1998). The cells retain the morphology of younger cells (*Figure 2d*) and do not express SA β-gal (*Figure 2e*), unlike vector controls which undergo replicative senescence (*Figures 2f* and *g*). Immortalization of cells by telomerase does not appear to confer any changes associated with malignancy (Jiang *et al.*, 1999; Morales *et al.*, 1999). They have intact cell cycle checkpoints, are karyotypically normal, become quiescent at high density and under conditions of serum starvation, fail to grow in soft agar or induce tumours *in vivo*. However in some cases telomerase expression in fibroblasts induces a senescence-like phenotype due to over-extended telomeres (C.J. Jones, unpublished data). One possible explanation is that there are insufficient telomere associated proteins available to bind these extended $(TTAGGG)_n$ tracts.

9. Future prospects

Expression of hTERT alone is not sufficient to immortalize all cell types. For example mammary epithelial cells require another event that knocks out the function of the Rb/p16 pathway (Kiyono *et al.*, 1998). Further investigation of the mechanisms underlying senescence are likely to uncover other factors controlling the lifespan of cells. An indication of how the telomerase gene is upregulated in cells is illustrated by the observation that activation of the hTERT promoter requires the function of c-myc (Greenberg *et al.*, 1999). Abrogation of telomerase function in mice induces certain features of ageing (Rudolph *et al.*, 1999). It is important to determine if telomerase reactivation in cells derived from progeroid diseases such as Werner's syndrome can reverse some of the symptoms of these disorders (Kipling and Faragher, 1999). The identity of the signal that activates p53 at senescence remains unknown. Loss of TRF2 function overstimulates a p53-mediated DNA damage response and induces apoptosis (Karlseder *et al.*, 1999). Replicative senescence is not a cell death mechanism and there is probably a more subtle activation of p53. Proteins involved in the repair of double-strand breaks in *S. cerevisiae* are required for normal telomere function (Critchlow and Jackson, 1998). The human protein DNA-PK is activated by strand breaks and phosphorylates p53 (Woo *et al.*, 1998), but there is no evidence to link these two observations.

It is likely that cell biologists and biochemists will benefit from the immortalizing ability of telomerase by allowing large-scale production of essentially normal cultures. It is hoped that telomerase will be a useful tool for clinicians as a diagnostic marker of malignancy and make a significant contribution to healthcare.

Acknowledgements

I wish to thank Julia Skinner for excellent technical support and Anne Jones for helpful comments on the manuscript.

References

Allsopp, R.C. and Harley, C.B. (1995) Evidence for a critical telomere length in senescent human fibroblasts. *Exp. Cell Res.* **219**: 130–136.

Allsopp, R.C., Vaziri, H., Patterson, C., Goldstein, S., Younglai, E.V., Futcher, A.B., Greider, C.W. and Harley, C.B. (1992) Telomere length predicts replicative capacity of human fibroblasts. *Proc. Natl Acad. Sci. USA* **89**: 10114–10118.

Avilion, A.A., Piatyszek, M.A., Gupta, J., Shay, J.W., Bacchetti, S. and Greider, C.W. (1996) Human telomerase RNA and telomerase activity in immortal cell lines and tumor tissues. *Cancer Res.* **56**: 645–650.

Bennett, V. (1992) Ankyrins. Adaptors between diverse plasma membrane proteins and the cytoplasm. *J. Biol. Chem.* **267**: 8703–8706.

Bianchi, A., Smith, S., Chong, L., Elias, P. and de Lange, T. (1997) TRF1 is a dimer and bends telomeric DNA. *EMBO J.* **16**: 1785–1794.

Bilaud, T., Brun, C., Ancelin, K., Koering, C.E., Laroche, T. and Gilson, E. (1997) Telomeric localization of TRF2, a novel human telobox protein. *Nature Genet.* **17**: 236–239.

Blasco, M.A., Lee, H.W., Hande, M.P., Samper, E., Lansdorp, P.M., DePinho, R.A. and Greider, C.W. (1997) Telomere shortening and tumor formation by mouse cells lacking telomerase RNA. *Cell* **91**: 25–34.

Bodnar, A.G., Ouellette, M., Frolkis, M., Holt, S.E., Chiu, C.P., Morin, G.B., Harley, C.B., Shay, J.W., Lichtsteiner, S. and Wright, W.E. (1998) Extension of life-span by introduction of telomerase into normal human cells. *Science* **279**: 349–352.

Broccoli, D., Smogorzewska, A., Chong, L. and de Lange, T. (1997) Human telomeres contain two distinct Myb-related proteins, TRF1 and TRF2. *Nature Genet.* **17**: 231–235.

Campisi, J. (1997) The biology of replicative senescence. *Eur. J. Cancer* **33**: 703–709.

Canman, C.E., Lim, D.S., Cimprich, K.A., Taya, Y., Tamai, K., Sakaguchi, K., Appella, E., Kastan, M.B. and Siliciano, J.D. (1998) Activation of the ATM kinase by ionizing radiation and phosphorylation of p53. *Science* **281**: 1677–1679.

Chong, L., Vansteensel, B., Broccolli, D., Erdjument-Bromage, H., Hanish, J., Tempst, P. and de Lange, T. (1995) A human telomeric protein. *Science* 270: 1663–1667.

Cooper, J.P., Nimmo, E.R., Allshire, R.C. and Cech, T.R. (1997) Regulation of telomere length and function by a Myb-domain protein in fission yeast. *Nature* 385: 744–747.

Counter, C.M., Hahn, W.C., Wei, W., Caddle, S.D., Beijersbergen, R.L., Lansdorp, P.M., Sedivy, J.M. and Weinberg, R.A. (1998) Dissociation among *in vitro* telomerase activity, telomere maintenance, and cellular immortalization. *Proc. Natl Acad. Sci. USA* 95: 14723–14728.

Critchlow, S.E. and Jackson, S.P. (1998) DNA end-joining: from yeast to man. *Trends Biochem. Sci.* 23: 394–398.

Dimri, G.P., Lee, X.H., Basile, G., Acosta, M., Scott, C., Roskelley, C., Medrano, E.E., Linskens, M., Rubelj, I., Pereira-Smith, O. *et al.* (1995) A biomarker that identifies senescent human-cells in culture and in aging skin in-vivo. *Proc. Natl Acad. Sci. USA* 92: 9363–9367.

Faragher, R.G. and Kipling, D. (1998) How might replicative senescence contribute to human ageing? *Bioessays* 20: 985–991.

Feng, J., Funk, W.D., Wang, S.-S., Weinrich, S.L., Avilion, A.A., Chiu, C.-P., Adams, R.R., Chang, E., Allsopp, R.C., Yu, J. *et al.* (1995) The RNA component of human telomerase. *Science* 269: 1236–1241.

Gire, V. and Wynford-Thomas, D. (1998) Reinitiation of DNA synthesis and cell division in senescent human fibroblasts by microinjection of anti-p53 antibodies. *Mol. Cell. Biol.* 18: 1611–1621.

Greenberg, R.A., O'Hagan, R.C., Deng, H., Xiao, Q., Hann, S.R., Adams, R.R., Lichtsteiner, S., Chin, L., Morin, G.B. and DePinho, R.A. (1999) Telomerase reverse transcriptase gene is a direct target of c-Myc but is not functionally equivalent in cellular transformation. *Oncogene* 18: 1219–1226.

Harley, C.B., Futcher, A.B. and Greider, C.W. (1990) Telomeres shorten during ageing of human fibroblasts. *Nature* 345: 458–460.

Hayflick, L. (1965) The limited *in vitro* lifetime of human diploid cell strains. *Exp. Cell. Res.* 37: 614–636.

Jiang, X.R., Jimenez, G., Chang, E., Frolkis, M., Kusler, B., Sage, M., Beeche, M., Bodnar, A.G., Wahl, G.M., Tlsty, T.D. and Chiu, C.P. (1999) Telomerase expression in human somatic cells does not induce changes associated with a transformed phenotype. *Nature Genet.* 21: 111–114.

Karlseder, J., Broccoli, D., Dai, Y., Hardy, S. and de Lange, T. (1999) p53- and ATM-dependent apoptosis induced by telomeres lacking TRF2. *Science* 283: 1321–1325.

Kilian, A., Bowtell, D.D., Abud, H.E., Hime, G.R., Venter, D.J., Keese, P.K., Duncan, E.L., Reddel, R.R. and Jefferson, R.A. (1997) Isolation of a candidate human telomerase catalytic subunit gene, which reveals complex splicing patterns in different cell types. *Hum. Mol. Genet.* 6: 2011–2019.

Kim, N.W. (1997) Clinical implications of telomerase in cancer. *Eur. J. Cancer* 33: 781–786.

Kim, N.W., Piatyszek, M.A., Prowse, K.R., Harley, C.B., West, M.D., Ho, P.L.C., Coviello, G.M., Wright, W.E., Weinrich, S.L. and Shay, J.W. (1994) Specific association of human telomerase activity with immortal cells and cancer. *Science* 266: 2011–2015.

Kipling, D. (1997a) Mammalian telomerase: catalytic subunit and knockout mice. *Hum. Mol. Genet.* 6: 1999–2004.

Kipling, D. (1997b) Telomere structure and telomerase expression during mouse development and tumorigenesis. *Eur. J. Cancer* 33: 792–800.

Kipling, D. and Faragher, R.G. (1999) Telomeres. Ageing hard or hardly ageing? *Nature* 398: 191, 193.

Kiyono, T., Foster, S.A., Koop, J.I., McDougall, J.K., Galloway, D.A. and Klingelhutz, A.J. (1998) Both Rb/p16INK4a inactivation and telomerase activity are required to immortalize human epithelial cells. *Nature* 396: 84–88.

Kolquist, K.A., Ellisen, L.W., Counter, C.M., Meyerson, M., Tan, L.K., Weinberg, R.A., Haber, D.A. and Gerald, W.L. (1998) Expression of TERT in early premalignant lesions and a subset of cells in normal tissues. *Nature Genet.* **19**: 182–186.

Lansdorp, P.M., Verwoerd, N.P., van de Rijke, F.M., Dragowska, V., Little, M.T., Dirks, R.W., Raap, A.K. and Tanke, H.J. (1996) Heterogeneity in telomere length of human chromosomes. *Hum. Mol. Genet.* **5**: 685–691.

Lee, H.W., Blasco, M.A., Gottlieb, G.J., Horner, J.W., 2nd, Greider, C.W. and DePinho, R.A. (1998) Essential role of mouse telomerase in highly proliferative organs. *Nature* **392**: 569–574.

Lengauer, C., Kinzler, K.W. and Vogelstein, B. (1998) Genetic instabilities in human cancers. *Nature* **396**: 643–649.

Makarov, V.L., Hirose, Y. and Langmore, J.P. (1997) Long G tails at both ends of human chromosomes suggest a C strand degradation mechanism for telomere shortening. *Cell* **88**: 657–666.

Marcand, S., Gilson, E. and Shore, D. (1997) A protein-counting mechanism for telomere length regulation in yeast. *Science* **275**: 986–990.

Martens, U.M., Zijlmans, J.M., Poon, S.S., Dragowska, W., Yui, J., Chavez, E.A., Ward, R.K. and Lansdorp, P.M. (1998) Short telomeres on human chromosome 17p. *Nature Genet.* **18**: 76–80.

Meyerson, M., Counter, C.M., Eaton, E.N., Ellisen, L.W., Steiner, P., Caddle, S.D., Ziaugra, L., Beijersbergen, R.L., Davidoff, M.J., Liu, Q. *et al.* (1997) hEST2, the putative human telomerase catalytic subunit gene, is up-regulated in tumor cells and during immortalization. *Cell* **90**: 785–795.

Morales, C.P., Holt, S.E., Ouellette, M., Kaur, K.J., Yan, Y., Wilson, K.S., White, M.A., Wright, W.E. and Shay, J.W. (1999) Absence of cancer-associated changes in human fibroblasts immortalized with telomerase. *Nature Genet.* **21**: 115–118.

Nakamura, T.M., Morin, G.B., Chapman, K.B., Weinrich, S.L., Andrews, W.H., Lingner, J., Harley, C.B. and Cech, T.R. (1997) Telomerase catalytic subunit homologs from fission yeast and human. *Science* **277**: 955–959.

Norsgaard, H., Clark, B.F. and Rattan, S.I. (1996) Distinction between differentiation and senescence and the absence of increased apoptosis in human keratinocytes undergoing cellular aging *in vitro. Exp. Gerontol.* **31**: 563–570.

Olovnikov, A.M. (1973) A theory of marginotomy: the incomplete copying of template margin in enzymatic synthesis of polynulceotides and biological significance of the problem. *J. Theor. Biol.* **41**: 181–190.

Ouellette, M.M., Aisner, D.L., Savre-Train, I., Wright, W.E. and Shay, J.W. (1999) Telomerase activity does not always imply telomere maintenance. *Biochem. Biophys. Res. Commun.* **254**: 795–803.

Rudolph, K.L., Chang, S., Lee, H.W., Blasco, M., Gottlieb, G.J., Greider, C. and DePinho, R.A. (1999) Longevity, stress response, and cancer in aging telomerase-deficient mice. *Cell* **96**: 701–712.

Shay, J.W. and Bacchetti, S. (1997) A survey of telomerase activity in human cancer. *Eur. J. Cancer* **33**: 787–791.

Shay, J.W., Pereira-Smith, O.M. and Wright, W.E. (1991) A role for both RB and p53 in the regulation of human cellular senescence. *Exp. Cell Res.* **196**: 33–39.

Smith, S., Giriat, I., Schmitt, A. and de Lange, T. (1998) Tankyrase, a Poly(ADP-ribose) polymerase at human telomeres. *Science* **282**: 1484–1487.

van Steensel, B. and de Lange, T. (1997) Control of telomere length by the human telomeric protein TRF1. *Nature* **385**: 740–743.

van Steensel, B., Smogorzewska, A. and de Lange, T. (1998) TRF2 protects human telomeres from end-to-end fusions. *Cell* **92**: 401–413.

Vaziri, H., West, M.D., Allsopp, R.C., Davison, T.S., Wu, Y.S., Arrowsmith, C.H., Poirier, G.G. and Benchimol, S. (1997) ATM-dependent telomere loss in aging human diploid fibroblasts and DNA damage lead to the post-translational activation of p53 protein involving poly(ADP-ribose) polymerase. *EMBO J.* **16**: 6018–6033.

Waga, S. and Stillman, B. (1994) Anatomy of a DNA replication fork revealed by reconstitution of SV40 DNA replication *in vitro*. *Nature* **369**: 207–212.

Weinrich, S.L., Pruzan, R., Ma, L., Ouellette, M., Tesmer, V.M., Holt, S.E., Bodnar, A.G., Lichtsteiner, S., Kim, N.W., Trager, J.B. *et al.* (1997) Reconstitution of human telomerase with the template RNA component hTR and the catalytic protein subunit hTRT. *Nature Genet.* **17**: 498–502.

Woo, R.A., McLure, K.G., Lees-Miller, S.P., Rancourt, D.E. and Lee, P.W. (1998) DNA-dependent protein kinase acts upstream of p53 in response to DNA damage. *Nature* **394**: 700–704.

Wright, W.E., Tesmer, V.M., Huffman, K.E., Levene, S.D. and Shay, J.W. (1997) Normal human chromosomes have long G-rich telomeric overhangs at one end. *Genes Dev.* **11**: 2801–2809.

10

Back from the brink: plant senescence and its reversibility

Howard Thomas and Iain Donnison

1. Many ways to get dead

All cells are born in more or less the same way, and die in more or less different ways. Leaf cells, with which this chapter is primarily concerned, are typical of plant cells in their origins and ultimate fates. A leaf cell faces the prospect of one of at least five distinct modes of death.

1. It may be programmed to die during normal development as part of processes that create complex organ shapes and specialized cell types. For instance, the pinnate (fan-like) leaves of palms arise by death of strips of cells (lorae) that divide the lamina into long thin leaflets (Kaplan *et al.*, 1982).
2. Environmental deviations beyond the adaptive limits of the cell usually result in necrotic cell death (Pennell and Lamb, 1997).
3. One form of disease resistance in plants is the hypersensitive response, a type of programmed cell death (PCD; Beers, 1997).
4. The tissue may be consumed by a predator, an extremely common fate for the foliage of many species, and one that triggers cell death processes with far-reaching ecological implications (see Chapter 22).
5. The cell may undergo senescence, a component of normal development intimately associated with, but significantly different from, autolytic and/or pathological cell death. This chapter is concerned mainly with senescence.

2. Organ senescence has a cellular basis

2.1 *Physiological and pathological modes of cell death*

Senescence has a particular meaning in plant science. It refers to the terminal phase in the development of leaves and homologous organs (flower parts, fruit tissues for example). Senescence follows organ maturity and generally occurs without growth or

---

morphogenesis. Death may follow senescence though, as discussed here, this is by no means inevitable. Conversely, death can occur without the intervention of senescence. *Figure 1* summarizes the interrelationships and semi-independence of the *physiological* (senescence, cell specialization) and *pathological* (biotic and abiotic stress) routes to plant cell death.

2.2 *Senescence of cell and tissue types*

Senescence of a whole organ usually means senescence of the dominant cell or tissue type. Thus changes in the green photosynthetic cells of mesophyll tissue signal the progress of foliar senescence. Some leaf cells senesce early – the xylem elements of the veins, for example (Fukuda, 1996; see also Chapter 14). At the other extreme, stomatal guard cells remain green almost indefinitely, as examination of leaves collected from the ground in a deciduous wood in autumn will show (Zeiger and Schwartz, 1982). It is not known how the specificity of cell death is regulated. There may be a 'die now' signal constantly present to which a given cell type becomes competent to respond in due time as its individual developmental programme is read out. Alternatively, there may be several or many cell-specific 'die now' stimuli that invoke essentially similar senescence/death programmes in

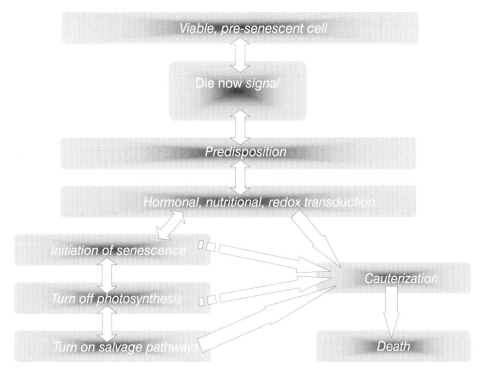

Figure 1. *Pathways leading to death of mesophyll cells. The senescence route and the pathological course are presented as operating either in series (where the reversible phase of senescence naturally progresses into the terminal stage and the asset-stripped cell is despatched according to the scorched-earth principle) or in parallel (as occurs in the hypersensitive response, for example). In both cases there are triggering and signal transduction events, and cells have to be competent to respond, in part because some of the machinery of death must already be in place. It remains to be established how much of the regulatory and metabolic machinery is common to the two pathways.*

their targets. Genetic variants, e.g. stay-greens, necrotic hybrids and disease lesion mimics, behave as if they have mutations in the genes regulating the timing or location of normal senescence and programmed cell death in organ development.

2.3 *Cell senescence is a symptom of organelle modifications*

Subcellular changes in senescence are dominated by the alterations undergone by the major (usually pigmented) organelles. The integrity of cell compartments is preserved until extreme senescence. The chloroplasts of green immature fruits redifferentiate into new organelles (chromoplasts) during ripening (Camara *et al.*, 1995). Likewise, plastids of senescing leaf cells are not degenerate but have been remodelled into geron-toplasts (Parthier, 1988). Another example of organelle redifferentiation during senes-cence is the transition of peroxisomes into glyoxysomes (Nishimura *et al.*, 1993). As will be seen, these subcellular modifications are often reversible. Reversibility and organelle differentiation are characteristics that suggest quite fundamental mecha-nistic differences between senescence and PCD.

2.4 *Senescing cells are metabolically active*

Cells undergoing senescence turn on new metabolic pathways and turn off others (*Figure 2*). Salvage and redistribution of metabolites and structural materials are characteristic of

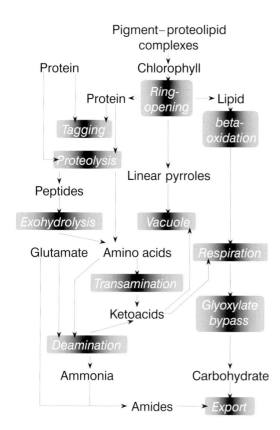

Figure 2. *Metabolism of senescing mesophyll cells. The fates of pigments, proteins and lipids of the dominant organelle, the plastid, are illustrated.*

senescing cells (Feller and Fischer, 1994). Yellowing of leaf cells is symptomatic of declining photosynthesis, which in turn means that the energy metabolism of such cells becomes increasingly heterotrophic. Some novel pathways activated in senescence have characteristics of detoxification mechanisms (Tommasini *et al.*, 1998). Changes in secondary compounds and other constituents that commonly occur in senescing cells may be related to interactions with pests or with beneficial organisms. For example, antibiotic compounds are often accumulated to prevent opportunist pathogens from using senescing tissue as a potential infection route (Harborne, 1993). Furthermore, pigments of ripening fruits, senescing foliage and floral organs attract seed-dispersing or pollinating animals (e.g. Stiles, 1982).

3. How do senescing cells change colour?

3.1 *Chlorophyll*

Even though the yellowing of leaves is a process that defines a whole season of the year and is spectacularly visible from space, the biochemical mechanism of chlorophyll breakdown is still incompletely understood. Catabolism begins with conversion of chlorophyll *a* to the green macrocyclic porphyrin phaeophorbide *a*, catalysed by chlorophyllase and magnesium dechelatase. It is believed that chlorophyll *b* is also degraded by first being converted to phaeophorbide *a* (Matile *et al.*, 1996). The critical step that opens the macrocycle to generate a colourless straight-chain tetrapyrrole requires two enzymes. Phaeophorbide *a* oxygenase (PaO) catalyses the ferredoxin-dependent reaction between phaeophorbide and O_2 resulting in the formation of a bright red bilin compound, RCC (Hörtensteiner *et al.*, 1998). RCC is immediately further metabolized by RCC reductase, a soluble enzyme which is also ferredoxin-dependent and catalyses reduction of a double bond in the pyrrole system of RCC to produce a colourless tetrapyrrole, FCC (Rodoni *et al.*, 1997). Chlorophyllase, PaO and (probably) dechelatase activity are associated with the gerontoplast envelope (Matile *et al.*, 1997). Chlorophyll is likely to be transported from the thylakoid membrane to the catabolic machinery in the envelope by some kind of carrier. FCC secretion from the gerontoplast requires ATP. Catabolites finally accumulate in the cell vacuole by the same ATP-driven tonoplast carrier system that deposits toxins and other xenobiotics in that cell compartment (Tommasini *et al.*, 1998). *En route* to the vacuole, FCC may be malonylated, glycosylated or otherwise modified (Hörtensteiner, 1998), resulting in numbers and structures of end-products that differ in different species.

3.2 *Carotenoids*

Chlorophyll loss from leaves and fruits often unmasks underlying carotenoids, which provide a yellow or orange background against which new pigments may accumulate. The activities of key enzymes in the isoprenoid pathway are enhanced during fruit ripening (Cunningham and Gantt, 1998). Carotenoids biosynthesized in ripening fruits concentrate in the fibrils, crystals or globules of chromoplasts. Fibrillins are specific proteins associated with the globules of chromoplasts and gerontoplasts (Pozueta-Romero *et al.*, 1997). They may be important in senescence by playing a role in lipid storage, in transporting newly synthesized or modified pigments or in stabilizing plastid structure.

3.3 *Phenylpropanoids*

Many of the red, purple and yellow pigments of autumnal foliage are cyanins and flavanoids, water-soluble phenylpropanoid derivatives that accumulate in cell vacuoles. Phenolics, tannins, phytoalexins, lignins and salicylate are other phenylpropanoids that are typically abundant in senescing tissues. Wounding, ethylene treatment, exposure to ozone and other stimuli that invoke senescence- or PCD-like pathological responses commonly increase the rate and alter the pattern of phenolic product synthesis (Booker and Miller, 1998; Tomas-Barberan *et al*., 1997). Phenylalanine ammonia lyase, the link between primary and phenylpropanoid metabolism, increases in activity with increasing leaf age (Fukasawa-Akada *et al*., 1996; Howles *et al*., 1996). The actions of individual regulatory genes, such as maize Lc and C1-B, are thought to be responsible for age-related changes in the flux into different branches of secondary metabolism (Bradley *et al*., 1998).

4. Senescence is an energy-demanding oxidative process

4.1 *Oxidative metabolism in senescence*

Transport and maintenance activities, pigment metabolism, protein breakdown and other senescence processes are energy-demanding. Their demands need to be satisfied for senescence to proceed correctly. But senescing tissues become increasingly carbon-starved as photosynthesis declines and sugars continue to be actively exported (see Chapter 11 and Brouquisse *et al*., 1998). Under these circumstances, amino acids derived from proteolysis are an important source of carbon skeletons (Feller and Fischer, 1994). In some (but not all) fruits, ripening is associated with the climacteric, a burst of respiration associated with production of, and responses to, ethylene (Rhodes, 1980).

4.2 *Gluconeogenesis*

The lipid-to-sugar pathway characteristic of germinating oilseeds (Falk *et al*., 1998) is also turned on during leaf senescence. Gluconeogenesis in senescence is associated with functional transformation of peroxisomes into glyoxysomes (Nishimura *et al*., 1993). Consistent with the reworking of protein complement within the organelle is the observation of specific senescence-enhanced endo- and exo-peptidases (Distefano *et al*., 1999).

4.3 *Degradation of nucleic acids during senescence*

Much of the organic phosphorus in leaves is in the form of nucleic acids. Ribonuclease (RNase) activity and gene expression increase in senescence (Lers *et al*., 1998). Some RNases are extracellular; others have endoplasmic reticulum and/or vacuole targeting motifs (Green, 1994). Just as for protein mobilization (see below), the senescence-enhanced RNases we know about seem to be in the wrong place. Phosphatases are ubiquitous and especially abundant during senescence. The nucleoside products of nuclease and phosphatase action are cleaved into sugars, purines and pyrimidines. Xanthine oxidase and uricase are strongly activated during leaf senescence. These enzymes are located in glyoxysomes and important in the redox functions of that organelle (del Rio *et al.*, 1998).

4.4 *Redox regulation in senescence?*

Redox regulation of transcriptional activators occurs in animal cells (Sun and Oberley, 1996) and something similar may happen in senescence. Expression of the genes encoding fibrillins involves a signalling pathway in which superoxide anion production is a key step (Chen *et al.*, 1998). Fluxes in the ascorbate/glutathione cycle change significantly during senescence (del Rio *et al.*, 1998; Kingston-Smith *et al.*, 1997). Excess reactive oxygen is catastrophic and is unlikely to contribute to anything but the most terminal phase of normal senescence.

5. Proteins are remobilized in cell senescence – but how?

5.1 *Proteolysis*

Most of the mobilizable protein in green cells is in the chloroplasts, e.g. RUBISCO (ribulose bisphosphate carboxylase-oxygenase) and chlorophyll-binding light-harvesting proteins. Proteolysis, or at least the critical first stages, is believed to occur within the plastid (Feller and Fischer, 1994). Senescence-enhanced cysteine proteases are closely related to proteases expressed during seed germination (Griffiths *et al.*, 1997). Cysteine endopeptidases from senescing tissues have structures that show them to be destined for the endoplasmic reticulum or the vacuole and not for the plastid. Very many plastid protease activities have been described (see, for example, Andersson and Aro, 1997), but there is little evidence of a function in senescence for any of them.

5.2 *How is the location of proteolysis controlled?*

If the major senescence-enhanced proteases are in the wrong place in the cell, how does plastid protein breakdown happen? There are several ideas about this. For example, could it be that the plastid envelope becomes modified in senescence so that, to the cell's protein trafficking machinery, it looks like the tonoplast? This seems unlikely but it is worth testing. More in tune with developing knowledge of cell death processes in animal systems is the possibility of proteolytic cascades (see Chapters 2 and 5). The major senescence-related proteases of plants are not in the same structural group as the caspases. However, Smart *et al.* (1995) cloned *See2*, a senescence up-regulated cDNA from maize leaves with sequence similarity to genes encoding proteases of the legumain type (Hara-Nishimura, 1998). *Figure 3* shows the general structural features of *SEE2* protein based on sequence comparisons with other legumains (Donnison *et al.*, 1999). It is significant that there are two pro-domains, indicating that the enzyme is likely to be activated by proteolysis. Furthermore, the legumain of the blood parasite *Schistosoma*, which was formerly considered to be a haemoglobinase (Meanawy *et al.*, 1990), is now known not to degrade haemoglobin but rather to activate another protease zymogen that does (Hara-Nishimura, 1998). By analogy, *See2* may represent the first glimpse of a protease cascade operating in senescing leaf cells. Such a cascade could finally carry the protein mobilization signal of senescence into the plastid. This looks to be a fruitful direction for future biochemical studies.

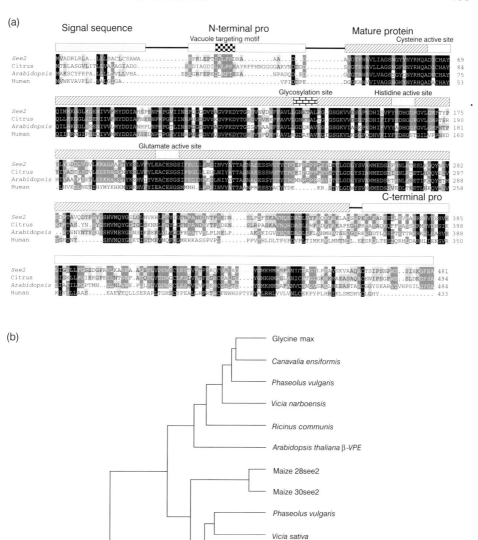

Figure 3. *(a) Protein sequence alignment of the maize senescence up-regulated protease SEE2 with other representative legumains. Conserved structural motifs are shown. Presumptive N- and C-terminal pro-domains are present, strongly suggesting that SEE2 is synthesized as a zymogen and activated by proteolysis. (b) Dendrogram showing sequence relationships between SEE2 and legumains from a range of other organisms. Similarity to the legumain from the blood parasite* Schistosoma *is significant because this enzyme is known to activate a protease which in turn degrades haemoglobin. By analogy, SEE2 may act in a proteolytic chain or cascade (Donnison* et al., *1999).*

6. Senescence is programmed

6.1 *Senescence mutants and other genetic variants*

Gregor Mendel analysed senescence well over 100 years ago. The recessive allele of the gene that Mendel called *B* (now *I*) is expressed in mature peas as green cotyledons. Cotyledons are modified leaves. It turns out that yellowing of senescing foliage of *ii* peas is also delayed (Thomas *et al.*, 1996). Such leaf senescence variants are often called 'stay-green' (Thomas and Smart, 1993). A recessive gene in the grass *Festuca* specifies a pattern of stay-green similar to that of Mendel's green peas. In such stay-green grasses and legumes the biochemical lesion is known to be a deficiency in the chlorophyll-degrading enzyme PaO (section 3.1; Thomas *et al.*, 1996; Vicentini *et al.*, 1995). Several mutants with altered colour development during fruit ripening are known, some of which have pleiotropic effects on leaves – 'greenflesh' in tomato, for example (Akhtar *et al.*, 1999). Stay-green is an economically important character in cereals and other grasses (Elings *et al.*, 1997; Thomas and Smart, 1993; Van Oosterom *et al.*, 1996). Greenness is easily screened and therefore highly convenient for genetic mapping by the procedures of marker tagging, chromosome painting and quantitative trait analysis (Thomas *et al.*, 1997; D. Thorogood and M. O. Humphreys, unpublished data). Three independently inherited stay-green mutations of *Arabidopsis* have been described (Oh *et al.*, 1997). Once they are classified and mapped, the powerful molecular genetic tools available for this species will be able to assign molecular and physiological functions to the corresponding genes.

The variant genes that modify senescence fall into two broad classes. Members of one class encode the individual components of the senescence syndrome, such as the enzymes of metabolic pathways activated in senescence. The *Festuca stay-green* gene is an example. The second type of gene regulates the initiation and/or rate of progress of the entire syndrome. The major *stay-green* loci of sorghum (Van Oosterom *et al.*, 1996) are examples. Each class of senescence gene is biologically and practically important in its own way. For a full understanding of senescence and how to exploit it, genes controlling the entire syndrome and those specifying individual components are of complementary significance.

6.2 *Cloned senescence genes*

Isolation of specific genes supports the evidence from classical genetics that senescence is programmed. Genes expressed during leaf senescence (*Sees*) have been cloned from *Arabidopsis, Brassica*, maize, cucumber, asparagus, tomato, rice, barley and a few other species (Buchanan-Wollaston, 1997; Nam, 1997; Smart, 1994). Ripening-associated genes from fruit such as tomato and banana include homologues of senescence-associated genes (Davies and Grierson, 1989; King and O'Donoghue, 1995; Medina-Suárez *et al.*, 1997). The same is true of gene expression in other senescence-like processes such as death of flower parts and of unpollinated fruit tissues. Around 50 cDNAs have been assigned possible functions in senescence on the basis of sequence homology (see Buchanan-Wollaston, 1997), and perhaps an equal number are known only as anonymous expressed sequence tags. Some *Sees* are expressed early in senescence, some late, some appear in natural senescence of attached tissue but not in detached tissue and vice versa. Some *Sees* are organ-specific whilst others, such as those encoding enzymes of gluconeogenesis, come on in leaves, cotyledons and germinating seeds. Expression of some *Sees* is enhanced by stresses such as drought, whereas others are suppressed in such conditions.

7. Senescence is regulated at the transcriptional and post-transcriptional level

7.1 *Senescence promoters*

Ethylene-responsive sequences in the regulatory regions of genes expressed in flower senescence and fruit ripening have been defined by promoter deletion analysis and trans-acting factors have been identified by DNase I footprinting (Deikman, 1997). The promoter of *sen1*, an *Arabidopsis See*, is responsive to dark and ABA (Chung *et al.*, 1997). The promoters of the genes encoding the gluconeogenic enzymes malate synthase and *iso*-citrate lyase of cucumber have sugar-responsive elements, which are distinct from the regions conferring germination specificity (DeBellis *et al.*, 1997). Dark induction of *sen1* is also suppressed by physiological concentrations of sucrose or hexoses but not by non-metabolizable carbohydrate analogues (Chung *et al.*, 1997). The promoter of GS1 (an important nitrogen mobilization enzyme) is highly responsive to the ratio of cellular glutamine to glutamate (Watanabe *et al.*, 1997). Carbon 'starvation' and amide production are characteristic of senescing tissues (Brouquisse *et al.*, 1998; Feller and Fischer, 1994) and are clearly important factors in the coordinated expression of a number of *Sees*.

7.2 *Post-transcriptional regulation of metabolic activity in senescing cells*

Metabolic control beyond the level of transcription and translation has some special features in senescing cells. Key proteins must retain structural and functional integrity in a cellular environment where proteolysis and other lytic processes predominate (see also Chapter 11). How? A number of mechanisms are employed. In many cases different enzymes are equally exposed to degradation but genes are selectively activated so that some are continuously replenished. As described in the example of chlorophyll degradation, subcellular compartmentation is also important. Another factor is protein stability in the presence of substrates, cofactors and allosteric effectors. For instance, chlorophyll–protein complexes are resistant to proteases but, without pigments, apoproteins cannot fold properly and are susceptible to proteolytic attack (Thomas, 1997). Chlorophyll degradation via PaO is necessary before proteins of the thylakoid complexes can be broken down in senescence. PaO-less mutants retain chlorophyll and have much reduced breakdown of light-harvesting and other pigment-binding proteins (Thomas *et al.*, 1999).

8. Senescence is reversible

8.1 *Cytokinins*

Exogenous application of cytokinins can cause delayed senescence in most tissues (see Chapter 12). A gene from *Agrobacterium tumefaciens* encodes iso-pentenyl transferase (*ipt*), which catalyses a limiting step in cytokinin biosynthesis. In plants transformed with *ipt* fused to a conditional promoter, increased levels of cytokinin where the promoter is active are associated with delayed senescence (Smart *et al.*, 1991). Gan and

Amasino (1995) fused *ipt* with the promoter of *SAG12*, a gene encoding a cysteine endopeptidase of senescing *Arabidopsis* leaf tissue. Tobacco plants transformed with this construct have an extreme stay-green phenotype as a consequence of auto-regulated cytokinin production.

8.2 *Re-greening*

Cytokinins act at two levels – at a distance by promoting differentiation and strong sink activity, and locally in senescing cells by mediating in the launch of the senescence programme (Kulaeva *et al.*, 1996; Tournaire *et al.*, 1996). The dual action of cytokinins is seen most dramatically in re-greening (*Figure 4*). At flowering, the lowest, oldest leaves of a mature plant of *Nicotiana rustica* will be almost completely yellow. If the shoot is cut off just above the lowest node and the plant is kept in dim light, the leaf

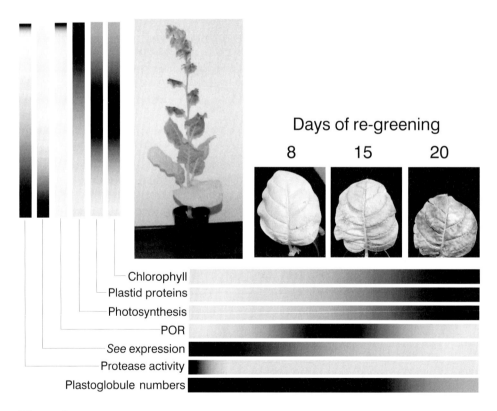

Figure 4. *Regreening in* Nicotiana rustica. *Leaves of* Nicotiana *plants that have begun to flower show a clear age-gradient from top to bottom of the shoot. If the shoot is cut off at the lowest node and the almost completely yellow leaf is treated with cytokinin, it will turn green again over the following 3 weeks in dim light. The abundances or activities of a number of leaf components along the age gradient from top to bottom of the shoot and over the re-greening period are summarized as intensity of shading. POR is NADPH-protochlorophyllide oxidoreductase, a key enzyme of chlorophyll biosynthesis. In situ immunoquantification of POR, together with counts of plastid and plastoglobule numbers, provide direct evidence that plastids reappear during greening by a reversal of the chloroplast-to-gerontoplast transition of senescence (Zavaleta-Mancera* et al.*, 1999a, 1999b).*

will gradually regain its green colour. This process is greatly accelerated if the leaf is treated with cytokinin solution. During re-greening, expression of *Sees* is suppressed, genes for plastid assembly are turned on, the gerontoplasts of the yellow leaf redifferentiate into chloroplasts and photosynthetic activity returns (Zavaleta-Mancera *et al.*, 1999a, 1999b; *Figure 4*).

9. Post-mortem

Being potentially reversible to an advanced stage makes senescence distinct from other PCD processes. But cell death and senescence share important features too. First, reactive oxygen seems to have a central role. Second, new gene expression works with pre-existing, often latent, factors to propagate the inductive stimulus. Third, by programming cell senescence and death, plants bow to the inevitability of ageing while exploiting mortality on their own, adaptively advantageous, terms (Thomas, 1994).

Acknowledgements

We thank the many colleagues, especially our co-authors in the listed references, who have worked with us on the biochemistry and genetics of senescence. Support from, among other sources, BBSRC, EU Framework IV, Advanta and Germinal Holdings is gratefully acknowledged.

References

Akhtar, M.S., Goldschmidt, E.E., John, I., Rodoni, S., Matile, P. and Grierson, D. (1999) Altered patterns of senescence and ripening in *gf*, a stay-green mutant of tomato (*Lycopersicon esculentum* Mill.) *J. Exp. Bot.* **50**: 1115–1122.

Andersson, B. and Aro, E.M. (1997) Proteolytic activities and proteases of plant chloroplasts. *Physiol. Plant.* **100**: 780–793.

Beers, E.P. (1997) Programmed cell death during plant growth and development. *Cell Death Differ.* **4**: 649–661.

Booker, F.L. and Miller, J.E. (1998) Phenylpropanoid metabolism and phenolic composition of soybean [*Glycine max* (L.) Merr.] leaves following exposure to ozone. *J. Exp. Bot.* **49**: 1191–1202.

Bradley, J.M., Davies, K.M., Deroles, S.C., Bloor S.J. and Lewis, D.H. (1998) The maize *Lc* regulatory gene up-regulates the flavonoid biosynthetic pathway of *Petunia*. *Plant J.* **13**: 381–392.

Brouquisse, R., Gaudillère, J.-P. and Raymond, P. (1998) Induction of a carbon-starvation-related proteolysis in whole maize plants submitted to light/dark cycles and to extended darkness. *Plant Physiol.* **117**: 1281–1291.

Buchanan-Wollaston, V. (1997) The molecular biology of leaf senescence. *J. Exp. Bot.* **48**: 181–199.

Camara, B., Hugueney, P., Bouvier, F., Kuntz, M. and Moneger, R. (1995) Biochemistry and molecular-biology of chromoplast development. *Int. Rev. Cytol.* **163**: 175–247.

Chen, H.C., Klein, A., Xiang, M.H., Backhaus, R.A. and Kuntz, M. (1998) Drought- and wound-induced expression in leaves of a gene encoding a chromoplast carotenoid-associated protein. *Plant J.* **14**: 317–326.

Chung, B.C., Lee, S.Y., Oh, S.A., Rhew, T.H., Nam, H.G. and Lee, C.H. (1997) The promoter activity of *sen1*, a senescence-associated gene of *Arabidopsis*, is repressed by sugars. *J. Plant Physiol.* **151**: 339–345.

Cunningham, F.X. and Gantt, E. (1998) Genes and enzymes of carotenoid biosynthesis in plants. *Annu. Rev. Plant Physiol. Plant Mol. Biol.* **49**: 557–583.

Davies, K.M and Grierson, D. (1989) Identification of cDNA clones for tomato (*Lyopersicon esculentum* Mill.) mRNAs that accumulate during fruit ripening and leaf senescence in response to ethylene. *Planta* **179**: 73–80.

DeBellis, L., Ismail, I., Reynolds, S.J., Barrett, M.D. and Smith, S.M. (1997) Distinct *cis*-acting sequences are required for the germination and sugar responses of the cucumber isocitrate lyase gene. *Gene* **197**: 375–378.

Deikman, J. (1997) Molecular mechanisms of ethylene regulation of gene transcription. *Physiol. Plant.* **100**: 561–566.

del Rio, L.A., Pastori, G.M., Palma, J.M., Sandalio, L.M., Sevilla, F., Corpas, F.J., Jiménez, A., López-Huertas, E. and Hernández, J.A. (1998) The activated oxygen role of peroxisomes in senescence. *Plant Physiol.* **116**: 1195–1200.

Distefano, S., Palma, J.M., McCarthy, I. and del Rio, L.A. (1999) Proteolytic cleavage of plant proteins by peroxisomal endoproteases from senescent pea leaves. *Planta* **209**: 308–313.

Donnison, I.S., Griffiths, C.M., Thomas, A., Hosken, S.E., Bridges, I. and Thomas, H. (1999) Characterisation of *See2*, a senescence enhanced cDNA from maize with homology to legumain. *Plant Mol. Biol.* (in press).

Elings, A., White, J.W. and Edmeades, G.O. (1997) Options for breeding for greater maize yields in the tropics. *Eur. J. Agron.* **7**: 119–132.

Falk, K.L., Behal, R.H., Xiang, C. and Oliver, D.J. (1998) Metabolic bypass of the tricarboxylic acid cycle during lipid mobilization in germinating oilseeds. Regulation of NAD^+-dependent isocitrate dehydrogenase versus fumarase. *Plant Physiol.* **117**: 473–481.

Feller, U. and Fischer, A. (1994) Nitrogen metabolism in senescing leaves. *Crit. Rev. Plant Sci.* **13**: 241–273.

Fukasawa-Akada, T., Kung, S.D. and Watson, J.C. (1996) Phenylalanine ammonia-lyase gene structure, expression, and evolution in *Nicotiana. Plant Mol. Biol.* **30**: 711–722.

Fukuda, H. (1996) Xylogenesis: initiation, progression, and cell death. *Annu. Rev. Plant Physiol. Plant Mol. Biol.* **47**: 299–325.

Gan, S. and Amasino, R.M. (1995) Inhibition of leaf senescence by autoregulated production of cytokinin. *Science* **270**: 1986–1988.

Green, P.J. (1994) The ribonucleases of higher plants. *Annu. Rev. Plant Physiol. Plant Mol. Biol.* **45**: 421–445.

Griffiths, C.M., Hosken, S.E., Oliver, D., Chojecki, J. and Thomas, H. (1997) Sequencing, expression pattern and RFLP mapping of a senescence-enhanced cDNA from *Zea mays* with high homology to oryzain gamma and aleurain. *Plant Mol. Biol.* **34**: 815–821.

Hara-Nishimura, I. (1998) Schistosome legumain. In: *Handbook of Proteolytic Enzymes* (eds A.J. Barret, N.D. Rawlings and J.F. Woessner). Academic Press, New York, pp. 749–754.

Harborne, J.B. (1993) *Introduction to Ecological Biochemistry*, 4th edn. Academic Press, London.

Hörtensteiner, S. (1998) NCC Malonyltransferase catalyses the final step of chlorophyll breakdown in rape (*Brassica napus*). *Phytochemistry* **49**: 953–956.

Hörtensteiner, S., Wüthrich, K.L., Matile, P., Ongania, K.-H. and Kräutler, B. (1998) The key step in chlorophyll breakdown in higher plants. Cleavage of pheophorbide *a* macrocycle by a monooxygenase. *J. Biol. Chem.* **273**: 15335–15339.

Howles, P.A., Sewalt, V.J.H., Paiva, N.L., Elkind, Y., Bate, N.J., Lamb, C. and Dixon, R.A. (1996) Overexpression of L-phenylalanine ammonia-lyase in transgenic tobacco plants reveals control points for flux into phenylpropanoid biosynthesis. *Plant Physiol.* **112**: 1617–1624.

Kaplan, D.R., Dengler, N.G. and Dengler, R.E. (1982) The mechanism of plication inception in palm leaves: histogenic observations on the palmate leaf of *Rhapis excelsa. Can. J. Bot.* **60**: 2999–3016.

King, G.A. and O'Donoghue, E.M. (1995) Unravelling senescence: new opportunities for delaying the inevitable in harvested fruit and vegetables. *Trend Food Sci. Technol.* **6**: 385–389.

Kingston-Smith, A.H., Thomas, H. and Foyer, C.H. (1997) Chlorophyll *a* fluorescence, enzyme and antioxidant analysis provide evidence for the operation of an alternative electron sink during leaf senescence in a *stay green* mutant of *Festuca pratensis*. *Plant Cell Environ*. **20**: 1323–1337.

Kulaeva, O.N., Karavaiko, N.N., Selivankina, S.Y., Moshkov, I.E., Novikova, G.V., Zemlyachenko, Y.V., Shipilova, S.V. and Orundgev, E.M. (1996) Cytokinin signalling systems – from whole-plant to the molecular-level. *Plant Growth Reg*. **18**: 29–37.

Lers, A., Khalchitski, A., Lomaniec, E., Burd, S. and Green, P.J. (1998) Senescence-induced RNases in tomato. *Plant Mol. Biol*. **36**: 439–449.

Matile, P., Hörtensteiner, S., Thomas, H. and Kräutler, B. (1996) Chlorophyll breakdown in senescent leaves. *Plant Physiol*. **112**: 1403–1409.

Matile, P., Schellenberg, M. and Vicentini, F. (1997) Localization of chlorophyllase in the chloroplast envelope. *Planta* **201**: 96–99.

Matile, P., Hörtensteiner, S. and Thomas, H. (1999) Chlorophyll degradation. *Annu. Rev. Plant Physiol. Plant Mol. Biol*. **50**: 67–95.

Meanawy, M.A.E., Toshiki, A.J.I., Phillips, N.F.B., Davis, R.E., Salata, R.A., Malhotra, I., McClain, D., Aikawa, M. and Davis, A.H. (1990) Definition of the complete *Schistosoma mansoni* hemoglobinase mRNA sequence and gene expression in developing parasites. *Am. J. Trop. Med. Hyg*. **43**: 67–78.

Medina-Suárez, R., Manning, K., Fletcher, J., Aked, J., Bird, C.R. and Seymour, G.B. (1997) Gene expression in the pulp of ripening bananas. Two-dimensional sodium dodecyl sulfate-polyacrylamide gel electrophoresis of in vitro translation products and cDNA cloning of 25 different ripening-related mRNAs. *Plant Physiol*. **115**: 453–461.

Nam, H.G. (1997) The molecular genetic analysis of leaf senescence. *Curr. Opin. Biotechnol*. **8**: 200–207.

Nishimura, M., Takeuchi, Y., DeBellis, L. and Haranishimura, I. (1993) Leaf peroxisomes are directly transformed to glyoxysomes during senescence of pumpkin cotyledons. *Protoplasma* **175**: 131–137.

Oh S.A., Park J.-H., Lee G.I., Paek K.H., Park S.K. and Nam H.G. (1997) Identification of three genetic loci controlling leaf senescence in *Arabidopsis thaliana*. *Plant J*. **12**: 527–535.

Parthier, B. (1988) Gerontoplasts – the yellow end in the ontogenesis of chloroplasts. *Endocytobio. Cell Res*. **5**: 163–190.

Pennell, R.I. and Lamb, C. (1997) Programmed cell death in plants. *Plant Cell* **9**: 1157–1168.

Pozueta-Romero, J., Rafia, F., Houlne, G., Cheniclet, C., Carde, J.P., Schantz, M.L. and Schantz, R. (1997) A ubiquitous plant housekeeping gene, PAP, encodes a major protein component of bell pepper chromoplasts. *Plant Physiol*. **115**: 1185–1194.

Rhodes, M.J.C. (1980) The maturation and ripening of fruits. In: *Senescence in Plants* (ed. K.V. Thimann). Boca Raton, FL, CRC Press, pp. 157–205.

Rodoni, S., Vicentini, F., Schellenberg, M., Matile, P. and Hörtensteiner, S. (1997) Partial purification and characterization of RCC reductase, a stroma protein involved in chlorophyll breakdown. *Plant Physiol*. **115**: 677–682.

Smart, C.M. (1994) Gene expression during leaf senescence. *New Phytol*. **126**: 419–448.

Smart, C.M., Scofield, S.R., Bevan, M.W. and Dyer, T.A. (1991) Delayed leaf senescence in tobacco plants transformed with tmr, a gene for cytokinin production in *Agrobacterium*. *Plant Cell* **3**: 647–656.

Smart, C.M., Hosken, S., Thomas, H., Greaves, J., Blair, B. and Schuch, W. (1995) The timing of maize leaf senescence and characterisation of senescence-related cDNAs. *Physiol. Plant*. **93**: 673–682.

Stiles, E.W. (1982) Fruit flags – 2 hypotheses. *Am. Nat*. **120**: 500–509.

Sun, Y. and Oberley, L.W. (1996) Redox regulation of transcriptional activators. *Free Rad. Biol. Med*. **21**: 335–348.

Thomas, H. (1994) Ageing in the plant and animal kingdoms – the role of cell death. *Rev. Clin. Gerontol*. **4**: 5–20.

Thomas, H. (1997) Chlorophyll: a symptom and a regulator of plastid development. *New Phytol*. **136**: 163–181.

Thomas H and Smart, C.M. (1993) Crops that stay green. *Ann. Appl. Biol*. **123**: 193–219.

Thomas, H., Schellenberg, M., Vicentini, F. and Matile, P. (1996) Gregor Mendel's green and yellow pea seeds. *Bot. Acta* **109**: 3–4.

Thomas, H., Evans, C., Thomas, H.M., Humphreys, M.W., Morgan, W.G., Hauck, B. and Donnison, I. (1997) Introgression, tagging and expression of a leaf senescence gene in *Festulolium. New Phytol*. **137**: 29–34.

Thomas H., Morgan W.G., Thomas, A.M and Ougham, H.J. (1999) Expression of the stay-green character introgressed into *Lolium temulentum* Ceres from a senescence mutant of *Festuca pratensis. Theor. Appl. Genet* **99**: 92–99.

Tomas-Barberan, F.A., Loaiza-Verlade, J., Bonfanti, A. and Saltveit, M.E. (1997) Early wound- and ethylene-induced changes in phenylpropanoid metabolism in harvested lettuce. *J. Am. Soc. Hort. Sci*. **122**: 399–404.

Tommasini, R., Vogt, E., Fromenteau, M., Hörtensteiner, S., Matile P., Amrhein, N. and Martinoia, E. (1998) An ABC-transporter of *Arabidopsis thaliana* has both glutathione-conjugate and chlorophyll catabolite transport activity. *Plant J*. **13**: 773–780.

Tournaire, C., Kushnir, S., Bauw, G., Inzé, D., de la Serve, B.T. and Renaudin, J.P. (1996) A thiol protease and an anionic peroxidase are induced by lowering cytokinins during callus growth in *Petunia. Plant Physiol*. **111**: 159–168.

Van Oosterom, E.J., Jayachandran, R. and Bidinger, F.R. (1996) Diallel analysis of the stay-green trait and its components in sorghum. *Crop Sci*. **36**: 549–555.

Vicentini, F., Hörtensteiner, S., Schellenberg, M., Thomas, H. and Matile, P. (1995) Chlorophyll breakdown in senescing leaves: identification of the biochemical lesion in a *stay-green* genotype of *Festuca pratensis* Huds. *New Phytol*. **129**: 247–252.

Watanabe, A., Takagi, N., Hayashi, H., Chino, M. and Watanabe, A. (1997) Internal Gln/Glu ratio as a potential regulatory parameter for the expression of a cytosolic glutamine synthetase gene of radish in cultured cells. *Plant Cell Physiol*. **38**: 1000–1006.

Zavaleta-Mancera, H.A., Franklin, K.A, Ougham, H.J., Thomas, H. and Scott, I.M. (1999a) Regreening of *Nicotiana* leaves. I. Reappearance of NADPH-protochlorophyllide oxidore-ductase and light-harvesting chlorophyll a/b-binding protein. *J. Exp. Bot*. **50**: 1677–1682.

Zavaleta-Mancera, H.A., Thomas, B.J., Thomas, H. and Scott, I.M. (1999b) Regreening of *Nicotiana* leaves. II. Redifferentiation of plastids. *J. Exp. Bot*. **50**: 1683–1689.

Zeiger, E. and Schwartz, A. (1982) Longevity of guard cell chloroplasts in falling leaves: implication for stomatal function and cellular aging. *Science* **218**: 680–682.

11

Senescence and cell death in *Brassica napus* and *Arabidopsis*

Vicky Buchanan-Wollaston and Karl Morris

1. What is senescence in plants?

A field of wheat turning golden in the August sunshine is a dramatic example of the role and the importance of plant senescence. Previously these plants were green, actively photosynthesizing to fix carbon from the air and taking up nitrogen from the soil, storing both in the developing leaves. Even under ideal growth conditions, these plants are programmed to initiate the senescence process at a particular time. Photosynthetic activity decreases and a highly controlled dismantling of the cellular components commences. Degradation of protein, lipids and nucleic acids results in the release of nitrogen, phosphorus and carbon as well as other minerals which can be mobilized from the senescing cells. As a result, a large proportion of the constituents of previously green tissue is transported from the senescing material and eventually is stored in the developing seeds. The efficiency of senescence is therefore important for the success of subsequent generations (Feller and Keist, 1986).

Another striking example of plant senescence is seen every autumn when the leaves of deciduous trees turn from green through an often spectacular colour display of red, yellow or brown. The purpose of this type of senescence is different from that occurring in annual plants such as wheat but the processes that take place within the leaves are probably very similar. The materials mobilized from the leaves of deciduous trees are not stored in developing seeds but are transferred into specialized parenchyma cells in the trunk of the tree from where mobilization can occur to provide nutrients for the growth of new leaves in the next season (Clausen and Apel, 1991).

Therefore, the senescence process that occurs as part of the developmental programme of the plant can be characterized primarily as a mobilization process. The function of senescence is to transfer as many nutrients as possible from the senescing tissue to another part of the plant, either for storage or for new growth. The final part of the senescence process is necrotic cell death but this is actively delayed until maximum mobilization has occurred.

Programmed Cell Death in Animals and Plants, edited by J.A. Bryant, S.G. Hughes and J.M. Garland.
© 2000 BIOS Scientific Publishers Ltd, Oxford.

Senescence is a highly regulated process and is genetically controlled (see also Chapter 10). A number of mutants in different plants that show delayed or altered patterns of senescence have been identified (Thomas and Smart, 1993). Expression of novel genes is required for senescence to take place and in the past few years many of these senescence-enhanced genes have been isolated from a number of different plants (reviewed in Buchanan-Wollaston, 1997; Smart, 1994). The identification of genes which are expressed during senescence has already increased our understanding of the enzymatic events that occur during this complex process and the availability of such genes allows a closer examination of the regulation of senescence.

2. Senescence and programmed cell death

Many types of programmed cell death (PCD) occur in developing plants (reviewed by Greenberg, 1996). These include the cell death that occurs in developing flowers of monoecious plants such as maize, to initiate single sexed flowers (Cheng and Pareddy, 1994), the cell death that occurs in developing xylem tissue to form the xylem vessels (Fukuda, 1997; see also Chapter 14) and the cell death that occurs in developing leaves of plants such as the Swiss cheese plant, *Monstera*, which results in the holes in the leaves. Another type of cell death is initiated in the hypersensitive response to pathogen infection where invaded cells die to prevent further pathogen spread (Morel and Dangl, 1997). The PCD that occurs in these situations may have similarities to apoptosis in animals in that DNA fragmentation, characteristic membrane changes and the formation of apoptotic bodies have been detected in some cases (Jones and Dangl, 1996; Pennell and Lamb, 1997).

Senescence has also been described as a programmed cell death in plants although occurrence of the characteristic signs of apoptosis has not been convincingly demonstrated during normal leaf senescence (Nooden and Guiamet, 1996). It is true that senescing cells do eventually die and therefore that death is pre-programmed, but one important difference between senescence and other types of programmed cell death may be that the senescing cell has a very specific role to play before it dies. Senescence is a well-controlled process that takes place over a number of days; the integrity of the nuclear DNA must be maintained to allow sequential biochemical pathways to be activated to complete the process (Makrides and Goldthwaite, 1981). The plant takes steps to ensure the maximum lifespan of the senescing cell before it dies. For example, chlorophyll is degraded only to remove this potentially toxic compound from the cell to prevent premature cell death (Matile *et al.*, 1989). None of the components of chlorophyll are mobilized out of the cell but the breakdown products are stored in the vacuoles of senescing cells (Matile *et al.*, 1988). Also, increased transcription of a number of genes encoding enzymes involved in removal of reactive oxygen species has been detected during senescence (Buchanan-Wollaston and Ainsworth, 1997). These 'precautions' would not be necessary if death was the primary role of the senescing cell.

It is likely that the different programmed cell death pathways that occur in plants have some distinctive and some common features (Heath, 1998; Jones and Dangl, 1996). As is described below there are clear links in the pathways that lead to senescence and to the hypersensitive-induced programmed cell death that occurs in the response to pathogens. There appear to be common genes involved in the hypersensitive response and senescence and equally there may be similarities in genes expressed in other types of cell death.

3. When is senescence induced?

Senescence can be initiated by a wide variety of different factors, both internal and external (Gan and Amasino, 1997; Nam, 1997). Senescence is characterized as the final stage of development of a leaf. The timing of the induction of developmental senescence depends on the plant species. For example, leaves on the short-lived *Arabidopsis* plant start to senesce almost immediately after they reach full expansion (Hensel *et al.*, 1993). Leaves on a wheat plant senesce when seed development signals the requirement for mobilized nutrients; leaves on a deciduous tree start to senesce when environmental conditions indicate the onset of winter and leaves (needles) on a bristle–cone pine tree senesce after about 30 years (Woolhouse, 1982). In all these situations senescence is a pre-programmed event.

As well as being an essential part of plant development, senescence in leaves is also induced prematurely by a number of different environmental stresses (Gan and Amasino, 1997; Smart, 1994). Plants cannot move away from adverse environmental conditions; senescence is one mechanism that they have evolved to cope with such problems. Nutrient stress, such as insufficient nitrogen, phosphorus or water, induces premature senescence in the deprived plants (Crafts-Brandner *et al.*, 1998; Smart, 1994). It is likely that the plant is responding to a severe limitation by mobilizing nutrients from the developed leaves to initiate seed production. Plants that have been exposed to oxidative chemicals such as ozone, or damaging radiation such as UV-B, also show senescence-like symptoms and localized cell death (Jansen *et al.*, 1998; Pell *et al.*, 1997). Ozone enters the cell via the plasma membrane where active oxygen species are generated. In acute cases, at high levels of ozone, rapid cell death occurs and this is seen as necrotic spots on the leaves. This response appears to have some similarity to the hypersensitive response seen with incompatible pathogens (Sanderman *et al.*, 1998). At lower levels of ozone, the increased oxidative stress leads to reduced transcription of genes encoding proteins involved in photosynthesis; chlorophyll and protein degradation occurs and early senescence symptoms are visible (Pell *et al.*, 1997). Plants exposed to supplementary UV-B radiation also show increased levels of active oxygen species, reduced transcription of genes encoding the proteins involved in photosynthesis and increased protein and chlorophyll degradation (reviewed in Jordan, 1996; A-H-Mackerness *et al.*, 1997). The irradiated leaves show both senescence-like symptoms and early necrosis.

Infection by pathogens can result in senescence-like symptoms. Yellowing of infected tissue often occurs (Matthews, 1991) and it is not clear whether this is a negative effect caused by pathogen growth or whether it is a defensive response from the plant by which nutrients are removed from the invading pathogen to restrict its development. Senescence and abscission of the infected leaf is a mechanism that could be useful to reduce the spread of the pathogen to the rest of the plant. Some pathogens attempt to inhibit this senescence by the production of cytokinins resulting in the 'green island' phenomenon that is seen in some pathogen-infected tissue (Coghlan and Walters, 1992). Instead of causing senescence-like symptoms, infection with incompatible pathogens can result in a different type of cell death. A common feature of disease resistance is the rapid development of cell death at the site of infection. A resistant plant responds to infection by an incompatible pathogen by the induction of the hypersensitive response where the invaded cells are rapidly killed to prevent the growth of the pathogen (reviewed in Morel and Dangl, 1997).

Senescence-like events also occur in plants after harvest (Huber, 1987). In particular, vegetable crops that are harvested while they are still growing, e.g. asparagus spears, broccoli florets or leafy crops such as lettuce and cabbage are subjected to considerable stress when they are separated from their nutrient, water and hormone supply (King *et al.*, 1990). These tissues rapidly show senescence-like symptoms such as chlorophyll loss, protein degradation and membrane deterioration and it is likely that some of the same biochemical processes are involved (King and Morris, 1994). Few genes involved in this process have been characterized as yet, but induced expression of asparagine synthetase has been detected in asparagus spears post-harvest as well as in senescing leaves (King *et al.*,1995). Also, genes encoding ethylene biosynthetic enzymes have been shown to increase in expression in broccoli post-harvest (Pogson *et al.*, 1995).

Senescence-like symptoms are seen, therefore, in response to a number of different stresses as well as during development and after harvest. The similarities and differences between stress-induced senescence and developmental senescence remain to be identified. The interesting questions are (a) how similar are the enzymatic events that occur in these different situations? (b) are the same genes involved? (c) are the same regulatory factors required? The identification of senescence-enhanced genes from *Brassica napus* and *Arabidopsis* is allowing us to start to answer some of these questions.

4. What happens during senescence?

It can be seen that many common events occur in the different types of senescence described above. The most obvious visible sign of senescence is yellowing of the tissue which is due to degradation of chlorophyll and retention of carotenoid pigments (Matile, 1992). Our study of gene expression during developmental senescence has indicated that visible yellowing is a relatively late symptom and many senescence-enhanced genes show increased expression well before the tissue is visibly yellow (Buchanan-Wollaston and Ainsworth, 1997). The earliest symptom of senescence is probably a reduction in photosynthesis and this is seen in all types of senescence. Protein degradation also starts early in senescence and the most obvious victim of this is the RUBISCO (ribulose-bisphosphate carboxylase–oxygenase) protein which is degraded rapidly and accounts for a large proportion of the protein mobilized from the senescing cell (Crafts-Brandner *et al.*, 1990). Dismantling of the chloroplast results in the degradation of the thylakoid membranes and the lipid components are probably recycled to fuel the senescence process (Matile, 1992). The degradation of macromolecules, especially lipids, is an oxidative process and the levels of reactive oxygen species increase within the senescing tissue. The rise in reactive oxygen species is concurrent with a decrease in the antioxidant status of the leaf. This causes an increase in membrane damage and eventually this reaches a point where the cell is no longer viable and cell death occurs.

5. Analysis of leaf senescence in *B. napus*

Many different senescence-enhanced genes have been isolated from *B. napus* leaves by a variety of techniques and we now have identified over 40 different genes encoding a number of different types of proteins (Buchanan-Wollaston, 1997; V. Buchanan-Wollaston, unpublished data). Expression analysis of these genes has shown that although they show increased expression during senescence, the induction of expression does not necessarily occur at the same stage of senescence. Therefore, a number of different regulatory pathways must be involved to control the expression of these genes. The northern analysis shown in *Figure 1* illustrates the expression patterns for a selection of the *B. napus* genes in leaves during development.

The enzymatic functions of many of the proteins encoded by these genes can be deduced from their DNA sequence and these can be then grouped according to their potential role in senescence. Some of these groups are shown in *Figure 1* and include genes that encode proteases (three different cysteine proteases and an aspartic protease show increased expression during senescence), enzymes that may be involved in mobilization such as glutamine synthetase (which probably converts amino acids for transport,) enzymes involved in gluconeogenesis (such as pyruvate orthophosphate dikinase) and enzymes involved in sulfur metabolism (such as ATP sulfurylase and APS reductase). Another group contains genes that encode proteins which may be involved in protecting the senescing plant cell from oxidative stress. These include a catalase gene, two different metallothionein genes, a superoxide dismutase and genes involved in glutathione metabolism. Other genes may have a role in protecting the senescing tissue from opportunistic pathogen infection. A group of particular interest contains genes that encode potential regulatory factors. We have identified several genes with similarity to different *Arabidopsis* protein kinases and transcription factors that show increased expression during senescence (*Figure 1*). These proteins may be

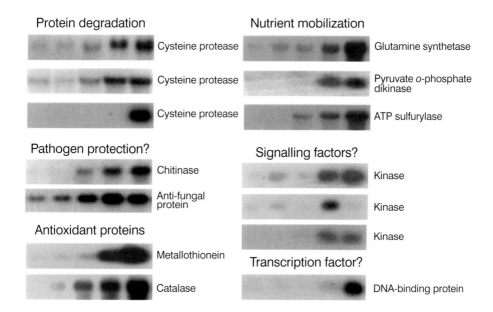

Figure 1. Genes expressed during leaf senescence in Brassica napus. *This figure shows either northern blots carrying total RNA isolated from leaves at five different stages of development, or virtual northern blots carrying PCR amplified cDNA made from the same RNA. Each blot has been hybridized with* [32]*P-labelled cDNA representing a senescence-enhanced gene. Samples are loaded from left to right as follows: (1) young green leaf, (2) mature green leaf, (3) late mature green leaf showing no loss of chlorophyll, (4) early senescent leaf (90% chl), (5) mid senescent leaf (50% chl). Only the last stage of development (5) was showing visible chlorophyll loss.*

involved in signalling pathways leading to senescence and in activating the expression of senescence-enhanced genes. A number of genes show no similarities to any genes currently in the databases and these may encode proteins that have not been characterized as yet. For example, no genes encoding enzymes involved in chlorophyll degradation have been identified yet.

We are currently using the technique of cDNA-AFLP (Bachem *et al.*, 1996; Money *et al.*, 1996) to clone more genes encoding regulatory factors. This is a very sensitive method that can be used to identify genes expressed at low levels during a particular phase of plant development. We are screening for cDNA clones that represent genes expressed at a very early stage of leaf senescence which may be candidates for key regulatory factors that control early phases of senescence.

A number of other groups have identified additional senescence-enhanced genes from *Arabidopsis* (Hensel *et al.*, 1993; Lohman *et al.*, *1994*; Park *et al.*, 1998). *B. napus* and *Arabidopsis* are both members of the Cruciferae family. The relationship between these two plants means that genes isolated from one species are very similar, at the DNA level, to the homologous gene in the other species. We have found that most of the genes that we have identified as exhibiting enhanced expression during senescence in *B. napus* also show enhanced expression in senescing *Arabidopsis* leaves.

6. Links between different types of senescence

One way to analyse the links between the different types of senescence that occur in different circumstances is to analyse the expression patterns of a range of senescence-enhanced genes. The obvious sign of senescence is yellowing, which occurs in all types of senescence. Therefore, it is likely that the genes involved in the control of chlorophyll degradation are expressed during all types of senescence. Other groups of genes may not be expressed if their roles are specific to a particular type of senescence. Comparison of gene expression patterns in normal developmentally regulated senescence, stress-induced senescence and post-harvest senescence will allow us to classify the senescence-enhanced genes into groups.

Recently, the expression of a number of senescence-enhanced *Arabidopsis* genes, in response to different stress or hormone treatments, has been analysed (Park *et al.*, 1998; Weaver *et al.* 1998). These studies showed that, although all genes tested were induced during natural developmental senescence, they showed different expression patterns after the stress treatments. For example, about half of the genes investigated by Weaver *et al.* (1998) were activated by ABA treatment and these, in general, were not the same genes as those induced when intact plants were placed in the dark. The expression patterns of the four genes investigated by Park *et al.* (1998) were all different from each other in response to treatments with the senescence-enhancing hormones ethylene and jasmonic acid.

In similar experiments, we have recently discovered that many of the *B. napus* genes that we have identified as showing enhanced expression during developmental senescence are also induced in broccoli heads post-harvest (T. Page and V. Buchanan-Wollaston, unpublished results). Genes encoding degradative, mobilization and protective functions are all induced during post-harvest deterioration indicating that a controlled dismantling of cellular components may be taking place in this tissue.

These results indicate that the senescence symptoms that are apparent when plants are exposed to different stress treatments are not necessarily caused by expression of the same groups of genes. There are likely to be common elements in the pathways that control the expression of these genes during these different types of senescence. Analysis of the expression of the regulatory factors that we are currently isolating may allow us to assign these to particular pathways leading to expression of different groups of senescence-enhanced genes.

7. Expression of the senescence-enhanced metallothionein gene, LSC54

An alternative way to investigate the expression pattern of senescence-enhanced genes is to fuse the gene promoter with a reporter gene and to analyse gene expression in transgenic plants. Using this method, we have examined the expression of three different senescence-enhanced promoters in response to a number of different stresses.

The gene *LSC54* encodes a metallothionein protein and is expressed at high levels during leaf senescence in *B. napus* (*Figure 1*; Buchanan-Wollaston, 1994). The role(s) of this protein during leaf senescence is not known; it may function to remove or store metal ions released during protein breakdown, or it may function as a protection against free radicals generated when cellular components are degraded. We have cloned and characterized the promoter region from the *LSC54* gene, fused it to the *GUS* gene and analysed the expression in transgenic *Arabidopsis* plants (Butt *et al.*, 1998). As was expected, the promoter fusion showed increased GUS expression during senescence. However, when seedlings from the transformed *Arabidopsis* plants were exposed to a range of different *Peronospora parasitica* isolates, expression of the *GUS* gene was found to be regulated in an isolate-dependent manner. In a compatible interaction, with *P. parasitica* isolates that could successfully infect and sporulate on the host, no expression of the *GUS* gene was observed until sporulation was well advanced (at this stage the cotyledons were yellowing, indicating that pathogen-induced senescence was underway). However, in an incompatible interaction, where infection is halted at an early stage, expression of the promoter was detectable in the cotyledon within 24 hours of infection. Microscopic analysis of these infected cotyledons indicated high levels of gene expression only in those cells that had been penetrated by the invading haustoria and which were undergoing the hypersensitive reaction. Therefore this senescence-enhanced gene is also expressed in cells undergoing hypersensitive cell death. The effect of infection with a necrotic pathogen, *Pseudomonas syringae*, on the expression of this gene was also tested. Early expression of the *GUS* gene was detected in the hypersensitive response to an incompatible isolate of *Pseudomonas* in a similar response to that with *Peronospora*. However, in contrast, in the compatible interaction which results in necrotic cell death as the bacteria grow within the leaf tissue, rather than the senescence like symptoms seen in the compatible *Peronospora* interaction, GUS expression was detected in the dying cells.

Using the *LSC54* promoter: *GUS* fusion, we have also shown that this gene shows increased expression in a number of abiotic stresses. Spraying plants with chemicals known to induce oxidative stress, such as silver nitrate, copper sulfate and 3-amino-1,2,4-triazole (a catalase inhibitor), resulted in increased expression of the *GUS* gene. These treatments also result in cell death in the treated tissue. The *LSC54* gene also shows increased expression in response to UV-B irradiation, abscisic acid treatment and in response to wounding.

Thus, the *LSC54* gene is expressed in response to a number of different stress situations. The gene is expressed in senescing cells (in both stress-induced and developmental senescence) and in cells undergoing programmed cell death (hypersensitive response) and necrotic cell death. The role of the metallothionein protein remains to be elucidated but it is likely to have some protective role. In the hypersensitive reaction the gene is expressed in the cells that are destined to die but it is also expressed in cells surrounding the dying cells. This could indicate a role of this gene in limiting the spread of the lesion or in maintaining viability. Its role in senescence is also likely to be in maintaining the oxidative balance of the cell to prevent premature cell death.

8. Expression of other senescence-enhanced genes

The expression of two other senescence-enhanced genes in response to pathogen infection has been analysed. The promoter of the senescence-enhanced chitinase gene, *LSC222* (*Figure 1*; Hanfrey *et al.*, 1996) has been fused to *GUS* and transformed into *Arabidopsis*. Expression patterns of this promoter in response to infection with compatible and incompatible *Peronospora* isolates were similar to those seen with the *LSC54* promoter except that the gene was expressed in a more extensive area around the infection site, not confined to a few neighbouring cells.

In contrast, expression driven by the *SAG12* promoter was completely different. No expression was seen in response to incompatible pathogen infection indicating that this gene is not expressed in the cell death that occurs in the hypersensitive reaction. In the compatible interaction, no GUS expression was observed 7 days after infection despite the visible yellowing of the cotyledons (K. Morris and V. Buchanan-Wollaston, unpublished data). The expression of the *SAG12* gene appears to be specific to a late stage of developmental senescence as this gene was not induced by a number of different stress treatments (Weaver *et al.*, 1998).

The expression of a senescence-enhanced gene *sen1* has been investigated using a *GUS* fusion with the promoter of this gene (Oh *et al.*, 1996). As well as being induced during developmental senescence, this gene was also activated in response to dark, ethylene and ABA treatment.

Therefore, these experiments add to those described above and provide a further indication that the expression patterns of different senescence-enhanced genes are not necessarily the same in the different treatments that induce senescence or cell death. An extension of this type of analysis may also help to identify and characterize the groups of genes involved in the different processes of senescence.

9. The screen for mutants

In order to elucidate fully the complexities of the signalling pathways that lead to the expression of the senescence syndrome it will be necessary to identify and characterize a number of the regulatory factors that are involved in the signalling pathways that lead to expression of senescence-enhanced genes. The key regulatory factors that control senescence may not be regulated at the transcriptional level but may encode repressor proteins or may be transcriptional activators that respond to the presence of a signal to become active. These genes will not be identified in a differential screen using AFLP cDNA as described above. Therefore, in addition to isolating cDNA clones, we have also initiated a mutant screen to identify senescence regulatory genes.

We have shown that senescence-enhanced genes such as *LSC54* are induced in many of the situations where senescence-like symptoms are observed. We have also shown that *LSC54* is expressed in a number of different types of cell death as well as during natural senescence. This indicates that there must be interconnections between the signalling pathways that lead to the senescence response. We are currently exploiting the *LSC54* expression pattern to try to identify mutants in genes encoding components of the numerous signalling pathways that lead to expression of the *LSC54* gene. Seed of *Arabidopsis* carrying the *LSC54* promoter *GUS* fusion has been mutagenized and is being screened for mutants that no longer express the *GUS* gene in response to a number of signals. We are screening for mutants that (a) fail to express GUS in the hypersensitive reaction, (b) show no expression in pathogen-induced senescence, (c) show no expression in response to oxidative stress, (d) show early expression in response to a compatible *Peronospora* isolate and (e) show constitutive GUS expression. Identification and characterization of these mutants should help to elucidate any links in the different pathways that lead to senescence and cell death. For example, mutants that no longer express GUS in the hypersensitive response may or may not also be affected in the expression of the *LSC54* gene in senescence. Pleiotropic effects of the mutations on senescence and other cell death pathways will be assessed by analysis of the phenotype of the mutant plants.

10. Conclusions

Plant senescence is a highly complex process that is controlled by the regulated expression of specific genes. The fact that so many different developmental and environmental signals can lead to the initiation of senescence means that the pathways leading to induced expression of senescence-enhanced genes are likely to be inter-linked with many other pathways that lead to more specific gene expression in response to each stress. *Figure 2* illustrates the complexities of the pathways that could

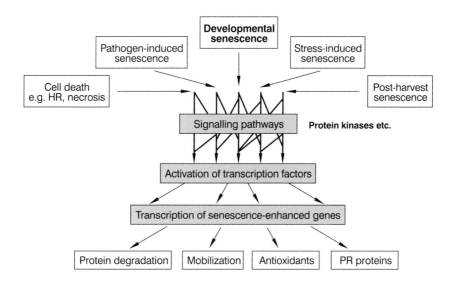

Figure 2. The regulation of senescence-enhanced gene expression. Expression of senescence-enhanced genes is induced during development, in response to biotic and abiotic stress and, also, in harvested leaves and other vegetables. Inter-linking signalling pathways leading to activation of transcription must be involved in the control of gene expression in these different situations.

be involved. Signals from a number of different stresses induce gene expression via signalling pathways. The groups of genes that are activated may differ depending on the initial stress and the signalling pathways that are involved. Common expression of these genes results in symptoms of senescence.

Our research is now aimed at the identification and characterization of genes that have a role in the regulation of senescence. If we can isolate a number of regulatory genes that are of key importance in the different pathways leading to senescence and cell death in response to different signals, we can start to investigate the links between these signalling pathways. Also, we will identify genes that function in the transcriptional regulation of the different groups of genes that take part in the processes of senescence. Once these genes are identified and isolated, this should enable future manipulation of senescence in agricultural and horticultural crops with consequent benefits in product quality, yield and shelf-life.

References

A-H-Mackerness, S., Jordan, B.R. and Thomas, B. (1997) UV-B effects on the expression of genes encoding proteins involved in photosynthesis. In: *Plants and UV-B: Responses to Environmental Change* (ed. P.J. Lumsden). Cambridge University Press, Cambridge, pp. 113–134.

Bachem, C., Hoeven, R., de Bruijn, S., Vreugdenhil, D., Zabeau, M. and Visser, R. (1996) Visualisation of differential gene expression using a novel method of RNA fingerprinting based on AFLP: Analysis of gene expression during potato tuber development. *Plant J.* 9: 745–753.

Buchanan-Wollaston, V. (1994) Isolation of cDNA clones for genes that are expressed during leaf senescence in *Brassica napus*. *Plant Physiol.* 105: 839–846.

Buchanan-Wollaston, V. (1997) The molecular biology of leaf senescence. *J. Exp. Bot.* 48: 181–199.

Buchanan-Wollaston, V. and Ainsworth, C. (1997) Leaf senescence in *Brassica napus*: cloning of senescence related genes by subtractive hybridisation. *Plant Mol. Biol.* 33: 821–834.

Butt, A., Mousley, C., Morris, K., Beynon, J., Can, C., Holub, E., Greenberg, J.T. and Buchanan-Wollaston, V. (1998) Differential expression of a senescence-enhanced metallothionein gene in *Arabidopsis* in response to isolates of *Peronospora parasitica* and *Pseudomonas syringae*. *Plant J.* 16: 209–221.

Cheng, P.C. and Pareddy, D.R. (1994) Morphology and development of the tassel and ear. In: *The Maize Handbook* (eds M. Freeling and V. Walbot) Springer, New York, pp. 37–47.

Clausen, S. and Apel, K. (1991) Seasonal changes in the concentration of the major storage protein and its mRNA in xylem ray cells of poplar trees. *Plant Mol. Biol.* 17: 669–678.

Coghlan, S.E. and Walters, S.R. (1992) Photosynthesis in green islands on powdery mildew-infected barley leaves. *Physiol. Mol. Plant Pathol.* 40: 31–38.

Crafts-Brandner, S.J., Salvucci, M.E. and Egli, D.B. (1990) Changes in ribulosebisphosphate carboxylase/oxygenase and ribulose 5-phosphate kinase abundances and photosynthetic capacity during leaf senescence. *Photosynthesis Res.* 23: 223–230.

Crafts-Brandner, S.J., Holzer, R. and Feller, U. (1998) Influence of nitrogen deficiency on senescence and the amounts of RNA and proteins in wheat leaves. *Physiol. Plant.* 102: 192–200.

Feller, U. and Keist, M. (1986) Senescence and nitrogen metabolism in annual plants. In: *Fundamental, ecological and agricultural aspects of nitrogen metabolism in higher plants* (eds H. Lambers, J.J. Neeterson and I. Stulen). Martinus Nijhoff Publishers, Dordrecht, pp. 219–234.

Fukuda, H. (1997) Tracheary element differentiation. *Plant Cell* 9: 1147–1156.

Gan, S. and Amasino, R.M. (1997) Making sense of senescence. Molecular genetic regulation and manipulation of leaf senescence. *Plant Physiol*. **113**: 313–319.

Greenberg, J.T. (1996) Programmed cell death: A way of life for plants. *Proc. Natl Acad. Sci. USA* **93**: 12094–12097.

Hanfrey, C., Fife, M. and Buchanan-Wollaston, V. (1996) Leaf senescence in *Brassica napus*: expression of genes encoding pathogenesis-related proteins. *Plant Mol. Biol*. **30**: 597–609.

Heath, M.C. (1998) Apoptosis, programmed cell death and the hypersensitive response. *European J. Plant Path*. **104**: 117–124.

Hensel, L.L., Grbic, V., Baumgarten, D.A. and Bleecker, A.B. (1993) Developmental and age-related processes that influence the longevity and senescence of photosynthetic tissues in *Arabidoposis. Plant Cell* **5**: 553–564.

Huber, D.J. (1987) Postharvest senescence: An introduction to the Symposium. *Hort. Sci*. **22**: 853–854.

Jansen, M.A.K., Gaba, V. and Greenberg, B.M. (1998) Higher plants and UV-B radiation: balancing damage, repair and acclimation. *Trends Plant Sci*. **3**: 131–135.

Jones, A.M. and Dangl, J.L. (1996) Logjam at the Styx: programmed cell death in plants. *Trends Plant Sci*. **1**: 114–119.

Jordan, B.R. (1996) The effects of UV-B radiation on plants: A molecular prospective. In: *Advances in Botanical Research* (ed. J.A. Callow). Academic Press, Boca Raton, FL., pp. 97–162.

King, G.A. and Morris, S.C. (1994) Early compositional changes during postharvest senescence of broccoli. *J. Amer. Soc. Hort. Sci*. **119**: 1000–1005.

King, G.A., Woollard, D.C., Irving, D.E. and Borst, W.M. (1990) Physiological changes in asparagus spear tips after harvest. *Physiol. Plant* **80**: 393–400.

King, G.A., Davies, K.M., Stewart, R.J. and Borst, W.M. (1995) Similarities in gene-expression during the postharvest-induced senescence of spears and natural foliar senescence of asparagus. *Plant Physiol.* **108**: 125–128.

Lohman, K.N., Gan, S., John, M.C. and Amasino, R.M. (1994) Molecular analysis of natural leaf senescence in *Arabidopsis thaliana*. *Physiol. Plant*. **92**: 322–328

Makrides, S.C. and Goldthwaite, J. (1981) Biochemical changes during bean leaf growth, maturity and senescence. Contents of DNA, polyribosomes, ribosomal RNA, protein and chlorophyll. *J. Exp. Bot*. **32**: 725–735.

Matile, P. (1992) Chloroplast senescence. In: *Crop photosynthesis: spatial and temporal determinants* (eds N.R. Baker and H. Thomas). Elsevier, Amsterdam, pp. 413–440.

Matile, P., Ginsburg, S., Schellenberg, M. and Thomas, H. (1988) Catabolites of chlorophyll in senescing barley leaves are localised in the vacuoles of mesophyll cells. *Proc. Natl Acad. Sci. USA* **85**: 9529–9532.

Matile, P., Duggelin, T., Schellenberg, M., Rentsch, D., Bortlik, K., Peisker, C. and Thomas, H. (1989) How and why is chlorophyll broken down in senescent leaves? *Plant Physiol. Biochem*. **27**: 595–604.

Matthews, R. (1991) Disease symptoms and effects on metabolism. In: *Plant Virology*. 3rd edn. Academic Press, London, pp. 380–422.

Money, T., Reader, S., Qu, L.J., Dunford, R.P. and Moore, G. (1996) AFLP-based mRNA fingerprinting. *Nucleic Acids Res*. **24**: 2616–2617.

Morel, J. and Dangl, J. (1997). The hypersensitive response and the induction of cell death in plants. *Cell Death Diff.* **4**: 671–683.

Nam, H.G. (1997) The molecular genetic analysis of leaf senescence. *Curr. Opin. Biotech*. **8**: 200–207.

Nooden, L.D. and Guiamet, J.J. (1996) Genetic control of senescence and ageing in plants. In: *Handbook of the Biology of Ageing*. 4th Edn (eds E.L. Schneider, W. Rowe and J. Orland). Academic Press, London, pp. 94–118.

Oh, S.A., Lee, S.Y., Chung, I.K., Lee, C.H. and Nam, H.G. (1996) A senescence-associated gene of *Arabidopsis thaliana* is distinctively regulated during natural and artificially induced leaf senescence. *Plant Mol. Biol*. **30**: 739–754.

Park, J.H., Oh, S.A., Kim, Y.H., Woo, H.R. and Nam, H.G. (1998) Differential expression of senescence-associated mRNAs during leaf senescence induced by different senescence-inducing factors in *Arabidopsis*. *Plant Mol. Biol.* **37**: 445–454.

Pell, E.J., Schlagnhaufer, C.D. and Arteca, R.N. (1997) Ozone-induced oxidative stress: Mechanisms of action and reaction. *Physiol. Plant.* **100**: 264–273.

Pennell, R. and Lamb, C. (1997). Programmed cell death in plants. *Plant Cell* **9**: 1157–1168.

Pogson, B.J., Downs, C.G. and Davies, K.M. (1995). Differential expression of two 1-amino-cyclopropane-1-carboxylic acid oxidase genes in broccoli after harvest. *Plant Physiol.* **108**: 651–657.

Sanderman, H., Ernst, D., Heller, W. and Langebartels, C. (1998) Ozone: an abiotic elicitor of plant defence reactions. *Trends Plant Sci.* **3**: 47–49.

Smart, C.M. (1994) Gene expression during leaf senescence. *New Phytol.* **126**: 419–448.

Thomas, H. and Smart, C.M. (1993) Crops that stay green. *Annu. Appl. Biol.* **123**: 193–219.

Weaver, L.M., Gan, S., Quirino, B. and Amasino, R.M. (1998) A comparison of the expression patterns of several senescence associated genes in response to stress and hormone treatment. *Plant Mol. Biol.* **37**: 455–469.

Woolhouse, H.W. (1982) Leaf senescence. In: *The Molecular Biology of Plant Development* (eds H. Smith and D. Grierson). University of California Press, Berkley, CA, pp. 257–281.

12

Cytokinin and its receptors

Richard Hooley

1. Introduction

Cytokinins were discovered through efforts to identify substances that would promote plant cells to divide in culture (reviewed by Chen, 1998). A wide range of cytokinins are now known to occur in plants (Auer, 1997). Most consist of an N^6-substituted adenine with a branched 5-carbon side-chain and various additional modifications to the adenine. In addition to their essential role in promoting cell division, cytokinins, acting both in synergy and antagonism with other plant hormones, influence a range of events during plant growth and development (*Figure 1*). In view of

CYTOKININS 'PROMOTE':
•Cell division
•Differentiation
•Auxiliary meristem growth
•Cotyledon expansion
•Nutrient mobilization
•Chloroplast development
•Stomatal opening
•Root hair growth
•Anthocyanin accumulation
•Flowering

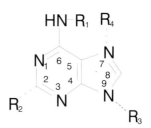

CYTOKININS 'INHIBIT':
•Senescence
•Root elongation
•Lateral root initiation
•Etiolation
•Flowering

Figure 1. *Cytokinins influence plant growth and development. The majority of plant cytokinins consist of an N^6-substituted adenine with a branched 5-carbon side-chain (benzyl; dihydrozeatin; isopentenyl; or zeatin) at R_1 and additional modifications at R_3 (H; glucosyl; ribosyl; or ribotide) and R_4 (H; or glucosyl). Events 'promoted' and 'inhibited' by cytokinins are listed.*

Programmed Cell Death in Animals and Plants, edited by J.A. Bryant, S.G. Hughes and J.M. Garland.
© 2000 BIOS Scientific Publishers Ltd, Oxford.

the scarcity of cytokinin biosynthesis and sensitivity mutants (Deikman, 1997) it is very likely that this group of hormones are essential for plant survival. Cytokinins play a role in senescence which is a large-scale cell-death event in plants. For this reason it is pertinent to consider our current understanding of the molecular basis of cytokinin signalling since it may cast light on some of the processes involved in the regulation of cell death in plants.

2. Cytokinins and leaf senescence

Leaf senescence is a phase of programmed cell death that marks the terminal stage of leaf development. It is subject to complex regulation by a number of factors (see Chapters 10 and 11; Nooden et al., 1997) that include the plant hormones ethylene (see Chapter 13) and cytokinin.

Application of cytokinins to both attached and detached leaves has long been known to delay leaf senescence symptoms in many plant species (Richmond and Lang, 1957; van Staden et al., 1988). In fact, inhibition of leaf senescence was one of the traditional bioassays for cytokinins. Does this mean that leaf senescence is normally regulated by changes in endogenous cytokinin concentration? A range of studies have shown an inverse correlation between leaf senescence and the amount of active cytokinins in leaf tissue (van Staden et al., 1988).

Although cytokinin metabolism is not particularly well understood (Prinsen et al., 1997) it is generally thought that developing roots are the main site of cytokinin biosynthesis and that cytokinins are transported to the rest of the plant through the xylem. The levels of the cytokinins zeatin riboside and dihydrozeatin riboside in xylem sap have been observed to decline coincident with the onset of leaf senescence in soybean suggesting a possible causal relationship (Nooden et al., 1990). The concept that root-to-shoot movement of cytokinins has a major influence on leaf senescence is attractive though, as argued by Jackson (1993), the evidence for it is not completely convincing.

In the Arabidopsis mutant amp1, endogenous levels of the cytokinins N^6-isopentenyladenine, dihydrozeatin, zeatin riboside and zeatin are between two and seven times higher than in the wild type (Chaudhury et al., 1993; Chin-Atkins et al., 1996). The discovery of a mutant that overproduces cytokinins is useful to help elaborate the role of this class of hormones in plant development. Leaves of the amp1 mutant show delayed leaf senescence compared with wild type plants, adding further support to the hypothesis that endogenous cytokinins do inhibit this process. However, amp1 plants have a range of other alterations in their morphology and development that make direct comparison with the wild type complicated. Essentially the same conclusion can be drawn from experiments in which the Agrobacterium tumefaciens isopentenyl transferase (IPT) gene, which catalyses the first step in the cytokinin biosynthesis pathway, is constitutively over-expressed in transgenic plants leading to elevated levels of endogenous cytokinins (Smart et al., 1991).

One of the main criticisms of experiments in which hormones are applied to plants, or of mutants and transgenics that have grossly altered levels of a hormone, is that they are so removed from physiological concentrations that deductions about the normal function of the hormone are meaningless. In an attempt to overcome this problem, and approach the question of what role cytokinins play in leaf senescence, Gan and Amasino (1995) put the IPT gene under the control of the promoter for a senescence-specific

Arabidopsis gene, *SAG12*. The *SAG12* gene encodes a protease enzyme and is only expressed in senescing leaves. Tobacco plants transformed with the *SAG12* promoter-driven *IPT* gene do not show leaf senescence and are indistinguishable from wild type plants in all other aspects of their growth and development (Gan and Amasino, 1995). The rational for the experiment is that at the onset of senescence the *IPT* gene will be expressed under the control of the *SAG12* promoter and cause an elevation in cytokinins in the senescing leaves. The increase in cytokinin will inhibit senescence and thereby attenuate expression of the *IPT* gene. Thus, cytokinin levels will be autoregulated rather than grossly elevated (*Figure 2*). Although Gan and Amasino did not confirm this by measuring cytokinin concentrations in the leaves directly they did use a *SAG12* promoter-driven *GUS* gene construct to report *SAG12* promoter activity. Their observations strongly suggest that cytokinins were not over-produced and that senescence-specific regulation of cytokinin levels can suppress leaf senescence.

Taken together these observations indicate that gross and subtle elevations of cytokinins in leaves inhibit the large-scale cell death associated with senescence. Whether or not cytokinins are natural regulators of leaf senescence, and the role of root-to-shoot cytokinin transport in the initiation of leaf senescence are still open to question.

3. Cytokinin receptors

Compared with other classes of plant hormones, the molecular mechanism of cytokinin action is only poorly understood. One of the reasons for this is that there are very few cytokinin sensitivity mutants (Deikman, 1997) and therefore cytokinin signalling has not seen the same advances that molecular genetics has brought to the

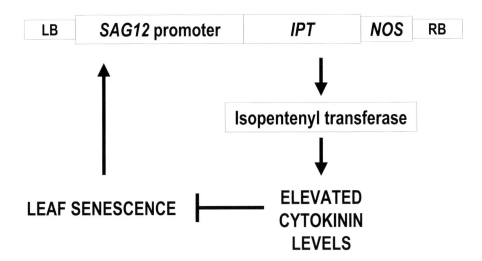

Figure 2. Auto-regulated cytokinin production inhibits tobacco leaf senescence. The rational for this classic experiment is that the onset of leaf senescence stimulates transcription from the senescence-specific SGA12 promoter leading to production of isopentenyl transferase enzyme from the Agrobacterium IPT gene. This presumably causes an elevation in cytokinin levels which inhibits leaf senescence and expression of the SAG12:IPT construct. Diagram is adapted from Gan and Amasino (1995).

understanding of other hormone signalling pathways. Cytokinin inhibition of leaf senescence clearly involves the inhibition of transcription of the *SAG12* promoter (Gan and Amasino, 1995). It is also clear that the expression of a number of other genes is regulated by cytokinins (Hare and van Staden, 1997). Another target of cytokinin signalling is a p34[cdc2]-like H1 histone kinase involved in regulation of the cell cycle at mitosis. Cytokinin promotes tyrosine dephosphorylation and activation of this enzyme (Zhang *et al.*, 1996).

The nature of the signal transduction pathways that are involved in cytokinin regulation of gene transcription and cell cycle control is however largely a mystery and it has been difficult to assign signalling intermediates or events to them with any degree of confidence. Various attempts have been made to identify cytokinin receptors and these have concentrated either on the isolation of cytokinin binding proteins or on molecular and genetic approaches.

3.1 *Cytokinin binding proteins*

Cytokinin binding proteins have been detected in extracts from a range of plant tissues using binding assays, affinity chromatography, anti-idiotypic antibodies and photoaffinity labelling (Brinegar, 1994). Of these, only a few have been characterized in any detail and in no case has there been a clear demonstration of a role for any of these proteins in a defined cytokinin response.

The best characterized cytokinin binding protein, CBF-1, is a 165 kDa homotrimeric protein present in wheat embryos (Brinegar, 1994). CBF-1 is an abundant vicilin-type seed storage protein present in protein bodies within the embryo. Although it binds the photoaffinity probe 2-azido-benzyladenine, and other active cytokinin derivatives, it is clearly not a cytokinin receptor (Brinegar, 1994). The discovery of CBF-1 highlights the technical difficulties in seeking low abundance, high affinity receptors, against a background of highly abundant binding proteins. A phenylurea-based photoaffinity probe, that is a powerful cytokinin agonist, has been used to label and characterize a glutathione S-transferase (Gonneau *et al.*, 1998). Similarly, this enzyme is very unlikely to be involved in cytokinin signalling and this is confirmed by its cytokinin binding properties which are not correlated with their biological activity (Gonneau *et al.*, 1998).

A cytokinin binding protein complex of 130 kDa has been purified from tobacco leaves and one of the protein components of it identified and cDNA cloned. This CBP57 subunit has high sequence similarity with *S*-adenosyl-L-homocysteine (SAH) hydrolase (Mitsui *et al.*, 1996). In other organisms SAH hydrolase is known to bind adenosine and cAMP and it is thought that certain purines may regulate SAH hydrolase activity. Thus, it has been proposed (Mitsui *et al.*, 1996; Mitsui *et al.*, 1997) that in plants cytokinins might regulate the activity of this enzyme and thereby influence DNA, protein or other methylation reactions that could be involved in cytokinin signalling or action. This is an interesting hypothesis, the validity of which should become clearer with further investigations.

Another potential target of cytokinin signalling is a stimulation of the general protein synthesis capacity of plant cells. This concept is logical in view of some of the physiological effects of cytokinins, for example, stimulation of cell division and inhibition of senescence. It is supported by the observation that kinetin stimulates steady-state levels of accurately initiated rRNA transcripts when applied to *Arabidopsis* leaves and an increase in the level of nascent rRNA transcripts in nuclei isolated from the treated leaves (Gaudino and Pikaard, 1997). How kinetin stimulates initiation of transcription by RNA polymerase I is not known, although it is intriguing that a soluble cytokinin binding protein isolated from barley leaves has, in the presence of active *trans*-zeatin, been reported to stimulate the elongation of nascent RNA transcripts in both a chromatin-associated RNA-polymerase I system, and in isolated nuclei (Kulaeva *et al.*, 1995).

3.2 *A two-component receptor involved in cytokinin signalling*

Recent molecular and genetic approaches have identified two different types of integral membrane receptor protein that have characteristics which suggest a role in cytokinin perception or signal transduction (Kakimoto, 1996; Plakidou-Dymock *et al.*, 1998). The first to be discovered, CKI1, was identified in an activation tagging screen in *Arabidopsis* because it conferred a cytokinin-independent phenotype in callus. Over-expression of CKI1 also gave a constitutive cytokinin response phenotype in callus (Kakimoto 1996). One interpretation of these observations is that CKI1 is a cytokinin receptor and that over-expression enables the cells to detect and respond to their endogenous cytokinins. There are other obvious interpretations of the data too and we must await identification of the ligand or regulation mechanism for CKI1 before a function can be ascribed with confidence. CKI1 has sequence similarity to histidine kinase two-component response regulators and has conserved motifs found in both the histidine kinase and response-regulator domains (*Figure 3*).

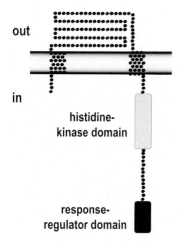

out

in

histidine-
kinase domain

response-
regulator domain

• **Activation tagged CKI1 callus are cytokinin autonomous**

• **Confers cytokinin-independent phenotype when over-expressed in callus cultures**

• **Belongs to same family as ethylene receptors**

Figure 3. CKI1 *is a candidate cytokinin receptor. The predicted structure of the CKI1 protein is shown. The N-terminus is cytoplasmic, two putative transmembrane spanning domains create an extracellular region that bears a number of potential glycosylation sites. The protein contains both the histidine kinase and response-regulator domains characteristic of the two-component regulators. Characteristics of CKI1 are listed (Kakimoto, 1996).*

Based on current knowledge of two-component systems (Chang and Stewart, 1998), CKI1 will presumably be activated either by binding a ligand such as cytokinin, or by interacting with a transport or receptor protein. This will lead to autophosphorylation of the conserved histidine residue in the sensory kinase domain, and transfer of the phosphate to the conserved aspartate in the response regulator or receiver domain. The downstream partners are not currently known. However, CKI1 is a member of the same family of receptors as the ethylene receptor ETR1 and this is considered most likely to signal through a MAP kinase cascade (Bleecker and Schaller, 1996). Perhaps CKI1 will signal through a similar route. Alternatively, a family of *Arabidopsis*-derived response-regulator proteins, and a maize-derived response-regulator protein might be recipients of the phosphotransfer reaction (Brandstatter and Kieber, 1998; Sakakibara *et al.*, 1998; Taniguchi *et al.*, 1998). Genes encoding these proteins are known to be up-regulated in leaves by cytokinins and by nitrate supplied to roots of nitrogen-starved plants, a treatment that probably elevates foliar cytokinin concentration.

3.3 *A seven transmembrane domain receptor involved in cytokinin signalling*

The other candidate cytokinin receptor is GCR1, the first plant homologue of the G protein-coupled receptor (GPCR) superfamily (Plakidou-Dymock *et al.*, 1998) (*Figure 4*). The *GCR1* cDNA encodes a 326 amino acid polypeptide that has up to 23% amino acid identity (53% similarity) to known GPCRs, and a number of amino acid motifs that are conserved in some classes of GPCRs. Hydropathy analysis indicates that GCR1 has seven potential transmembrane spanning domains, and membrane topology prediction algorithms support a structure characteristic of GPCRs. *GCR1* is a single copy gene in *Arabidopsis* and is expressed at very low

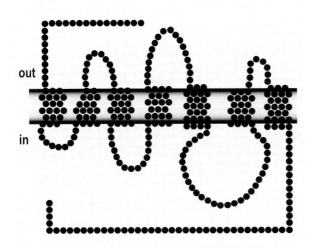

- **Antisense-suppression of *GCR1* expression confers reduced sensitivity to benzyl adenine**

- **Homologue of the superfamily of G protein-coupled receptors**

Figure 4. GCR1 is a candidate cytokinin receptor. The predicted structure of the GCR1 protein is shown. The polypeptide is predicted to have an extracellular N-terminus, seven transmembrane-spanning domains connected by alternate intracellular and extracellular loops and an intracellular C-terminus. Characteristics of GCR1 are listed (Plakidou-Dymock et al., 1998).

concentrations. Using RT-PCR, transcript can be detected in 4-day-old seedlings and in roots, and vegetative tissues of 2–7-week-old plants. Transgenic *Arabidopsis* containing antisense *GCR1* behind the constitutive CaMV*35S* promoter have reduced sensitivity to the cytokinin benzyl adenine in roots and shoots, yet respond normally to all other plant hormones. This suggests a functional role for *GCR1* in cytokinin signal transduction.

GCR1 may be a cytokinin receptor; alternatively it might be a downstream component in cytokinin signalling, or a receptor for a different ligand, the signalling pathway for which interacts with cytokinin signalling. While the downstream signalling partners for GCR1 are unknown it is likely, though by no means certain, that ligand activated GCR1 will interact with a heterotrimeric G protein (Hooley, 1999).

Further research is needed to establish the function of, and relationship between, GCR1 and CKI1, and the extent to which either or both of them are involved in cytokinin inhibition of cell death associated with leaf senescence.

Acknowledgements

IACR receives grant-aided support from the Biotechnology and Biological Sciences Research Council (BBSRC) of the United Kingdom. This work was supported by the BBSRC Integration in Cellular Responses Programme and European Union Framework IV Programme.

References

Auer, C.A. (1997) Cytokinin conjugation: recent advances and patterns in plant evolution. *Plant Growth Reg.* **23**: 17–32.

Bleecker, A.B. and Schaller, E.G. (1996) The mechanism of ethylene perception. *Plant Physiol.* **111**: 653–660.

Brandstatter, I. and Kieber, J.J. (1998) Two genes with similarity to bacterial response regulators are rapidly and specifically induced by cytokinin in arabidopsis. *Plant Cell* **10**: 1009–1019.

Brinegar, A.C. (1994) Cytokinin binding proteins and receptors. In: *Cytokinins: Chemistry, Activity, and Function* (eds D.W.S. Mok and M.C. Mok). CRC Press, Boca Raton, pp. 217–232.

Chang, C. and Stewart, R.C. (1998) The two component system. Regulation of diverse signalling pathways in prokaryotes and eukaryotes. *Plant Physiol.* **117**: 723–731.

Chaudhury, A.M., Letham, S., Craig, S. and Dennis, E.S. (1993) *amp1*-a mutant with high cytokinin levels and altered embryonic pattern, faster vegetative growth, constitutive photomorphogenesis and precocious flowering. *Plant J.* **4**: 907–916.

Chen, C.-M. (1998) The discovery of cytokinins. In: *Discoveries in Plant Biology Volume 1* (eds S.-D. Kung and S.-F. Yang). World Scientific Publishing Co, Singapore, pp. 1–15.

Chin-Atkins, A.N., Craig, S., Hocart, C.H., Dennis, E.S. and Chaudhury, A.M. (1996) Increased endogenous cytokinin in the *Arabidopsis amp1* mutant corresponds with de-etiolation responses. *Planta* **198**: 549–556.

Deikman, J. (1997) Elucidating cytokinin response mechanisms using mutants. *Plant Growth Reg.* **23**: 33–40.

Gan, S. and Amasino, R.M. (1995) Inhibition of leaf senescence by autoregulated production of cytokinin. *Science* **270**: 1986–1988.

Gaudino, R.J. and Pikaard, C.S. (1997) Cytokinin induction of RNA polymerase I transcription in *Arabidopsis thaliana*. *J. Biol. Chem.* **272**: 6799–6804.

Gonneau, M., Mornet, R. and Laloue, M. (1998) A *Nicotiana plumbaginifolia* protein labeled with an azido cytokinin agonist is a glutathione S-transferase. *Physiol. Plant.* **103**: 114–124.

Hare, P.D. and van Staden, J. (1997) The molecular basis of cytokinin action. *Plant Growth Reg.* **23**: 41–78.

Hooley, R. (1999) A role for G proteins in plant hormone signalling? *Plant Physiol. Biochem.* **37**: 393–402.

Jackson, M.B. (1993) Are plant hormones involved in root to shoot communications? *Adv. Bot. Res.* **19**: 103–186.

Kakimoto, T. (1996) CKI1, a histidine kinase homolog implicated in cytokinin signal transduction. *Science* **274**: 982–985.

Kulaeva, O.N., Karavaiko, N.N., Selivankina, S.Y., Zemlyachenko, Y.V. and Shipilova, S.V. (1995) Receptor of *trans*-zeatin involved in transcription activation by cytokinins. *FEBS Lett.* **366**: 26–28.

Mitsui, S., Wakasugi, T. and Sugiura, M. (1996) A cytokinin-binding protein complex from tobacco leaves. The 57 kDa subunit has high homology to S-adenosyl-L-homocysteine hydrolase. *Plant Growth Reg.* **18**: 39–43.

Mitsui, S., Wakasugi, T., Hanano, S. and Sugiura, M. (1997) Localization of a cytokinin-binding protein CBP57/S-adenosyl-L-homocysteine hydrolase in a tobacco root. *J. Plant Phsyiol.* **150**: 752–754.

Nooden, L.D., Singh, S. and Letham, D.S. (1990) Correlation of xylem sap cytokinin levels with monocarpic senescence in soybean. *Plant Physiol.* **93**: 33–39.

Nooden, L.D., Guiamet, J.J. and Isaac, J. (1997) Senescence mechanisms. *Physiol. Plant.* **101**: 746–753.

Plakidou-Dymock, S., Dymock, D. and Hooley, R. (1998) A higher plant seven-transmembrane receptor that influences sensitivity to cytokinins. *Curr. Biol.* **8**: 315–324.

Prinsen, E., Kamfnek, M. and van Onckelen, H.A. (1997) Cytokinin biosynthesis: a black box? *Plant Growth Reg.* **23**: 3–15.

Richmond, A.E. and Lang, A. (1957) Effect of kinetin on protein content and survival of detached *Xanthium* leaves. *Science* **125**: 650–651.

Sakakibara, H., Suzuki, M., Takei, K., Deji, A., Taniguchi, M. and Sugiyama, T. (1998) A response-regulator homologue possibly involved in nitrogen signal transduction mediated by cytokinin in maize. *Plant J.* **14**: 337–344.

Smart, C.M., Scofield, S.R., Bevan, M.W. and Dyer T.A. (1991) Delayed leaf senescence in tobacco plants transformed with *tmr*, a gene for cytokinin production in *Agrobacterium*. *Plant Cell* **3**: 647–656.

Taniguchi, M., Kiba, T., Sakakibara, H., Ueguchi, C., Mizuno, T. and Sugiyama, T. (1998) Expression of *Arabidopsis* response regulator homologs is induced by cytokinins and nitrate. *FEBS Lett.* **429**: 259–262.

van Staden, J., Cook, E.L. and Nooden, L.D. (1988) Cytokinins and senescence. In: *Senescence and Aging in Plants.* (eds L.D. Nooden and A.C. Leopold). Academic Press, San Diego, pp. 281–328.

Zhang, K., Letham, D.S. and John, P.C.L. (1996) Cytokinin controls the cell cycle at mitosis by stimulating the tyrosine dephosphorylation and activation of p34[cdc2]-like H1 histone kinase. *Planta* **200**: 2–12.

13

Ethylene-triggered cell death during aerenchyma formation in roots

Malcolm C. Drew, Chuan-Jiu He and Page W. Morgan

1. Introduction

All higher plants, like animals, are aerobic organisms, fully dependent on a supply of molecular oxygen for cytochrome oxidase in oxidative phosphorylation, and for other oxidases. Even plants that are adapted to live in O_2-deficient wetland conditions, retain their dependence on O_2, and at the vegetative stage can usually only tolerate brief periods of anoxia (Drew, 1997). How do plants in such environments get enough O_2 to remain aerobic? Wetland species typically form gas-filled spaces (air spaces) in their stems and roots. These spaces or lacunae, together with the surrounding living cells are known as aerenchyma. Interconnected lacunae form continuous channels from leaves or stems that are exposed to the air, to plant organs that are in an O_2-deficient environment, i.e. roots or rhizomes submerged in water or in anaerobic mud. In many species, not just wetland ones, aerenchyma formation takes place by the death of healthy cells in a highly selective and predictable manner, probably by programmed cell death. In this chapter, we review the occurrence, function and formation of aerenchyma, and suggest that programmed cell death is indeed involved, triggered by the gaseous hormone, ethylene.

2. Patterns of aerenchyma formation

Aerenchyma formation in the roots and stems of higher plants takes place either **shizogenously**, when cells separate at the middle lamella, and cavities develop during subsequent expansion growth, or **lysigenously**, when cells undergo cell death followed by autolysis, leaving gas-filled cavities. Among hydrophytes and other wetland species, both shizogenous and lysigenous aerenchyma are found (Bristow, 1975; Schussler and Longstreth, 1996).

In some species of wetland origin such as rice (*Oryza sativa*), lysigenous aerenchyma mostly forms constitutively (Jackson *et al.*, 1985a). Aerenchyma is seen

Programmed Cell Death in Animals and Plants, edited by J.A. Bryant, S.G. Hughes and J.M. Garland.
© 2000 BIOS Scientific Publishers Ltd, Oxford.

even in upland (dryland) varieties that do not experience flooding, and when these varieties are grown in the laboratory under well-oxygenated conditions (M.C. Drew, unpublished observations). For many species, however, aerenchyma forms in roots only as they grow into waterlogged or O_2-deficient environments (Justin and Armstrong, 1987), and even in some rice varieties, additional aerenchyma form under O_2-deficient conditions (Justin and Armstrong, 1991). In nodal maize roots in solution culture with an O_2 concentration of about 3–4% (balance, N_2) in the aeration gas stream to make the tissue hypoxic, aerenchyma begins to form in the zone about 10 mm from the tip, where cell expansion is nearly finished (Campbell and Drew, 1983). Cells in this zone were initiated in the apical meristem some 36 h earlier. Cells do not die synchronously however, so that aerenchyma formation is not complete until cells reach a zone 3–4 cm behind the tip (Drew et al., 1979) at which point 40–45% of all cortical cells have lysed. With a root extension rate of about 6 mm per day at 25°C under these conditions, that means that complete formation of aerenchyma takes place in 3 to 5 days. Individual cells collapse and autolyse in a much shorter time than that, possibly within 24 h. At the electron microscope level, cell death is first detectable by the collapse of cells in the mid-cortex with apparent loss of vacuolar integrity (Campbell and Drew, 1983). Dissolution of the protoplasm and much of the cell wall is complete within a further 12–24 h, leaving gas-filled spaces (lacunae) that inter-connect to the gas space system in the shoot.

Aerenchyma formation also takes place lysigenously in the roots of maize when subjected to mechanical impedance, under conditions that do not involve O_2-deficiency. The roots of young plants were allowed to grow inside a triaxial cell packed with incompressible particles of calcined clay. By applying pressure to the external membrane that contains the particles, the rigidity of the matrix, and the resistance it offers to growing plant organs that have to displace the particles, can be controlled. The matrix was kept moist, and air was circulated at normal ambient pressure through the cell so that O_2 was not limiting (He et al., 1996a; Sarquis et al., 1991). An applied pressure of 1 bar induced aerenchyma formation to about the same degree as did hypoxia, produced by passing a mixture of gases (4% O_2, balance N_2) through the cell.

Aerenchyma can also be induced by temporary shortage of a source of N (as NO_3^- or NH_4^+) or $H_2PO_4^-$ in the rooting medium (Drew et al., 1989; He et al., 1992).

3. Aerenchyma function

It has been generally accepted since Williams and Barber (1961) that aerenchyma formation serves a dual purpose under conditions of O_2-deficiency, namely reduction in the number of respiring, O_2-consuming cells and the simultaneous provision of a low resistance pathway for the movement of O_2 from aerial organs to O_2-deficient, submerged organs. Oxygen movement takes place in the gaseous phase by diffusion down a concentration gradient to the O_2-starved cells, and also by mass flow (convection) in response to pressure changes. Modelling of O_2 diffusion indicates that roots with a typical 'wetland anatomy' like rice could readily elongate 20 cm into completely O_2-deficient environments before the apical meristem at the root tip becomes O_2-limited (Armstrong, 1979; Armstrong et al., 1991). Factors that determine the limiting length of roots under these circumstances are diameter, respiration rate (and therefore temperature), the extent of aerenchyma formation (porosity), wall resistance to outward O_2 diffusion, and the extent to which the soil is a sink for O_2. It is notable that roots of wetland species have modifications to the wall structure to make them impermeable to the outward diffusion of gases (Clark and Harris, 1981). The advantage of this structural adaptation is that it helps conserve O_2 internally, so that more is available to diffuse to the root tip. Here, the apical meristem is highly demanding of O_2 for cell division and expansion. O_2 also leaks to the rhizosphere from the tip zone and detoxifies reduced chemical species so that unprotected cells in the apical meristem are not in direct contact with harmful molecules that would otherwise easily diffuse across primary cell walls and contact the plasma membrane.

The significance of aerenchyma formation in mechanically impeded roots is less certain. It may be that O_2-deficiency could occur in such roots because of the densely-packed soil particles that surround the growing root tip. Aerenchyma formation would therefore assist internal aeration. With nutrient deficiency, cell death followed by lysis may be of importance in making nutrients (N and P) available to the apical zone, where growth processes require a continuous supply of inorganic nutrients which are not directly available from the external environment.

4. Ethylene biosynthesis and induction of cell death

Aerenchyma formation was attributed for a long time to the direct effects of severe O_2-deficiency on the viability of maize root cortical cells (McPherson, 1939). However, working with stems of sunflower, aerenchyma formation was found by Kawase (1979; 1981a,b) to be promoted by exogenous ethylene. With adventitious (nodal) roots of maize, Drew et al. (1979) showed that low concentrations of ethylene (0.1–1.0 μl l^{-1} air) bubbled through nutrient solution, readily induced aerenchyma in the root cortex under conditions where no O_2-deficiency could be involved. The structure of these roots was indistinguishable from that induced by hypoxia. Hypoxic root tips contained higher concentrations of ethylene than aerobic ones, suggesting that ethylene biosynthesis might be accelerated under these conditions (Drew et al., 1979). Supporting that notion, higher concentrations of the ethylene precursor ACC (Atwell et al., 1988) and greater activities of ACC synthase and ACC oxidase have since been found in hypoxic maize root tips (He et al., 1994; 1996a). Furthermore, low, non-toxic concentrations of inhibitors of ethylene biosynthesis (AVG) or of ethylene action (Ag$^+$) effectively block aerenchyma formation in hypoxic roots (Drew et al., 1981; Jackson et al., 1985b; Konings, 1982). The inhibitory effect of AVG (which blocks ACC synthase activity) is reversed by simultaneous addition of ACC to the rooting medium (Jackson et al, 1985a). Such results present a strong case for believing that cell death in the maize root cortex is not a direct consequence of O_2-starvation, but is triggered by an enhanced rate of ethylene production, stimulated by hypoxia.

ACC synthase activity is enhanced in maize root tips by hypoxia (He et al., 1994; 1996a), but gradually declines under anoxia (He et al., 1994). There have been other reports of increases in ACC synthase activity in tomato roots (Wang and Arteca, 1992) or of ACC concentration in xylem sap, presumably reflecting a greater ACC synthase activity (Bradford and Yang, 1980). ACC synthase activity is also enhanced in submerged (O_2-depleted) internodes of rice (Cohen and Kende, 1987). In all these cases, it seems probable that the tissue was hypoxic rather than anoxic. However, ACC synthase is encoded by multigene families in Arabidopsis and tomato, and several studies have found increases in mRNA levels for some ACC synthase genes under conditions of O_2-deficiency, including anoxia (Shiu *et al.*, 1998).

Additional evidence for a role for ethylene in aerenchyma formation in maize roots comes from studies involving transient starvation of N or P. Under fully aerobic conditions, replacement of a normal nutrient solution with one lacking either element induces the formation of aerenchyma within a few days (Drew *et al.*, 1989), indistinguishable from that induced by hypoxia or exogenous ethylene. However, N or P starvation tends to depress ethylene biosynthesis, not accelerate it. Instead, nutrient starvation enhances the sensitivity of cortical cells to ethylene so that cell death occurs at lower ethylene concentrations, and more quickly, than in unstarved controls (He *et al.*, 1992). Other examples of hormonal regulation of plant development, involving changes in sensitivity to hormones, are reviewed by Bradford and Trewavas (1994). The ability of Ag^+ or AVG to block this promotion of cell death by transient nutrient starvation is consistent with the role of ethylene.

Because of the complete lysis of the protoplast and most of the cell wall following cell death in aerenchyma formation, it is reasonable to suppose that a wide array of cell-degrading enzymes is involved. Activity of cellulase increases about 16-fold in hypoxic root tips, or in N-starved root tips just prior to aerenchyma formation (He *et al.*, 1994), and this rise is blocked by AVG (and reversed again by simultaneous addition of ethylene), indicating the essential role of ethylene in cellulase induction. Levels of mRNA encoding xyloglucan *endo*-transglycosylase (XET) increase in maize roots made hypoxic or exposed to ethylene, which suggests that this wall-loosening enzyme might also have a role in wall degradation (Saab and Sachs, 1996). However, cell wall degradation is observed under the electron microscope to take place towards the end of the cell degeneration process (Campbell and Drew, 1983) and we have directed our research more toward understanding the initiation of cell death.

An overall model for the processes leading to cell death and degradation in maize roots is summarized in *Figure 1*.

5. Ethylene signal transduction pathways

Apart from aerenchyma formation, ethylene is known to regulate many developmental processes in higher plants, including leaf and flower senescence and abscission, fruit ripening, epinasty, stem elongation as well as response to pathogen invasion (Abeles *et al.*, 1992). Ethylene is also involved in many aspects of 'stress physiology' in higher plants (reviewed by Morgan and Drew, 1997). Several of the components in ethylene perception and signal transduction have recently been defined by various ethylene response mutants isolated in *Arabidopsis thaliana* (reviewed by Bleeker and Schaller, 1996; Ecker, 1995; Kieber, 1997; Zarembinski and Theologis, 1994; see also Chapter 12). Summarizing some points that are relevant to the present chapter, the *ETR1* gene has sequence homology with both components of the prokaryotic 'two-component'

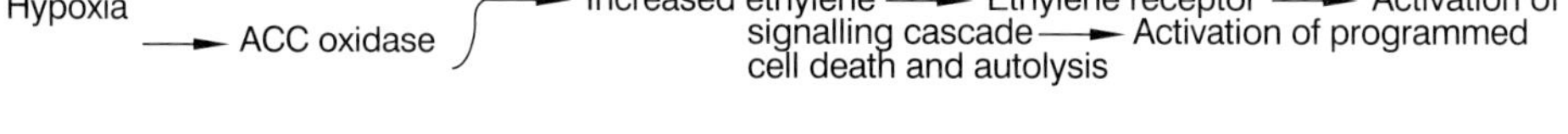

Figure 1. Model of activation by hypoxia of programmed cell death in cortical cells of maize roots.

receptor system (Chang *et al.*, 1993). It encodes a 179 kD protein dimer (Schaller *et al.*, 1995), which includes the ethylene binding site or receptor (Schaller and Bleecker, 1995). This implies that an early step, by analogy with the prokaryotic sensor–response regulator system, involves phosphorylation of a histidine kinase. The *CTR1* gene is downstream of *ETR1* in the signal transduction pathway, and encodes a serine/threonine protein kinase with homology with the Raf protein kinase family. Genes hybridizing to *CTR1* have been found in many plant species, including maize, suggesting that this pathway, or parts of it, may be conserved in plant species, and have features in common with animal cells. Clearly, protein phosphorylation via kinases occurs at several steps in the ethylene signal transduction in *Arabidopsis.* Other components of the ethylene response pathway in *Arabidopsis* have been characterized genetically (Roman *et al.*, 1995) and their sequence tentatively assigned on the basis of epistasis. Thus, *EIN2* is downstream of both *ETR1* and *CTR1*, but its function is not yet defined. *EIN3* acts downstream of *ETR1* and *CTR1*, and is known to encode a nuclear protein (Chao *et al.*, 1997). The ethylene signalling pathway in *Arabidopsis* has so far been examined in relation only to the 'triple response' of seedlings (curvature of apical hook; radial swelling of hypocotyl; inhibition of root and hypocotyl elongation).

Components of an ethylene signal transduction pathway have also been revealed by a pharmacological approach in a study of induction of chitinase and other pathogenesis-related (*PR*) genes in tobacco (Raz and Fluhr, 1992, 1993): reagents that chelated Ca^{2+} blocked the ethylene-dependent rise in gene expression, whereas drugs that artifically raised endogenous Ca^{2+} concentrations stimulated *PR* gene expression.

There is evidence that anoxia, independently of ethylene, also elicits a signal transduction pathway involving Ca^{2+}. The anoxic induction of alcohol dehydrogenase was blocked in maize roots by an inhibitor of organellar Ca^{2+} fluxes (Subbaiah *et al.*, 1994a), whereas a drug that raised cytosolic $[Ca^{2+}]$ caused expression of ADH in the absence of anoxia (Subbaiah *et al.*, 1994b). As ethylene production is blocked by strict anoxia, ethylene could not have been involved in the anoxic signalling pathway, but it would be interesting to know whether parts of the signal transduction pathways starting with anoxia or with ethylene, are shared.

Working with the nodal roots of intact maize plants, we have used a pharmacological approach to characterize an ethylene signal transduction pathway that results in cell death in the root cortex (He *et al.*, 1996b). All reagents were used at low concentrations that did not arrest root growth or boost ethylene biosynthesis, and appropriate controls were used throughout. Cell death could be induced in normoxic roots by reagents that activate G-proteins (GTP-γ-S), or that raise cytosolic Ca^{2+} (caffeine or thapsigargin). By contrast in hypoxic roots, cell death was blocked in the presence of antagonists of phospholipid metabolism (neomycin), or reagents that lowered cytosolic Ca^{2+} (EGTA or ruthenium red). Direct visualization of corresponding changes in cytosolic free Ca^{2+} confirmed that the application of agonists or antagonists produced the expected changes in $[Ca^{2+}]$ (Knight *et al.*, 1991; Subbaiah *et al.*, 1994b). Additionally, cell death was blocked in the cortex of hypoxic roots exposed to K252a, which is an inhibitor of protein kinases, especially protein kinase C. The use of okadaic acid, an inhibitor of protein phosphatases, promoted death of cortical cells in normoxic and hypoxic roots.

Our model for the ethylene signal transduction pathway in maize roots, based on the above evidence is summarized in *Figure 2*. Although these studies offer no clues as to the nature or position in the pathway of the kinases and phosphatases, it is likely

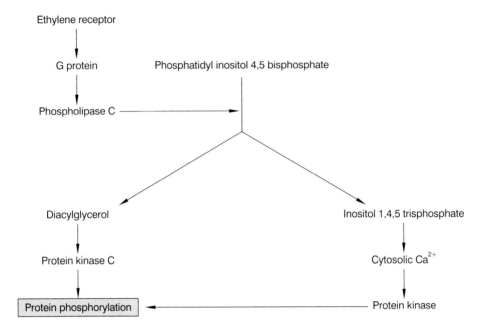

Figure 2. Proposed ethylene signal transduction pathway for cell death in cortex of maize roots.

that several steps could be affected simultaneously, perhaps at the site of ethylene perception and involving a histidine kinase, as well as downstream where Ca^{2+} or calcium–calmodulin (CaCM) might activate a specific protein kinase.

Building on the evidence obtained using various agonists/antagonists (*Figure 2*) we have recently looked for direct evidence of changes in the concentration or activity of some of the putative signal transduction components. There is an early increase in the activity of phospholipase C (PLC) in the plasma membrane fraction prepared from maize roots made hypoxic or when normoxic roots are exposed to ethylene. A doubling of activity in the root tips of maize occurs within the first 4 h of exposure to hypoxia, well ahead of the first signs of cell death, usually at 48 h. PLC enzyme activity in the plasma membrane can thus be regarded as an early marker of cell death. By contrast, PLC activity in the total cell extract (reflecting the enzyme activity in the cytoplasm as a whole) decreased during the same timespan. PLD appears an unlikely contender for signalling cell death in maize roots: activity in both the plasma membrane fraction and in the total cell extract either increased slowly, or increased at the time that cell death and degradation were already underway. PLD may thus be involved more in cell degradation than in signalling in aerenchyma formation in maize roots. We have examined the activity of PLC in roots treated with agonists or antagonists that are known to modify initiation of cell death (He *et al.*, 1996a). For example, in roots treated with neomycin which blocks cell death during hypoxia, PLC activity was strongly inhibited (*Figure 2*). Neomycin is recognized as an inhibitor of PLC. By contrast, EGTA which also blocks cell death, but acting downstream of PLC (by chelating intracellular Ca^{2+}) had little effect on PLC activity in the initial 24 h.

Although we cannot rule out changes in gene expression as part of a signal trans-duction pathway in maize roots, it seems more likely, by analogy to animal systems, and to recent work on higher plant cells (Munnik *et al.*, 1995), that rapid responses to ethylene will be through activation of a cascade of molecular events, without the requirement for synthesis of new proteins.

6. Programmed cell death

The highly selective death of cells in the root cortex, in response to ethylene, is reminiscent of programmed cell death or apoptosis in animal cells (Kerr and Harmon, 1991). Processes akin to programmed cell death have been identified in higher plants, as part of the hypersensitive reaction to pathogens (Greenberg, 1996, 1997; Mittler and Lam, 1995a), during death of cultured plant cells induced by exposure to host-selective fungal toxins (Wang *et al.*, 1996) and during maturation and death of xylem cells (Fukuda, 1996; Groover *et al.*, 1997; Groover and Jones, 1999; Mittler and Lam, 1995b). DNA fragmentation and formation of typical apoptotic bodies were found also in sloughed (dying) root cap cells (Wang *et al.*, 1996), and in salt-stressed barley root tips (Katsuhara and Kawasaki, 1996). Leaf senescence may also be regarded as a PCD phenomenon (Gan and Amasino, 1997; see also Chapters 10 and 11), and increases in cytosolic $[Ca^{2+}]$ are thought to be involved in the signal transduction pathway (Huang *et al.*, 1997). An apoptosis suppressor gene, *Dad-1*, cloned from *Arabidopsis*, is functional in hamster mutants (Gallois *et al.*, 1997). A homologue of the gene has been identified also in rice (Tanaka *et al.*, 1997) and in roots of maize (Finkelstein, 1998). Thus there may be conservation, from plants and animals of at least some components of a genetic pathway leading to PCD (however, see also Chapter 1).

In seedling roots of rice, in which aerenchyma formation is constitutional, the initiation of cell death occurs with greatest frequency in the middle of the root cortex (Kawai *et al.*, 1998). However, further death of cells is preferentially in a radial direction. This has led to the suggestion that the initial steps in aerenchyma formation in rice are by PCD, but that further spread may occur through transmission of a diffusible factor, causing necrotic death.

Very few studies with higher plant cells have been published yet, but in the development of tracheary elements *in vitro* it was found that the sequence of events in cell death did not fit the apoptotic model (Groover and Jones, 1999). By contrast, with tomato protoplasts in culture, and with intact onion root apices, cell death in response to contact with a fungal toxin followed the animal paradigm (Wang *et al.*, 1996). Protoplasts of tobacco undergo cell death in a manner that closely parallels that observed in mammalian cells (O'Brien *et al.*, 1998).

At present little is known about the details of cell death during lysigenous aerenchyma formation in roots of any species. For example, there are no published reports on whether there are characteristic signs of PCD such as DNA 'ladders' in gel electrophoresis, or TUNEL-positive nuclei (but see Chapter 1). Much research remains to be done to see whether cell death in aerenchyma formation has any steps in common with apoptosis in animal cells.

References

Abeles, F.B., Morgan, P.W. and Saltveit, M.E. (1992) *Ethylene in Plant Biology*, 2nd edn. Academic Press, New York.

Armstrong, W. (1979). Aeration in higher plants. *Adv. Bot. Res.* **7**: 225–331.

Armstrong, W., Beckett, P.M., Justin, S.H.F.W. and Lythe, S. (1991) Modelling and other aspects of root aeration. In: *Plant Life Under Oxygen Stress* (eds M.B. Jackson, D.D. Davies and H. Lambers). SPB Academic Publishing, The Hague, pp. 267–282.

Atwell, B.J., Drew, M.C. and Jackson, M.B. (1988) The influence of oxygen deficiency on ethylene synthesis, 1-aminocyclopropane-1-carboxylic acid levels and aerenchyma formation in roots of *Zea mays* L. *Physiol. Plant* 72: 15–22.

Bleecker, A.B. and Schaller, G.E. (1996) The mechanism of ethylene perception. *Plant Physiol.* 111: 653–660.

Bradford, K.J. and Trewavas, A.J. (1994) Sensitivity thresholds and variable time scales in plant hormone action. *Plant Physiol.* 105: 1029–1036.

Bradford, K.J. and Yang, S.F. (1980) Xylem transport of 1-aminocyclopropane-1-carboxylic acid, an ethylene precursor, in waterlogged tomato plants. *Plant Physiol.* 65: 322–326.

Bristow, J.M. (1975) The structure and function of roots in aquatic vascular plants. In: *The Development and Function of Roots* (eds J.G. Torrey and D.T. Clarkson). Academic Press, London, pp. 221–236.

Campbell, R. and Drew, M.C. (1983) Electron microscopy of gas space (aerenchyma) formation in adventitious roots of *Zea mays* L. subjected to oxygen shortage. *Planta* 157: 350–357.

Chang, C., Kwok, S.F., Bleecker, A.B. and Meyerowitz, E.M. (1993) *Arabidopsis* ethylene-response gene ETRI: similarity of product to two-component regulators. *Science* 262: 539–544.

Chao, Q., Rothenberg, M., Solano, R., Roman, G., Terzaghi, W. and Ecker, J.R. (1997) Activation of the ethylene gas response pathway in Arabidopsis by the nuclear protein ETHYLENE-INSENSITIVE 3 and related proteins. *Cell* 89: 1133–1144.

Clark, L.H. and Harris, W.H. (1981) Observations on the root anatomy of rice (*Oryza sativa* L.). *Amer. J. Bot.* 68: 154–161.

Cohen, E. and Kende, H. (1987) *In vivo* 1-aminocyclopropane-1-carboxylate synthase activity in internodes of deep water rice. Enhancement by submergence and low oxygen levels. *Plant Physiol.* 84: 282–286.

Drew, M.C. (1997) Oxygen deficiency and root metabolism: injury and acclimation under hypoxia and anoxia. *Annu. Rev. Plant Physiol. Plant Mol. Biol.* 48: 223–250.

Drew, M.C., Jackson, M.B. and Giffard, S. (1979) Ethylene-promoter adventitious rooting and development of cortical air spaces (aerenchyma) in roots may be adaptive responses to flooding in *Zea mays* L. *Plant* 147: 83–88.

Drew, M.C., Jackson, M.B., Giffard, S.C. and Campbell, R. (1981) Inhibition by silver ions of gas space (aerenchyma) formation in adventitious roots of *Zea mays* L. subjected to exogenous ethylene or to oxygen deficiency. *Planta* 153: 217–224.

Drew, M.C., He, C.J. and Morgan, P.W. (1989) Decreased ethylene biosynthesis, and induction of aerenchyma, by nitrogen- or phosphate-starvation in adventitious roots of *Zea mays* L. *Plant Physiol.* 91: 266–271.

Ecker, J.R. (1995) The ethylene signal transduction pathway in plants. *Science* 268: 667–675.

Finkelstein, D. (1998). The discovery of apoptotic genes, genes induced by hypoxia, and genes induced by mechanical impedance in the root tips of *Zea mays* L. PhD Dissertation, Texas A&M University, College Station Texas, USA.

Fukuda, H. (1996) Xylogenesis: initiation, progression, and cell death. *Annu. Rev. Plant Physiol. Plant Mol. Biol.* 47: 299–325.

Gallois, P., Makishima, T., Hecht, V., Despres, B., Laudie, M., Nishimoto, T. and Cooke, R. (1997) An *Arabidopsis thaliana* cDNA complementing a hamster apotosis suppressor mutant. *Plant J.* 11: 1325–1331.

Gan, S. and Amasino, R.M. (1997) Making sense of senescence. Molecular genetic regulation and manipulation of leaf senescence. *Plant Physiol.* 113: 313–319.

Greenberg, J.T. (1996) Programmed cell death: A way of life for plants. *Proc. Natl Acad. Sci. USA* 93: 12094–12097.

Greenberg, J.T. (1997) Programmed cell death in plant–pathogen interactions. *Annu. Rev. Plant Physiol. Plant Mol. Biol.* **48**: 525–546.

Groover, A. and Jones, A.M. (1999) Tracheary element differentiation uses a novel mechanism coordinating programmed cell death and secondary cell wall synthesis. *Plant Physiol.* **119**: 375–384.

Groover, A., DeWitt, N., Heidel, A. and Jones, A. (1997) Programmed cell death of plant tracheary elements differentiating *in vitro*. *Protoplasma* **196**: 197–211.

He, C.J., Morgan, P.W. and Drew, M.C. (1992) Enhanced sensitivity to ethylene in nitrogen- or phosphate-starved roots of *Zea mays* L during aerenchyma formation. *Plant Physiol.* **98**: 137–142.

He, C.J., Drew, M.C. and Morgan, P.W. (1994) Induction of enzymes associated with lysigenous aerenchyma formation in roots of *Zea mays* during hypoxia or nitrogen-starvation. *Plant Physiol.* **105**: 861–865.

He, C.J., Finlayson, S.A., Drew, M.C., Jordan, W.R. and Morgan, P.W. (1996a). Ethylene biosynthesis during aerenchyma formation in roots of *Zea mays* subjected to mechanical impedance and hypoxia. *Plant Physiol.* **112**: 1679–1685.

He, C.J., Morgan, P.W. and Drew, M.C. (1996b) Transduction of an ethylene signal is required for cell death and lysis in the root cortex of maize during aerenchyma formation induced by hypoxia. *Plant Physiol.* **112**: 463–472.

Huang, F.Y., Philosoph-Hadas, S., Meir, S., Callaham, D.A., Sabato, R., Zelcer, A. and Hepler, P.K. (1997) Increases in cytosolic Ca^{2+} in parsley mesophyll cells correlate with leaf senescence. *Plant Physiol.* **115**: 51–60.

Jackson, M.B., Fenning, T.M. and Jenkins, W. (1985a) Aerenchyma (gas-space) formation in adventitious roots of rice (*Oryza sativa* L.) is not controlled by ethylene or small partial pressures of oxygen. *J. Exp. Bot.* **36**: 1566–1572.

Jackson, M.B., Fenning, T.M., Drew, M.C. and Saker, L.R. (1985b) Stimulation of ethylene production and gas-space (aerenchyma) formation in adventitious roots of *Zea mays* L. by small partial pressures of oxygen. *Planta* **165**: 486–492.

Justin, S.H.F.W. and Armstrong, W. (1987) The anatomical characteristics of roots and plant response to soil flooding. *New Phytol.* **106**: 465–495.

Justin, S.H.F.W. and Armstrong, W. (1991) Evidence for the involvement of ethene in aerenchyma formation in adventitious roots of rice (*Oryza sativa* L.). *New Phytol.* **118**: 49–62.

Katsuhara, M. and Kawasaki, T. (1996) Salt stress induced nuclear and DNA degradation in meristematic cells of barley roots. *Plant Cell Physiol.* **37**: 169–173.

Kawai, M., Samarajeewa, P.K., Barrero, R.A., Nishiguchi, N. and Uchimiya, H. (1998) Cellular dissection of the degradation pattern of cortical cell death during aerenchyma formation of rice roots. *Planta* **204**: 277–287.

Kawase, M. (1979) Role of cellulase in aerenchyma development of sunflower. *Am. J. Bot.* **66**: 183–190.

Kawase, M. (1981a) Anatomical and morphological adaptation of plants to waterlogging. *HortiSci.* **16**: 30–34.

Kawase, M. (1981b) Effect of ethylene on aerenchyma development. *Am. J. Bot.* **68**: 651–658.

Kerr, J.F.R. and Harmon, B.V. (1991) Definition and incidence of apotosis: an historical perspective. In: *Apoptosis: the Molecular Basis of Cell Death* (eds L.D. Tomei and F.O. Lope). Cold Spring Harbor Laboratory Press, New York, pp. 5–29.

Kieber, J. (1997) The ethylene response pathway in *Arabidopsis*. *Annu. Rev. Plant Physiol. Plant Mol. Biol.* **48**: 277–296.

Knight, M.R., Campbell, A.K., Smith, S.M. and Trewavas, A.J. (1991) Transgenic plant aequorin reports the effect of touch and cold shock and elicitors on cytoplasmic calcium. *Nature* **352**: 524–526.

Konings, H. (1982) Ethylene-promoted formation of aerenchyma in seedling roots of *Zea mays* L. under aerated and non-aerated conditions. *Physiol. Plant.* **54**: 119–124.

McPherson, D.C. (1939) Cortical air spaces in the roots of *Zea mays* L. *New Phytol.* **38**: 190–202.

Mittler, R. and Lam, E. (1995a) Identification, characterization and purification of a tobacco endonuclease activity induced upon hypersensitive response cell death. *Plant Cell* **7**: 1951–1962.

Mittler, R. and Lam, E. (1995b) *In situ* detection of nDNA fragmentation during the differentiation of tracheary elements in higher plants. *Plant Physiol* **108**: 489–493.

Morgan, P.W. and Drew, M.C. (1997) Ethylene and plant responses to stress. *Physiol. Plant.* **100**: 620–630.

Munnik, T., Arisz, S.A., de Vrije, T. and Musgrave, A. (1995) G protein activation stimulates phospholipase D signaling in plants. *Plant Cell* **7**: 2197–2210.

O'Brien, I.E.W., Baguley, B.C., Murray, B.G., Morris, B.A.M. and Ferguson, I.B. (1998) Early stages of the apoptotic pathway in plant cells are reversible. *Plant J.* **13**: 803–814.

Raz, V. and Fluhr, R. (1992) Calcium requirement for ethylene-dependent responses. *Plant Cell* **4**: 1123–1140.

Raz, V. and Fluhr, R. (1993) Ethylene signal is transduced via protein phosphorylation events in plants. *Plant Cell* **5**: 520–530.

Roman, G., Lubarsky, B., Kieber, J.J., Rothenburg, M. and Ecker, Jr. (1995) Genetic analysis of ethylene signal transduction in *Arabidopsis thaliana*: five novel mutant loci integrated into a stress response pathway. *Genetics* **139**: 1393–1409.

Saab, I.N. and Sachs, M.M. (1996) A flooding-induced xyloglucan endo-transglycosylase homolog in maize is responsive to ethylene and associated with aerenchyma. *Plant Physiol.* **112**: 385–391.

Sarquis, J.I., Jordan, W.R. and Morgan, P.W. (1991) Ethylene evolution from maize (*Zea mays* L.) seedling roots and shoots in response to mechanical impedance. *Plant Physiol.* **96**: 1171–1177.

Schaller, G.E. and Bleecker, A.B. (1995) Ethylene-binding sites generated in yeast expressing the *Arabidopsis* ETR1 gene. *Science* **270**: 1809–1811.

Schaller, G.E., Ladd, A.N., Lanahan, M.B., Spanbauer, J.M. and Bleecker, A.B. (1995) The ethylene response mediator ETRI from *Arabidopsis* forms a disulfide-linked dimer. *J. Biol. Chem.* **270**: 12526–12530.

Schussler, E. and Longstreth, D.J. (1996) Aerenchyma develops by cell lysis in roots and cell separation in leaf petioles in *Sagittaria lancifolia* (*Alismataceae*). *Am. J. Bot.* **83**: 1266–1273.

Shiu, O.Y., Oetiker, J.H., Yip, W.K. and Yang, S.F. (1998) The promoter of *LE-ACS7*, an early flooding-induced 1-aminocyclopropane-1-carboxylate synthase gene of the tomato, is tagged by a *Sol3* transposon. *Proc. Natl Acad. Sci. USA* **95**: 10334–10339.

Subbaiah, C.C., Zhang, J. and Sachs, M.M. (1994a) Involvement of intracellular calcium in anaerobic gene expression and survival of maize seedlings. *Plant Physiol.* **105**: 369–376.

Subbaiah, C.C., Bush, D.S. and Sachs, M.M. (1994b) Elevation of cytosolic calcium precedes anoxic gene expression in maize suspension-culture cells. *Plant Cell* **6**: 1747–1762.

Tanaka, Y., Makishima, T., Sasabe, M., Ichinose, Y., Shiraishi, T., Nishimeto, T. and Yamada, T. (1997) *dad-1*, a putative programmed cell death suppressor gene in rice. *Plant Cell Physiol.* **38**: 379–383.

Wang, H., Li, J., Bostock, R.M. and Gilchrist, D.G. (1996) Apoptosis: a functional paradigm for programmed plant cell death induced by a host-selective phytotoxin and invoked during development. *Plant Cell* **8**: 375–391.

Wang, T.W. and Arteca, R.N. (1992) Effects of low O_2 root stress on ethylene biosynthesis in tomato plants (*Lycopersicon esculentum* Mill cv Heinz 1350). *Plant Physiol* **98**: 97–100.

Williams, W.T. and Barber, D.A. (1961) The functional significance of aerenchyma in plants. In: *Mechanisms in Biological Competition*, Symposium 15, Society of Experimental Biology, London, pp. 132–144.

Zarembinski, T.I. and Theologis, A. (1994) Ethylene biosynthesis and action: a case for conservation. *Plant Mol. Biol.* **26**: 1579–1597.

14

Targeted cell death in xylogenesis

Maureen C. McCann, Nicola J. Stacey and Keith Roberts

1. Introduction

Programmed cell death (PCD) is an integral part of organismal development for both animals and plants. It is distinguished from necrotic death by cell-autonomous, active and ordered suicide in which specific proteases are recruited to destroy a limited number of key cellular proteins (Nicholson and Thornberry, 1997). Necrotic death from environmental insult, or triggered by neighbouring cells, is not an ordered process and cannot be blocked by protease inhibitors. In plants, PCD occurs in specific cells during development of floral organs, embryos, the vascular system, leaves and roots (Jones and Dangl, 1996). In this chapter, we will focus on a single plant cell type, the tracheary element, which undergoes PCD as a final step in cell specialization. We will review the progress made by a number of groups in identifying components of the cell death machinery in a model xylogenic system, the *Zinnia* mesophyll cell system. We then present some preliminary data using protease inhibitors in this model system to show that other differentiation-related events can proceed independently of PCD.

2. Plant programmed cell death

2.1 *Tracheary element formation in a model system*

The conductive tissues of plants comprise sieve tubes in the phloem and tracheids or vessels in the xylem. Sieve tubes are formed from sieve-tube elements and transport nutrients such as sucrose: these cells often lose their nucleus but are alive. In contrast, vessels and tracheids transport water and salts, and are built from dead tracheary elements (TEs). TEs are highly distinctive because of the formation of a secondary cell wall with annular, spiral, reticulate or pitted wall thickenings. As they mature, the cell contents are autolysed to produce hollow dead water-conducting tubes. The process of xylogenesis is difficult to study in intact plants as only a few cells are forming TEs at a given time. However, a model system exists in which freshly isolated mesophyll and

Programmed Cell Death in Animals and Plants, edited by J.A. Bryant, S.G. Hughes and J.M. Garland.

palisade cells from the leaves of *Zinnia elegans* are cultured *in vitro* in the presence of two phytohormones, auxin and cytokinin, and will trans-differentiate to TEs (Fukuda and Komamine, 1980) (*Figure 1*). About 60% of the living cells will form TEs by 96 h of culture relatively synchronously, making it an excellent system with which to study the biochemistry and molecular biology of PCD.

Prior to cell death, the TE becomes specialized for its function by depositing hoops of secondary wall material, predominantly cellulose and xylan, transverse to the long axis of the cell, to reinforce it from collapse under the negative tension in the xylem (Fukuda, 1997; McCann, 1997). The secondary thickenings are lignified to waterproof them and to contribute to cell strength. The patterning of secondary thickening is a hierarchical process – microtubule orientation patterns cellulose deposition and this in turn patterns lignification, but inhibition of any of these events does not prevent cell autolysis (Taylor *et al.*, 1992; Taylor and Haigler, 1993). However, autolysis does not proceed until thickenings have formed. This anomaly suggests that the timing of autolysis and secondary wall thickening are coupled by a mechanism that is not dependent upon cellulose deposition or lignification (McCann, 1997).

2.2 *Components of the cell death machinery are present in plants*

The detection of fragmented nuclear DNA in vessel elements of pea raised the possibility that TE formation might share features of apoptotic cell death (Mittler and Lam, 1995). However, other morphological features of apoptosis are not observed (Groover *et al.*, 1997). For example, in animals, macrophages can be recruited to remove the corpses of cell suicides, but plant PCD must be autophagic by necessity. Regardless of the morphology of death, it is now clear that there are many features of the machinery of cell death, in which the induction and action of specific proteins bring about the controlled disassembly of the cell, that are shared by plants and animals.

Animal PCD is mediated by a family of cysteinyl aspartate-specific proteinases, known as caspases, which are thought to act in a proteolytic cascade, cleaving key intracellular proteins (Nicholson and Thornberry, 1997; see also Chapters 1 and 2). There are at least two and possibly three subfamilies of caspases. Members of the ICE subfamily are related to mammalian interleukin-1β converting enzyme (also called caspase 1) and predominantly play a role in inflammation whereas members of the CED-3 subfamily are largely (if not exclusively) involved in apoptosis (Nicholson and Thornberry, 1997). In the nematode *C. elegans*, the *CED 3* gene product is absolutely required for the apoptotic death of all 131 cell deaths occurring during hermaphrodite development (Chapter 1). Caspases require an aspartic acid residue to the immediate left and at least four amino acids on the amino terminal side of the scissile bond. A positional scanning substrate combinatorial library has identified consensus sequences for cleavage of DExD for CED-3 family members and WEHD for ICE-family members. There is therefore a limited range of cellular proteins that are substrates for caspases. The net effect of cleavage of these proteins appears to be to (1) halt cell-cycle progression, (2) disable homeostatic and repair mechanics, (3) initiate cell detachment from surrounding tissues, (4) disassemble structural components of the cytoskeleton and nucleus, and (5) mark the dying cells for engulfment by other cells such as macrophages (Nicholson and Thornberry, 1997). Pro-caspases are activated by cleavage at sites that resemble their own substrate sites, suggesting that auto- or trans-activation can occur, leading to amplification of the cascade. A variety of factors can

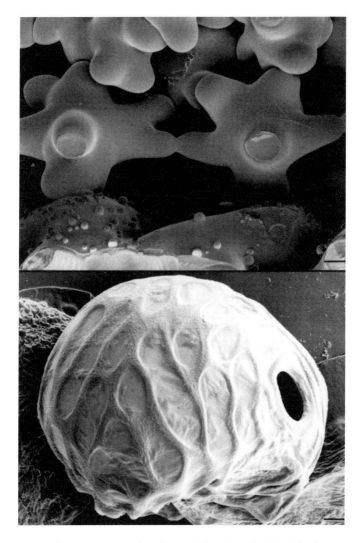

Figure 1. *Scanning electron micrographs of mesophyll cells in the* Zinnia *leaf prior to cell isolation (top) and an* in vitro *tracheary element formed after 96 h of culture of mesophyll and palisade cells in medium containing auxin and cytokinin (bottom). Micrographs courtesy of Kim Findlay.*

induce the cascade, including Fas ligands, tumour necrosis factors, a serine protease called granzyme B, and hormones; diverse signals can be transduced into a common mechanism of death.

Both caspase and serine proteases have been implicated in plant PCD. A well-documented form of plant PCD is the hypersensitive response (HR), in which plant cells die surrounding sites of pathogen ingress (Dangl *et al.*, 1996). Bacterial induction of cell death in tobacco leaf can be prevented by caspase inhibitors, Ac-YVAD-CMK and Ac-DEVD-CHO, but not by a general serine protease inhibitor, phenylmethylsulfonyl fluoride, and caspase-like protease activity was detected *in vitro* during the development of HR induced by tobacco mosaic virus (del Pozo and Lam, 1998). In contrast, HR elicited in a tobacco cell culture by a xylanase from *Trichoderma viride* was blocked by a serine protease inhibitor but caspase inhibitors had no effect (Yano *et al.*, 1999). A key feature of the HR is the oxidative burst, characterized by a dramatic increase in the level of reactive oxygen intermediates, including the superoxide anion, hydrogen peroxide and the hydroxyl radical produced by elicited plant cells (Dangl *et al.*, 1996). In both cases, the induction of the oxidative burst and other aspects of HR were not affected (del Pozo and Lam, 1998; Yano *et al.*, 1999).

TE PCD does not involve an oxidative burst. Treating mesophyll cells with exogenous hydrogen peroxide does not induce genes encoding specific cysteine and serine proteases and RNases (Groover *et al.*, 1997; Ye and Droste, 1996). The HR may represent a less specialized form of PCD than when death is an integral part of differentiation in a pre-programmed developmental pathway. The most appropriate parallel with animals may be in the differentiation of keratinocytes where the terminal stage is to die and form a layer of corpses (squames) on the surface of the skin (Jacobson *et al.*, 1997). Caspases are activated during normal human keratinocyte differentiation (Weil *et al.*, 1999), but the outcome is very different from that of classical apoptosis, since only the organelles are degraded.

Although the components of the death machinery are thought to be constitutively expressed in all cells, they remain inactive in most cells. Some genes are anti-apoptotic, for example, *CED-9* is required to keep the death programme off if a cell is to survive in the developing worm (Chapter 1). In plants, there are mutants that exhibit constitutive initiation of HR-like cell death in the absence of pathogen. Two classes of *Arabidopsis* mutants, *lsd* (lesions simulating disease resistance) and *acd* (accelerated cell death) (Dietrich *et al.*, 1997), a rice *dad-1* (defender against apoptotic death) gene (Tanaka *et al.*, 1997) and a maize *Lls1* (lethal leaf spot1) gene (Gray *et al.*, 1997) are all candidates for direct negative control of cell death.

2.3 *PCD in the* Zinnia *mesophyll cell system*

Because of the twin advantages of synchronicity and inducibility of TE differentiation in the *Zinnia* mesophyll cell system, the morphology and mechanisms of PCD in TE formation have been best characterized using this *in vitro* system (Fukuda, 1996). The key event in building a TE, the deposition of bands of secondary wall material between the plasma membrane and the primary cell wall, occurs at about 72 h of culture (McCann, 1997). About 6 h after the appearance of visible cell wall thickenings, there is a very rapid disruption of the tonoplast (occurring within 3 minutes), and this is succeeded by swelling and disruption of single-membraned organelles (ER and Golgi), followed by double-membraned organelles (chloroplasts, mitochondria, the nucleus) within hours after vacuole collapse (Groover *et al.*, 1997). Nuclear DNA is degraded and can be assayed in individual cells using TUNEL (Groover *et al.*, 1997; Groover and Jones, 1999).

Two cDNAs, p48h-17 (Ye and Varner, 1996) and ZCP4 (Minami *et al.*, 1996) have been isolated from cells in inductive medium and are associated with differentiating xylem in stems. p48h-17 encodes a cysteine protease with homology to papain, and has a pro-peptide sequence consistent with vacuolar residence. ZCP4 may encode a cysteine protease (30 kDa) that is present transiently in differentiating TEs in the *Zinnia* system before autolysis (Minami and Fukuda, 1995). The protease activity detected by Minami and Fukuda has a pH optimum of 5.5, that of vacuolar pH, and presumably the pH of the cytoplasm after vacuolar disintegration. Caspase inhibitors block entry into the late stages of TE formation, although inhibitors of ICE do not (Fukuda, 1997). The 30 kDa caspase activity is inhibited by E-64 (N-{N-(L-3-trans-carboxirane-2-carbonyl)-L-leucyl]-agmatine, a thiol protease inhibitor, and activated by DTT. Ye and Varner (1996) detected a caspase activity of 20 kDa by an activity gel assay with a pI of 5.7, which may be the processed product of the 23 kDa protein from the predicted sequence of p48h-17, and the 24 kDa activity detected by Beers and Freeman (1997). Beers and Freeman (1997) also detected a protease of 28 kDa which could be inhibited by thiol protease inhibitors, leupeptin and E-64.

Other hydrolytic activities have been detected including two serine proteases, one of 145 kDa (Beers and Freeman, 1997) and one of 40 kDa (Groover and Jones, 1999). Ye and Varner (1996) detected a 60 kDa protein and Beers and Freeman (1997) also detected a 59 kDa serine protease that was present in all culture conditions but strongly up-regulated during TE formation. There is an accompanying dramatic increase in the activity of intracellular hydrolytic enzymes including three RNases and a nuclease that degrades single-stranded DNA (Fukuda, 1997; McCann, 1997).

2.4 *Induction of PCD during TE formation*

Different hormones have been implicated in the induction of different examples of plant PCD; ethylene triggers aerenchyma formation, GA3 in aleurone cells and auxin in all examples of TE formation. Both exogenous auxin and cytokinin are absolutely required for the induction of TEs in the *Zinnia* mesophyll cell system. After 48 h in culture, the hormones are no longer required to be present for formation of TEs by 96 h; the mesophyll cells are committed to trans-differentiation by 48 h and will form TEs in the absence of the inducing stimulus (Church and Galston, 1988; Stacey *et al.*, 1995). However, the *in vitro* differentiation depends critically upon the cell concentration within the medium until a very late stage, at or after the deposition of secondary thickenings. At this point, cells may be diluted by 10- or 100-fold and still go on to complete TE formation. However, up to this point, there is a clear requirement for cell–cell interaction. Other hormones have been invoked as signalling molecules including ethylene, gibberellin, abscisic acid and brassinosteroids. These may act downstream of the initial inductive auxin and cytokinin but some or all may be involved in further commitment of the cell to a TE fate. Uniconazole, an inhibitor of brassinosteroid biosynthesis, blocks a late stage of TE formation, and prevents the expression of genes involved in secondary wall thickening and also in cell death, in particular, the *ZCP4* gene (Yamamoto *et al.*, 1997). In animal cells, death can apparently be suppressed by signals from other cells (Jacobson *et al.*, 1997). About half of all vertebrate neurons die as a consequence of having to compete for a limited supply of signals necessary to survive (Jacobson *et al.*, 1997). Thus, death results either from direct activation of the cell death machinery or from a failure to receive adequate death-suppressing signals.

Calcium has been implicated in the regulation both of secondary wall formation and of the final commitment to cell death. Membrane-associated Ca^{2+} is elevated during differentiation, and differentiation can be inhibited using Ca^{2+} channel blockers or calmodulin antagonists (Fukuda, 1996). Methylxanthines, which act to reduce intracellular sequestration of Ca^{2+}, inhibit differentiation prior to secondary wall formation (Roberts and Haigler, 1992). Additionally, secondary wall thickening correlates with the expression of two calmodulin-binding proteins and an increase in calmodulin levels (Kobayashi and Fukuda, 1994). Groover and Jones (1999) have shown that the collapse of the vacuole is preceded by a large influx of Ca^{2+} into the cell. Reducing calcium influx protects against nuclear DNA fragmentation. Addition of trypsin to the cell cultures initiates cell death suggesting that specific proteolysis of the extra-cellular matrix triggers Ca^{2+} influx, vacuole collapse, cell death and chromatin degra-dation (Groover and Jones, 1999). A 40 kDa serine protease is secreted during secondary cell wall synthesis (Groover and Jones, 1999). Soybean trypsin inhibitor inhibits both PCD in the *Zinnia* system and the activity of this serine protease. Interestingly, granzyme B is a serine protease delivered by cytotoxic T lymphocytes and mediates cell suicide by activating endogenous caspases within the target cell. Such a mechanism is consistent with the dependence on cell concentration of TE differenti-ation in the *Zinnia* system until a late stage.

3. Cell death is blocked by a cathepsin B inhibitor

In this section, we report our preliminary experiments using caspase inhibitors in the *Zinnia* mesophyll cell system. Fukuda (1997) has previously reported that E64 added before secondary wall thickening was effective in inhibiting TE formation, whereas no inhibition was observed at or just after thickenings are observed. However, lignification of the secondary thickenings may prevent uptake of E64 into the cells. In addition, Woffenden *et al.* (1998) have shown that inhibitors of proteasome activity inhibited TE differentiation by 96 h if added at time 0 but not at 48 h. However, TE formation was delayed with addition at 48 h, and treatment with carbobenzoxy-leucinyl-leucinyl-leucinal (LLL) resulted in incomplete autolysis, perhaps due to the ability of LLL to inhibit cysteine proteases as well as the proteasome.

We used two cell-permeable peptide caspase inhibitors based on benzyl-oxy-carboxyl-fluoromethyl ketones with different amino acids attached, ZVAD-fmk and BocD-fmk, and added at a final concentration of 200 μM in DMSO in the cell culture, and similarly with a cathepsin B inhibitor (ZFA-fmk). Each of these was added to *Zinnia* cells in inductive medium at two timepoints, 24 h and 48 h after the start of culture. The caspase inhibitors did not block TE formation by 96 h. Controls of DMSO alone added to the cultures, and a general protease inhibitor PMSF, also did not block TE formation. However, the cathepsin B inhibitor delayed differentiation by at least 48 h, with TE thickenings having a brown rather than dark grey appearance and still containing chloroplasts and other organelles (*Figure 2*).

Thus, caspase inhibitors are effective only at a late stage of differentiation as reported by Fukuda (1997), at a time corresponding to the transient expression of ZCP4 and p48h-17 shortly before secondary wall thickening commences. However, the cathepsin B inhibitor acts at an earlier stage to inhibit cell death but not secondary wall formation, and is effective from 24 h of culture.

Cathepsin B is a lysosomal cysteine protease of the papain family thought to be responsible for the degradation of endocytosed proteins (Mort and Buttle, 1997). It is processed to active form by acidification and it is interesting to note in this context that acidic pHs have been found to favour TE differentiation (Roberts and Haigler, 1994). In pathological conditions, cathepsin B may be inappropriately located: secreted as a pro-enzyme and activated extracellularly, or released from dead cells.

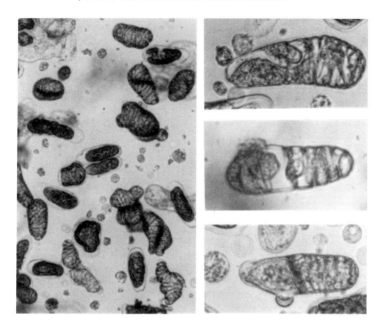

Figure 2. *The left-hand panel shows* Zinnia *cells cultured for 96 h, after addition of a caspase inhibitor, Boc-Asp(O-methyl)-fluoromethylketone (BocD-fmk), which have* trans-*differentiated to TEs. The right-hand panels show three divided cells in the presence of a cathepsin B inhibitor, benzyloxycarboxyl-Phe-Ala(O-methyl)-fluoromethylketone (ZFA-fmk), with cell contents still visible in the TEs.*

Lysosomal cathepsin B is secreted by some cancer cells, perhaps to increase their metastatic potential, as cathepsin B can hydrolyse many animal extracellular matrix components. In arthritic diseases, cathepsin B may activate pro-metalloproteinases intracellularly, which then attack proteoglycan and type II collagen, damaging cartilage extracellular matrix (Mort and Buttle, 1997). The mechanical isolation of *Zinnia* mesophyll and palisade cells makes it probable that some vacuolar enzymes are released from the dead or dying cells present in the culture at an early stage. Isolated cells may be regarded as wounded cells, and wounding is known to induce TE formation in intact plants. Such a mechanism in the *in vitro* system may underpin the synchronous induction of TE formation in 60% of the living cells, supposing that a substrate cleaved by cathepsin B may act as a signal to induce PCD of neighbouring cells.

4. Conclusion

The *Zinnia* mesophyll cell system provides an excellent model system in which to study plant PCD, by virtue of its semi-synchronous differentiation and as an *in vitro* system to which inhibitors can readily be added. PCD in TE formation involves homologues of the cysteine and serine proteases that are part of the cell death machinery in animal cells. Transcription and translation of some of these proteases are up-regulated shortly before secondary wall synthesis (Groover and Jones, 1999; Minami and Fukuda, 1995; Ye and Varner, 1996) whilst others are constitutively expressed (Beers and Freeman, 1997). The caspases are optimally active at vacuolar pH (Minami and

Fukuda, 1995) and autolysis occurs rapidly after vacuolar disruption, when other hydrolytic enzymes are also released to mix with the cytoplasm (Groover *et al.*, 1997). Secondary wall deposition and autolysis, the irreversible step of commitment to cell death, can be blocked by caspase or serine protease inhibitors (Fukuda, 1997; Groover and Jones, 1999). However, we have found that a cathepsin B inhibitor will block autolysis and cell death, but not secondary wall deposition, if present at an early stage of culture. This suggests that secondary wall deposition and other differentiation-related events can be uncoupled from cell death in TEs as it is in phloem cells *in planta*. In the HR, other aspects of the defence response are not inhibited by blocking PCD (del Pozo and Lam, 1998, Yano *et al.*, 1999). As yet, the upstream signals and down-stream targets of the cell death machinery remain to be elucidated in plants.

References

Beers, E.P. and Freeman, T.B. (1997) Proteinase activity during tracheary element differenti-ation in *Zinnia* mesophyll cultures. *Plant Physiol.* 113: 873–880.

Church, D.L. and Galston, A.W. (1988) Kinetics of determination in the differentiation of isolated mesophyll cells of *Zinnia elegans* to tracheary elements. *Plant Physiol.* 88: 92–96.

Dangl, J., Dietrich, R. and Richberg, M. (1996) Death don't have no mercy: cell death programs in plant–microbe interactions. *Plant Cell* 8: 1793–1807.

del Pozo, O. and Lam, E. (1998) Caspases and programmed cell death in the hypersensitive response of plants to pathogens. *Curr. Biol.* 8: 1129–1132.

Dietrich, R.A., Richberg, M.H., Schmidt, R., Dean, C. and Dangl, J.L. (1997) A novel zinc finger protein is encoded by the Arabidopsis LSD1 gene and functions as a negative regulator of plant cell death. *Cell* 88: 685–694.

Fukuda, H. (1996) Xylogenesis: initiation, progression, and cell death. *Annu. Rev. Plant Physiol. Plant Mol. Biol.* 47: 299–325.

Fukuda, H. (1997) Tracheary element differentiation. *Plant Cell* 9: 1147–1156.

Fukuda, H. and Komamine, A. (1980) Establishment of an experimental system for the tracheary element differentiation from single cells isolated from the mesophyll of *Zinnia elegans*. *Plant Physiol.* 52: 57–60.

Gray, J., Close, P.S., Briggs, S.P. and Johal, G.S. (1997) A novel suppressor of cell death in plants encoded by the Lls1 gene of maize. *Cell* 89: 25–31.

Groover, A. and Jones, A.M. (1999) Tracheary element differentiation uses a novel mechanism coordinating programmed cell death and secondary cell wall synthesis. *Plant Physiol.* 119: 375–384.

Groover, A., DeWitt, N., Heidel, A. and Jones, A. (1997) Programmed cell death of plant tracheary elements differentiating *in vitro*. *Protoplasma* 196: 197–211.

Jacobson, M.D., Weil, M. and Raff, M.C. (1997) Programmed cell death in animal development. *Cell* 88: 347–354.

Jones, A. and Dangl, J. (1996) Logjam at the Styx. *Trends Plant Sci.* 1: 114–119.

Kobayashi, H. and Fukuda, H. (1994) Involvement of calmodulin and calmodulin-binding proteins in the differentiation of tracheary elements in *Zinnia* cells. *Planta* 194: 388–394.

McCann, M.C. (1997) Tracheary element formation: building up to a dead end. *Trends Plant Sci.* 2: 333–338.

Minami, A. and Fukuda, H. (1995) Transient and specific expression of a cysteine endopro-teinase associated with autolysis during differentiation of *Zinnia* mesophyll cells into tra-cheary elements. *Plant Cell Physiol.* 36: 1599–1606.

Minami, A., Demura, T. and Fukuda, H. (1996) Specific and transient expression of gene for cysteine protease during tracheary element differentiation of Zinnia cells. *Plant Cell Physiol.* 37(suppl.): s123, No. 459.

Mittler, R. and Lam, E. (1995) *In situ* detection of nDNA fragmentation during the differentiation of tracheary elements in higher plants. *Plant Physiol.* **108**: 489–493.

Mort, J.S. and Buttle, D.J. (1997) Cathepsin B. *Int. J. Biochem. Cell Biol.* **29**: 715–720.

Nicholson, D.W. and Thornberry, N.A. (1997) Caspases: killer proteases. *Trends Biochem. Sci.* **22**: 299–306.

Roberts, A.W. and Haigler, C.H. (1992) Methylxanthines reversibly inhibit tracheary element differentiation in suspension cultures of *Zinnia elegans* L. *Planta* **186**: 586–592.

Roberts, A.W. and Haigler, C.H. (1994) Cell expansion and tracheary element differentiation are regulated by extracellular pH in mesophyll cultures of *Zinnia elegans* L. *Plant Physiol.* **105**: 699–706.

Stacy, N.J., Roberts, K., Carpita, N.C., Wells, B. and McCann, M.C. (1995) Dynamic changes in cell surface molecules are very early events in the differentiation of mesophyll cells from *Zinnia elegans* into tracheary elements. *Plant J.* **8**: 891—906.

Tanaka, Y., Makishima, T., Sasabe, M., Ichinose, Y., Shiraishi, T., Nishimoto, T. and Yamada, T. (1997) dad-1, a putative programmed cell death suppressor gene in rice. *Plant Cell Physiol.* **38**: 379–383.

Taylor, J.G. and Haigler, C.H. (1993) Patterned secondary cell-wall assembly in tracheary elements occurs in a self-perpetuating cascade. *Acta. Bot. Neer.* **42**: 153–163.

Taylor, J.G., Owen, T.P., Koonce, L. and Haigler, C.H. (1992) Dispersed lignin in tracheary elements treated with cellulose synthesis inhibitors provides evidence that molecules of the secondary cell wall mediate wall patterning. *Plant J.* **2**: 959–970.

Weil, M., Raff, M.C. and Braga, V.M.M. (1999) Caspase activation in the terminal differentiation of human epidermal keratinocytes. *Curr. Biol.* **9**: 361–364.

Woffenden, B.J., Freeman, T.B. and Beers, E.P. (1998) Proteasome inhibitors prevent tracheary element differentiation in *Zinnia* mesophyll cell cultures. *Plant Phys.* **118**: 419–430.

Yamamoto, R., Demura, T. and Fukuda, H. (1997) Brassinosteroids induce entry into the final stage of tracheary element differentiation in cultured *Zinnia* cells. *Plant Cell Physiol.* **38**: 980–983.

Yano, A., Suzuki, K. and Shinshi, H. (1999) A signaling pathway, independent of the oxidative burst, that leads to hypersensitive cell death in cultured tobacco cells includes a serine protease. *Plant J.* **18**: 105–109.

Ye, Z-H. and Droste, D.L. (1996) Isolation and characterization of cDNAs encoding xylogenesis – associated and wounding-induced ribonucleases in *Zinnia elegans*. *Plant Mol. Biol.* **30**: 697–709.

Ye, Z-H. and Varner, J.E. (1996) Induction of cysteine and serine proteases during xylogenesis in *Zinnia elegans*. *Plant Mol. Biol.* **30**: 1233–1246.

15

Abscission and dehiscence

Jeremy A. Roberts

1. Introduction

The ancient Greek word apoptosis literally means 'falling away from' and was originally applied to the shedding of petals or leaves (Kerr and Harmon, 1991). In modern
usage the term has become more synonymous with programmed cell death (PCD) and
in this chapter we will examine the extent to which PCD plays a role in the process of
abscission and the analogous phenomenon of dehiscence.

2. Events associated with abscission

2.1 *Role of ethylene*

While the precise sequence of events that initiates the abscission programme is
unknown it is evident that in leaves, flowers and fruit the phenomenon is associated
with the action of ethylene (Gonzalez-Carranza *et al.*, 1998). Under some circumstances it may be an increase in ethylene production that promotes organ shedding
while at other times it may be an increase in sensitivity to the gas that hastens the
process. If the action of ethylene is blocked, by silver ions or by a mutation in one of
the receptors that perceives the gaseous regulator then abscission is delayed (Lanahan
et al., 1994; Wilkinson *et al.*, 1997).

2.2 *Changes in enzyme activity*

Leaf, flower and fruit fall are the consequence of cell wall separation at the site of
abscission (Sexton and Roberts, 1982). There is a considerable body of evidence that
wall degradation is brought about by an increase in the activity of several hydrolytic
enzymes. One of these, cellulase or β-1,4-glucanase has been found to be ubiquitously
associated with abscission and the elevation in activity is primarily restricted to the site
of cell separation and occurs at the time when a decline in break-strength is detected
(Webb *et al.*, 1993). The other principal enzyme is polygalacturonase (PG) and the
activity of endo-acting PGs have been reported to increase in a spectrum of abscission
systems including leaves, flowers and fruit (Taylor *et al.*, 1993; Zanchin *et al.*, 1995).
The precise substrate for the action of β-1,4-glucanase is unclear as the enzyme is not

Programmed Cell Death in Animals and Plants, edited by J.A. Bryant, S.G. Hughes and J.M. Garland.
© 2000 BIOS Scientific Publishers Ltd, Oxford.

thought to degrade crystalline cellulose and anatomical studies have revealed that the first site of wall breakdown is the pectic-rich middle lamella. A detailed analysis of the changes in macromolecular constituents of the cell wall is difficult during abscission as many zones are composed of only a few layers of separating cells. However, the leaflet abscission zone in *Sambucus nigra* comprises in excess of 30 rows of cells and an analysis of the polyuronide composition during ethylene-promoted abscission has revealed that the size of the uronic acid fragments becomes more disperse and this would be in accord with the action of an enzyme such as endo-PG (Taylor *et al.*, 1993).

2.3 *Abscission-related gene expression*

The increase in activity of both β-1,4-glucanase and PG is associated with an up-regulation of gene expression and mRNA specifically accumulates in the abscission zone cells. Interestingly, northern analysis of the expression of the genes encoding these hydrolytic enzymes reveals that the mRNA often exhibits some degradation with the extent of the breakdown increasing as cell separation progresses (Taylor *et al.*, 1994; Roberts *et al.*, 1997). Other abscission-related mRNAs do not show this feature and this raises the possibility that either the messages encoding the hydrolytic enzymes are specifically targeted for degradation or that an abscission zone is made up of a composite of cell layers.

A spectrum of abscission-related genes have now been isolated as a consequence of strategies such as differential screening and differential display. In *S. nigra* a number of genes encoding pathogenesis-related (PR) proteins have been found to be up-regulated during ethylene-promoted cell separation of the leaflet tissue. These include a polyphenol oxidase, a PR-1 type protein, a PR-4 type protein similar to the potato *win* peptides, a proteinase inhibitor, and both an acidic and basic chitinase (Coupe *et al.*, 1997). These peptides have been proposed to protect the exposed fracture surface from pathogenic attack. There is no evidence that the expression of these genes is induced by invading pathogens; rather they are up-regulated as part of the abscission programme to pre-empt colonization of the separating cells by bacteria or fungi. The mRNA encoding a metallothionein-like protein has also been shown to accumulate in the abscission zone tissue. The role of this protein is unclear particularly as there is no evidence that the abscission zone accumulates heavy metal ions (Coupe *et al.*, 1995). An alternative function of the peptide might be to sequester free radicals and these highly toxic cellular constituents have been proposed to play a role in the induction of PR-proteins (Coupe *et al.*, 1997).

3. Abscission and PCD

From the previous discussion it can be seen that the events that are associated with abscission are in some ways reminiscent of the hypersensitive response in plants. In the hypersensitive response the invasion of a plant by a pathogen elicits a series of events that lead to PCD (Morel and Dangl, 1997; see also Chapters 10 and 11). Commonly these include the accumulation of PR-proteins and the pathogen may gain entry into the plant by secreting cell wall degrading enzymes. It is evident from ultra-structural studies that organ shedding is not precipitated by cell death but rather that metabolic activity increases prior to cell separation. However, evidence is accumulating that an abscission zone is not composed of layers of a single cell type. For instance the most likely interpretation of the observed differential degradation of abscission-related mRNAs is that their accumulation takes place in different cell layers. If this hypothesis is correct then it would be predicted that the line of cells immediately adjacent to the site of separation would be the origin of mRNAs encoding such enzymes as β-1,4-glucanase and PG and once these peptides have been secreted and commenced wall breakdown, cell death might take place (*Figure 1*). This event could be either a consequence of PCD or a result of desiccation of the cells associated with wall loosening and tissue shedding. The abscission-related accumulation of PR proteins would be predicted to occur in cells that remain viable and can continue to secrete peptides that would protect the exposed fracture face from pathogens. Localizing the expression of the abscission-related mRNAs or the peptides that they encode could readily test this hypothesis.

4. Dehiscence

Like abscission, dehiscence is also the consequence of a sequence of events that culminates in cell separation. Pod dehiscence is a phenomenon that has serious agricultural consequences and losses of oilseed rape seed have been reported to be as great as 50% in unfavourable weather conditions. Moreover, seeds that are shed and persist in the soil give rise to weed oilseed rape in the following season and this can result in contamination of the germplasm in neighbouring crops.

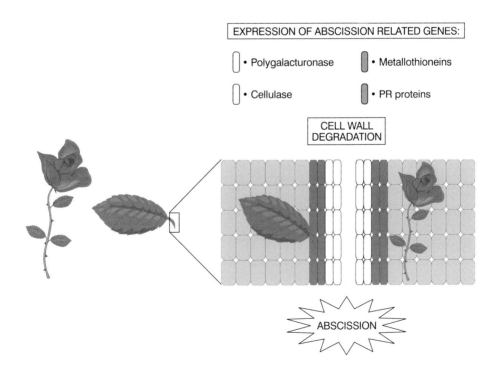

Figure 1. Schematic diagram of the abscission process showing the possible expression of wall degrading enzymes and pathogenesis-related proteins in different cell layers of the abscission zone (after Gonzalez-Carranza et al., 1998).

5. Events associated with dehiscence

Within 20 days after anthesis, oilseed rape pods have reached their maximum size and further visible development takes place primarily in the seeds. It is difficult to identify precisely the site at which pod opening will take place at this time as, unlike abscission zone cells, the layer that makes up the dehiscence zone is anatomically indistinguishable (Meakin and Roberts, 1990a). However, by 40 days after anthesis extensive lignification has taken place in the region adjacent to the vascular tissue along which splitting of the siliques take place. Two cell layers remain unlignified and these cells comprise the dehiscence zone. It is evident from this observation that the dehiscence zone cells are in some way exempt from the signal that drives lignification and, unlike their neighbours, do not go through a PCD programme at this stage (Gray and Johal, 1998). Indeed if the dehiscence zone cells became lignified pod shatter would not take place and this would be one strategy by which the problem might be alleviated.

5.1 Changes in enzyme activity

Increases in the activity of both polygalacturonase and β-1,4-glucanase have been reported to take place prior to pod opening in the dehiscence zone cells (Meakin and Roberts, 1990b; Peterson et al., 1996). This correlation does not prove that either of these enzymes contributes to dehiscence zone wall loosening; however the observation that an increase in the activity of both these enzymes is found in the dehiscence zone of young pods infected with *Dasineura brassicae* (pod midge) supports the assertion. The eggs of this insect are deposited in the young siliques and the larva consumes the seed before emerging from the maturing pods by prematurely triggering the cell separation process in a discrete region of the dehiscence zone. The increase in hydrolytic enzyme activity induced by the developing larva is restricted to those sites where dehiscence takes place (Meakin and Roberts, 1991).

5.2 Dehiscence-related gene expression

A number of strategies have been employed to identify dehiscence-related genes. Using degenerate PCR a fragment of nucleic acid was amplified and used to isolate a full-length PG cDNA from a *B. napus* pod dehiscence-related library. Northern analysis has revealed that the mRNA encoding this PG accumulates within the dehiscence zone tissues from 35 days after anthesis and expression cannot be detected in adjacent tissues of the silique (Jenkins et al., 1996; Peterson et al., 1996). The homologous gene has been isolated from *Arabidopsis* and the promoter fused to β-glucuronidase (GUS). Transformation of *B. napus* with this reporter construct has revealed that PG expression is restricted to three sites. The principal region is the two layers of cells in the pod that constitutes the dehiscence zone; however, GUS expression is also observed at the junction of the seed and the funiculus and in the stomium region of the anther. This discovery (Jenkins et al., 1999) confirms a role for PG in cell separation as wall loosening takes place at all three positions where the gene is expressed. Moreover, it is entirely consistent for seed abscission and pod dehiscence to operate via a common mechanism and for anther opening to be regulated in a similar manner.

Degenerate PCR is a powerful procedure when it is possible to predict the nature of the genes that may be up- or even down-regulated. Differential screening relies on no assumptions and has been used to isolate two dehiscence-related mRNAs. One of these is a proline-rich protein that has homology to a family of peptides in a range of species with diverse patterns of expression (Coupe *et al.*, 1993). These include regulation by low temperature, salt, ABA and developmental cues. Hydropathy profiles of these proteins indicate that they are secreted and the proline-rich C-terminal region has led to the proposal that they may anchor the plasma membrane to the cell wall and provide signals to the protoplast about the external environment (Goodwin *et al.*, 1996). The other dehiscence-related protein has some homology to a dehydrogenase but its function remains unknown (Coupe *et al.*, 1994).

Although there is overwhelming evidence that ethylene regulates the timing of leaf, flower and fruit abscission it seems unlikely that the gas plays a similar role during either pod or anther dehiscence or perhaps seed abscission. In support of this hypothesis the *etr1* mutant of *Arabidopsis* exhibits greatly delayed petal shedding while remaining fully fertile and shedding seed (Wilkinson *et al.*, 1997). Although the ligand(s) responsible for co-ordinating dehiscence remains unidentified, the recent isolation of a response regulator protein that is transitorily up-regulated in the dehiscence zone prior to the accumulation of PG, exhibiting homology to the C-terminal region of ETR1 (Chang *et al.*, 1993) and CKI1 (Kakimoto, 1996), raises the possibility that the ligand may interact with a histidine kinase type of peptide (Whitelaw *et al.*, 1999). The search for both the upstream and downstream components of this putative signalling cascade is now underway.

6. Dehiscence and PCD

From the foregoing discussion it is evident that dehiscence shares a number of features in common with abscission. The end product of both developmental processes is targeted cell separation and this is mediated, in part, by an increase in the activity of cell wall degrading enzymes regulated at the level of transcription. Furthermore, degradation of the mRNA encoding PG within the dehiscence zone is apparent and this raises the possibility that these cells may be exhibiting some form of PCD. Ultrastructural analysis of the dehiscence zone is consistent with this proposal and immediately prior to pod opening the cells in this region show signs of organellar and membrane breakdown (Meakin and Roberts, 1990a). If this hypothesis is correct it raises the question of the timing of dehiscence as the controlled synthesis and secretion of wall-degrading enzymes would not be possible in cells that were undergoing PCD. One possibility is that these cells accumulate hydrolytic enzymes but that they are not released into the apoplast until the plasma membrane degenerates. Certainly the PG peptide has a putative signal peptide that would make the lysis of the protoplast an unnecessary mechanism for release (Jenkins *et al.*, 1996). Alternatively it could be that the accumulation of pectolytic enzymes within the cytoplasm reaches a toxic level. Although there is no evidence that endogenous PGs are deleterious to plant cells it is well documented that fungal pectin-degrading enzymes readily precipitate cell death.

7. PCD and the manipulation of dehiscence

As stated previously, pod dehiscence has important implications for the agricultural industry and an ability to manipulate the process would be beneficial. The lignification of the dehiscence zone cell walls would be one approach; another is inducing a form of PCD or targeted ablation of the site at which shatter takes place. The identification of a promoter that is specifically turned on in the dehiscence zone cells has led to this strategy being attempted (Jenkins *et al.*, 1999). The PG promoter from *Arabidopsis* has been fused to the ribonuclease, barnase, and the construct used to transform *B. napus*. The resultant transgenic plants proved to be male sterile (*Figure 2*) due to the activity of the PG promoter in the stomium region of the anthers (Jenkins *et al.*, 1999). Pods arising from pollination using a wild type pollen donor showed increased resistance to pod shatter. These observations have revealed that the approach is worth pursuing although the promoter will have to be refined to prevent expression in the dehiscence zone of the anther whilst maintaining expression in the dehiscence zone of the pod.

8. Conclusions

In this review the relationship between apoptosis and cell separation during abscission and dehiscence has been explored. It is evident that PCD does not bring about cell separation in either event. Indeed the cells that separate must remain metabolically functional or wall loosening will not take place. This is especially the case during abscission, although evidence is accumulating to support the hypothesis that more than a single class of abscission zone cell might exist and that those cells that are immediately adjacent to the site of separation might lose their viability. However, the timing of this and the mechanism responsible is unknown. The ability to remove separating cells from the abscission zone of *S. nigra* in considerable quantities would enable DNA laddering and TUNEL assays (Havel and Durzan, 1996) to be carried out in an effort to resolve whether an apoptotic-like process is involved. During pod dehiscence PCD may contribute at two stages. Firstly the lignification of surrounding tissues may provide the necessary structure to transmit the tensional forces associated with silique desiccation (Spence *et al.*, 1996). Secondly the cells of the dehiscence zone show signs of cytoplasmic degradation prior to pod opening and the integrity of this layer during pod development remains unclear as does the mechanism of PG secretion into the cell wall. Further work will be necessary to ascertain whether some of the genes that have been reported to influence the timing of PCD are differentially expressed during either of these cell separation processes.

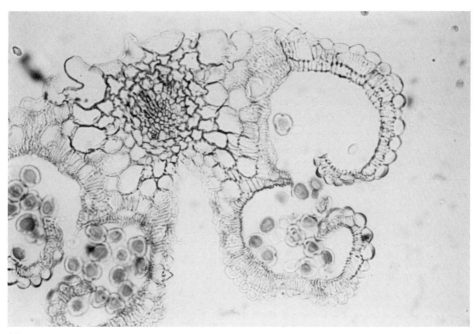

(a)

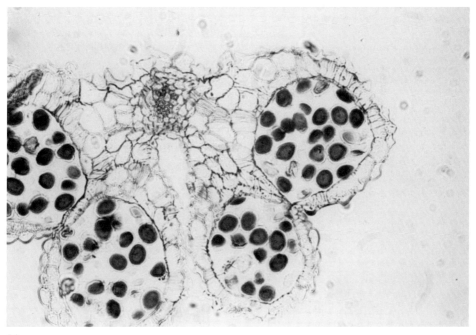

(b)

Figure 2. *Transverse section through a mature anther from (A) a wild-type* B. napus *(Westar) plant exhibiting dehiscence or (B) a non-dehiscent* B. napus *plant transformed with the* A. thaliana *PG promoter (1.408 kb) driving the barnase gene. Both sections were stained with Toluidine blue O.*

References

Chang, C., Kwok, S.F., Bleecker, A.B. and Meyerowitz, E.M. (1993) *Arabidopsis* ethylene response gene *ETR1*: similarity of products to two-component regulators. *Science* **262**: 539–544.

Coupe, S.A., Taylor, J.E., Isaac, P.G. and Roberts, J.A. (1993) Identification and characterisation of a proline-rich mRNA that accumulates during pod development in oilseed rape (*Brassica napus* L.). *Plant Mol. Biol.* **23**: 1223–1232.

Coupe, S.A., Taylor, J.E., Isaac, P.G. and Roberts, J.A. (1994) Characterization of a mRNA that accumulates during development of oilseed rape pods. *Plant Mol. Biol.* **24**: 223–227.

Coupe, S.A., Taylor, J.E. and Roberts, J.A. (1995) Characterisation of an mRNA encoding a metallothionein-like protein that accumulates during ethylene-promoted abscission of *Sambucus nigra* L. leaflets. *Planta* **197**: 442–447.

Coupe, S.A., Taylor, J.E. and Roberts, J.A. (1997) Temporal and spatial expression of mRNAs encoding pathogenesis-related proteins during ethylene-promoted leaflet abscission in *Sambucus nigra*. *Plant Cell Environ.* **20**: 1517–1524.

Gonzalez-Carranza, Z.H., Lozoya-Gloria, E. and Roberts, J.A. (1998) Recent developments in abscission: shedding light on the shedding process. *Trends Plant Sci.* **3**: 10–14.

Goodwin, W., Palls, J.A. and Jenkins, G.I. (1996) Transcripts of a gene encoding a putative cell wall-plasma membrane linker protein are specifically cold-induced in *Brassica napus*. *Plant Mol. Biol.* **31**: 771–781.

Gray, J. and Johal, G.S. (1998) Programmed cell death in plants. In: *Arabidopsis* (eds M.A. Anderson and J.A. Roberts). Sheffield Academic Press, Sheffield, UK, pp. 360–394.

Havel L. and Durzan, D.J. (1996) Apoptosis in plants. *Bot. Acta* **109**: 268–277.

Jenkins, E.S., Paul, W., Coupe, S.A., Bell, S.J., Davies, E.C. and Roberts, J.A. (1996) Characterization of an mRNA encoding a polygalacturonase expressed during pod development in oilseed rape (*Brassica napus* L.). *J. Exp. Bot.* **47**: 111–115.

Jenkins, E.S., Paul, W., Craze, M., Whitelaw, C.A., Weigand, A. and Roberts, J.A. (1999) Dehiscence-related expression of an *Arabidopsis thaliana* gene encoding a polygalacturonase in transgenic plants of *Brassica napus*. *Plant Cell Environ.* **22**: 159–167.

Lanahan, M.B., Yen, H-C., Giovannoni, J.J. and Klee, H.J. (1994) The *Never Ripe* mutation blocks ethylene perception in tomato. *Plant Cell* **6**: 521–530.

Kakimoto, T. (1996) CKI1, a histidine kinase homolog implicated in cytokinin signal transduction. *Science* **274**: 982–985.

Kerr, J.F.R. and Harmon, B.V. (1991) Definition and incidence of apoptosis. In: *Apoptosis: The molecular basis of cell death* (eds L.D. Tomei and F.O. Cope). Cold Spring Harbor Press, Cold Spring Harbor, NY, pp. 5–29.

Meakin, P.J. and Roberts, J.A. (1990a) Dehiscence of fruit in oilseed rape. 1. Anatomy of pod dehiscence. *J. Exp. Bot.* **41**: 995–1002.

Meakin, P.J. and Roberts, J.A. (1990b) Dehiscence of fruit in oilseed rape. 2. The role of cell wall degrading enzymes. *J. Exp. Bot.* **41**: 1003–1011.

Meakin, P.J. and Roberts, J.A. (1991) Induction of oilseed rape pod dehiscence by *Dasineura brassica*. *Ann. Bot.* **67**: 193–197.

Morel, J.-B. and Dangl, J.F. (1997) The hypersensitive response and the induction of cell death in plants. *Cell Death Diff.* **4**: 671–683.

Peterson, M., Sander, L., Child, R., Van Onckelen, H., Ulvskov, P. and Borkhardt, B. (1996) Isolation and characterization of a pod dehiscence zone-specific polygalacturonase from *Brassica napus*. *Plant Mol. Biol.* **31**: 517–527.

Roberts, J.A., Coupe, S.A., Taylor, J.E. and Whitelaw, C.A. (1997) Spatial and temporal expression of abscission-related genes during ethylene-promoted organ shedding. In: *Biology and Biotechnology of the Plant Hormone Ethylene*. NATO Series. (ed. A.K. Kanellis).

Sexton, R. and Roberts, J.A. (1982) Cell biology of abscission. *Annu. Rev. Plant Physiol.* **33**: 133–162.

Spence, J., Vercher, Y., Gates, P. and Harris, N. (1996) Pod shatter in *Arabidopsis thaliana*, *Brassica napus* and *B. juncea*. *J. Microscopy* **181**: 195–203.

Taylor, J.E., Webb, S.T.J., Coupe, S.A., Tucker, G.A. and Roberts, J.A. (1993) Changes in polygalacturonase activity and solubility of polyuronides during ethylene-stimulated leaf abscission in *Sambucus nigra*. *J. Exp. Bot.* **258**: 93–98.

Taylor, J.E., Coupe, S.A., Picton, S.J. and Roberts, J.A. (1994) Isolation and expression of a mRNA encoding an abscission-related β 1,4 glucanase from *Sambucus nigra*. *Plant Mol. Biol.* **24**: 961–964.

Webb, S.T.J., Taylor, J.E., Coupe, S.A., Ferrarese, L. and Roberts, J.A. (1993) Purification of β1,4 glucanase from ethylene-treated abscission zones of *Sambucus nigra*. *Plant Cell Environ.* **16**: 329–333.

Whitelaw, C.A., Paul, W., Jenkins, E.S., Taylor, V.M. and Roberts, J.A. (1999) A mRNA encoding a response regulator protein from *Brassica napus* is up-regulated during pod development. *J. Exp. Bot.* **50**: 335–341.

Wilkinson, J.Q., Lanahan, M.B., Clark, D.G., Bleecker, A.B., Chang, C., Meyerowitz, E.M. and Klee, H.J. (1997) A dominant mutant receptor from *Arabidopsis* confers ethylene insensitivity in heterologous plants. *Nature Biotech.* **15**: 444–447.

Zanchin, A., Marcato, C., Trainotti, L., Casadoro, G. and Rascio, N. (1995) Characterization of abscission zones in the flower and fruits of peach (*Prunus persica* L.). *New Phytol.* **129**: 345–354.

16

Regulation of cell cycle arrest and cell death – alternative responses to DNA damage

Chris Norbury

1. Introduction

Many anti-cancer therapies rely for their success on the induction of DNA damage. The responses of mammalian cells to such damage are complex, and encompass mechanisms that act on the one hand to delay cell cycle progression and promote DNA repair and on the other to promote cell death or long-term withdrawal from the cell cycle. Following DNA damage eukaryotic cells typically activate 'checkpoint' mechanisms that arrest the cell cycle (Hartwell and Weinert, 1989; Weinert and Hartwell, 1988), most strikingly in the G_2 phase though also in G_1 and S (*Figure 1*). Proteins

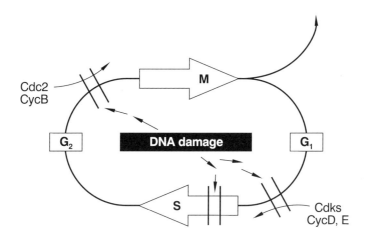

Figure 1. DNA damage checkpoint mechanisms constitute intracellular signalling pathways that serve to arrest the cell cycle following DNA damage. Cyclin dependent kinases (Cdks) are the ultimate targets for checkpoint pathways arresting the cell cycle in G_1 or G_2; the equivalent target(s) allowing temporary arrest in S phase are less well defined.

Programmed Cell Death in Animals and Plants, edited by J.A. Bryant, S.G. Hughes and J.M. Garland.
© 2000 BIOS Scientific Publishers Ltd, Oxford.

involved in these checkpoint pathways in human cells include p53 and its transcriptional target p21^{WAF1}, known to be required for efficient arrest in G$_1$ (Kuerbitz et al., 1992; Waldman et al., 1995), and the product of the ATM (ataxia-telangiectasia mutated) gene (Beamish et al., 1996; Savitsky et al., 1995a, 1995b), which appears to act upstream of p53 in the G$_1$ arrest pathway and which is also required for checkpoint arrest in S$_2$ and G$_2$.

Mutations in p53 are found in the majority of human tumours, and germline mutations in p53 or ATM give rise to increased cancer risk, suggesting that impairment of checkpoint controls could be an important step in tumorigenesis. The fact that DNA damage can also trigger apoptotic cell death in a relatively tumour-specific manner probably explains why DNA-directed therapies are ever effective, but the molecular mechanisms that couple DNA damage recognition to apoptosis and that underlie this tumour cell selectivity are not yet clearly defined. An exciting development in this area was the demonstration that cells lacking functional p53 can be selectively sensitized to combinations of DNA damage and agents such as caffeine, pentoxyfylline and 7-hydroxystaurosporine (UCN-01) that suppress checkpoint arrest (Fan et al., 1995; Powell et al., 1995; Russell et al., 1995; Wang et al., 1996; Yao et al., 1996). This suggests that p53 can modulate G$_2$ arrest and that the selectivity of some therapies could be improved by combining them with drugs of this sort that override checkpoints.

2. A role for p53 in the G$_2$ DNA damage checkpoint

Tumorigenesis in a wide variety of cell types is characterized by strong selective pressure for loss of function of the p53 tumour suppressor protein. Intensive study of p53 has revealed a number of biological roles for the wild-type protein that could explain why tumours so frequently lack it. Cells lacking p53 appear to have a less robust G$_2$ checkpoint response to DNA damage than wild type controls and enforced over-production of p53 can result in G$_2$ arrest in the absence of exogenous DNA damage (Agarwal et al., 1995). Cell cycle arrest in G$_2$ following DNA damage is in part dependent on inhibitory phosphorylation of the mitosis-promoting protein kinase Cdc2 (Figure 2; see also Norbury and Nurse, 1992; Nurse, 1990). Until recently, however, there was no indication of how activation of p53 might impact on G$_2$-M regulation to confer relative resistance to checkpoint-inhibitory drugs.

We have demonstrated that activation of p53 makes a significant contribution to inhibition of entry into mitosis following irradiation of human cells (Winters et al., 1998), and similar conclusions have since been drawn by others (Bunz et al., 1998). Cdc2 was found to be inhibited following p53 activation by a mechanism independent of inhibitory phosphorylation. Surprisingly, p53-mediated inhibition of Cdc2 was accompanied by its dephosphorylation and translocation from the cytoplasm to the nucleus. By contrast, the G$_2$ damage checkpoint in cells lacking p53 was found to be entirely reliant on inhibitory phosphorylation of Cdc2; the sensitivity of these cells to caffeine was correlated with Cdc2 dephosphorylation after administration of the drug (Table 1).

To allow us to study the effect of p53 activation on radiation-induced G$_2$ arrest we generated a pair of transfected cell lines derived from the p53-null human non-small cell lung cancer line NCI-H1299. One transfectant stably expresses a temperature-sensitive (Ala138→Val) human p53 which adopts the transcriptionally inactive, mutant conformation at 37–39°C, but behaves as wild-type at 32°C (Yamato et al., 1995). In

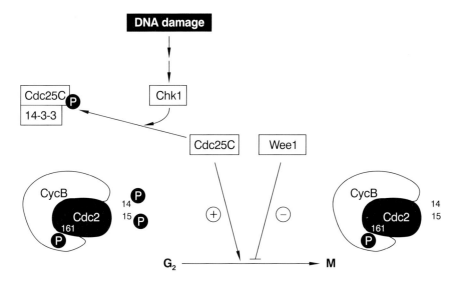

Figure 2. Phosphorylation events in the G_2 DNA damage pathway. Entry into mitosis from G_2 is controlled by the prototypical cyclin dependent kinase Cdc2. Phosphorylation of the Cdc2 catalytic subunit at residues threonine 14 and tyrosine 15 inhibits Cdc2 and is brought about by Wee1 and related inhibitory protein kinases. Dephosphorylation of Cdc2 threonine 14 and tyrosine 15 is performed by the Cdc25C phosphatase, which is in turn subject to inhibitory phosphorylation by the Chk1 protein kinase. Phosphorylated Cdc25C may be sequestered by 14-3-3- proteins. Thus activation of the Chk1 kinase in response to DNA damage results indirectly in the inhibition of Cdc2.

these cells, p53 is transcriptionally active only when the incubation temperature is shifted from 37°C to 32°C. Like the parental cell line, the control (empty vector) transfectant completely lacks p53.

Using this pair of human cell lines with and without functional p53 we examined the expression, activity and subcellular distribution of cell cycle regulators such as Cdc2, cyclin B1 and p21^{WAF1} in G_2-arrested cells in order to identify components that are influenced by p53 status. Cells induced to express functional p53 (by shift down to 32°C) after exposure to 6 Gy gamma radiation in early S phase, exhibited a protracted G_2 arrest that could not be overridden by checkpoint-suppressing drugs. In contrast,

Table 1. Two modes of G_2 arrest in human cells exposed to ionizing radiation (6 Gy) in the presence and absence of functional p53

	+ p53	−p53
G_2 delay	Protracted	Brief
Cdc2	Nuclear, dephosphorylated; associated with p21	Cytoplasmic, phosphorylated
Cyclin B	Nuclear	Cytoplasmic
Caffeine	Resistant	Sensitive

control populations treated in the same way but lacking p53 exhibited only a transient, caffeine-sensitive G_2 delay. This transient delay was characterized by tyrosine-15 and threonine-14 phosphorylation of Cdc2 which, with cyclin B1, was restricted to the cytoplasm in line with earlier studies of cells lacking p53. The more protracted G_2 delay seen after induction of p53 was associated with the accumulation of tyrosine-15- and threonine-14-dephosphorylated Cdc2 in the nucleus, an event conventionally associated with the onset of mitosis. A substantial fraction of the Cdc2 in these cells was associated with p21^{WAF1}, as judged by co-immunoprecipitation experiments, suggesting the possible involvement of this Cdk inhibitor in the mode of Cdc2 inhibition seen after p53 activation. In line with this interpretation, we found that transient transfection of a plasmid encoding p21 into cells lacking p53 was sufficient to drive Cdc2 into the nucleus and to promote Cdc2 dephosphorylation on threonine 14 and tyrosine 15. A strikingly similar role for p21 in the nuclear localization and inhibition of Cdk4/cyclin E complexes has been reported by others (LaBaer *et al.*, 1997).

In the light of these data, we believe it may be informative to examine expression and subcellular localization of Cdc2 and p21^{WAF1} in primary human breast cancer. In this way it should be possible to correlate the p53 status of tumours with expression of proteins involved in G_2 checkpoint arrest and to predict the likely degree of sensitization by caffeine-related checkpoint-suppressing drugs. Novel caffeine-related compounds having improved selectivity and specificity in checkpoint-suppression assays may be of clinical importance in the longer term. Our recent data suggest that it may be possible to use fission yeast (*Schizosaccharomyces pombe*) to identify checkpoint-overriding drugs, as caffeine has been shown to be capable of inhibiting the S–M checkpoint response in this simple eukaryote (Wang *et al.*, 1999). In addition, fission yeast genetics can be used to identify gene products that modulate checkpoint responses to caffeine, which may include the molecular targets of this class of drug.

Perhaps the best characterized function of p53 is its action as a transcriptional inducer of the *WAF1/CIP1* gene in response to DNA damage. *WAF1/CIP1* encodes the p21 Cdk inhibitor that is required for efficient p53-mediated arrest at the G_1 cell cycle checkpoint after DNA damage (*Figure 3*). A separable function of p53 is the promotion of apoptotic cell death, which provides a means for the elimination of cells in which DNA damage is beyond repair. In some cell types the apoptotic function of p53 involves transcriptional induction of the *BAX* gene. Thus p53 can participate in molecular mechanisms determining the relative probabilities of cell survival and cell death after DNA damage.

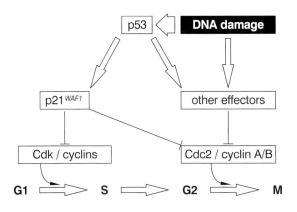

Figure 3. *p53 influences G_1 and G_2 checkpoint integrity after DNA damage. DNA damage detection activates p53-dependent and -independent checkpoint signalling pathways. Activation of p53 reinforces arrest in G_2 and is essential for arrest in G_1. The p53-mediated influence on G_2 arrest may be mediated in part by the p21*^{WAF1} *Cdk inhibitor, which is transcriptionally induced by p53 and which is capable of binding to and inhibiting Cdc2.*

The p53 protein is subject to complex post-translational modification by a variety of protein kinases, and a detailed analysis of p53 phosphorylation has revealed that residue serine 315 is the target for phosphorylation by Cdks *in vitro* and probably also *in vivo* (Bischoff *et al.*, 1990; Muller and Scheidtmann, 1995). The identity of the Cdk/cyclin responsible for the observed phosphorylation at serine 315 *in vivo* has yet to be established unambiguously, but it is perhaps significant that Cdk2/cyclin A and Cdc2/cyclin B can both phosphorylate the site with comparatively high efficiency *in vitro* (Bischoff *et al.*, 1990; Wang and Prives, 1995). An important regulatory role for serine 315 phosphorylation was strongly suggested by experiments in which Cdc2-mediated phosphorylation at this site was shown to increase the affinity of recombinant p53 for a specific subset of p53-responsive DNA elements *in vitro* (Wang and Prives, 1995). These and related experiments (Hecker *et al.*, 1996) together suggested that the serine 315-phosphorylated form of p53 has an elevated affinity for elements including that found in the WAF1/CIP1 promoter, but a reduced affinity for the BAX promoter element. In this case, phosphorylation of p53 by Cdc2 (or a related Cdk) might bias p53-mediated transcription in favour of cell cycle arrest-associated target genes such as *WAF1/CIP1* and against cell death-associated genes such as *BAX*. This would provide an explanation for our previous observation of elevated apoptosis on inhibition of Cdc2 in cells exposed to DNA damage (Ongkeko *et al.*, 1995).

Thus it is apparent that there is an intimate and bi-directional relationship between the key cell cycle regulator Cdc2 and p53, the 'guardian of the genome'; not only does p53 activation contribute to Cdc2 inhibition after DNA damage, but Cdc2 may also regulate p53 by direct phosphorylation at serine 315.

3. ATM and cellular responses to DNA damage

Ataxia telangiectasia (AT) is a recessive, heritable human disorder characterized clinically by progressive neurodegeneration, abnormal microvasculature, immune system defects, radiosensitivity and an increased incidence of tumours, particularly those of lymphoid origin (for reviews see Bundey (1994); Lavin and Shiloh (1996)). Cells derived from AT patients are also sensitive to ionizing radiation *in vitro*, though the fundamental basis for this radiosensitivity remains contentious. AT cells fail to display a number of the normal cell cycle checkpoint arrest responses to ionizing radiation (Ford *et al.*, 1984; Houldsworth and Lavin, 1980; Nagasawa and Little, 1983; Rudolph and Latt, 1989), but suffer in addition from DNA repair defects, hyper-recombination and progressive telomere shortening (Dar *et al.*, 1997; Foray *et al.*, 1997a, 1997b; Meyn, 1993).

The cloning of the *ATM* gene, mutations in which are responsible for this disease, identified the Atm protein as a member of a family of large lipid/protein kinases that also includes DNA-dependent protein kinase, required for the efficient repair of double-strand DNA breaks, and the Rad3/Mec1/Tel1 cell cycle checkpoint proteins (Greenwell *et al.*, 1995; Hartley *et al.*, 1995; Kato and Ogawa, 1994; Morrow *et al.*, 1995; Savitsky *et al.*, 1995a, 1995b; Seaton *et al.*, 1992). While this molecular evidence indicates a likely role for Atm in DNA repair and/or cell cycle checkpoint pathways, it has also been proposed that cells lacking functional Atm could suffer elevated levels of radiation-induced apoptotic cell death. Such an effect could either be direct, if Atm normally functions to suppress apoptosis (Meyn, 1995), or indirect, if the unscheduled cell cycle progression seen after irradiation of AT cells leads to secondary damage that in turn triggers cell death (Enoch and Norbury, 1995).

Atm-deficient mice generated by targeted gene disruption have been used to investigate the significance of Atm in the responses of specific tissues to irradiation. Perhaps surprisingly, thymocytes from *atm–/–* mice were found to be more resistant than wild-type thymocytes to gamma radiation-induced apoptosis (Westphal *et al.*, 1997a, 1997b; Xu and Baltimore, 1996). The favoured explanation for this resistance is that radiation-induced thymocyte apoptosis is largely dependent on the tumour suppressor protein p53, induction of which is impaired in cells lacking Atm (Kastan *et al.*, 1992; Lu and Lane, 1993). Interestingly, while a similar lack of radiation-induced death was also seen in lymphocytes and virally immortalized lymphoblasts from AT patients compared with normal controls, the basal level of spontaneous apoptosis was elevated in the AT cells (Duchaud *et al.*, 1996). As in AT patients, the intestinal epithelium (and in some cases skin) of *atm–/–* mice, but not wild-type controls, exhibited acute radiation-induced oedema (Barlow *et al.*, 1996; Westphal *et al.*, 1997a). It was not clear if this tissue toxicity was the result of excess apoptosis in the cells concerned, though any cell death that was involved was not dependent on p53 function (Westphal *et al.*, 1997a, 1997b).

AT patients and cell lines derived from them are hypersensitive to agents that induce double-strand DNA breaks, and many of the clinical features of AT could be explained in terms of excess apoptotic cell death (Enoch and Norbury, 1995). Previous studies of AT fibroblasts have generally used SV40-immortalized cells, the response of which to DNA damage is potentially influenced by the presence of virally-encoded T antigen (T-Ag). We have used time lapse videomicroscopy and *in situ* end-labelling (in collaboration with C. Gilbert and D. Lyon, ICRF Videomicroscopy Laboratory, London and with E. Capulas and C. Arlett, MRC Cell Mutation Unit, University of Sussex) to measure levels of spontaneous and ionizing radiation-induced apoptosis in primary and SV-40 T-Ag immortalized human fibroblasts, both from ataxia telangiectasia patients and from apparently normal subjects. In this way we have been able to explore further the significance of apoptosis in the AT radiosensitive phenotype. None of the primary cell populations exhibited significant levels of cell death either before or after exposure to a dose of gamma radiation sufficient to reduce viability of the AT cells at least tenfold. In contrast the T-Ag-immortalized fibroblasts displayed high levels of spontaneous apoptosis. Immortalized fibroblasts from normal individuals and from one of two AT patients studied did not exhibit any additional radiation-induced apoptosis.

Similar results were obtained using the topoisomerase-II-cleavable complex-stabilizing drug mitoxantrone, rather than irradiation, to induce DNA damage. In contrast to the earlier findings of others (Meyn *et al.*, 1994), we therefore conclude that the radiosensitivity of A-T fibroblasts is not explicable in terms of excess radiation-induced apoptosis and there is no indication that elevated rates of spontaneous apoptosis might contribute to the pathogenesis of this disease. Importantly, SV-40-mediated transformation, frequently used to establish human fibroblasts in culture, had a much more significant impact on the rate of apoptosis than did either low-dose (1–2 Gy) gamma irradiation or loss of Atm function. This powerful influence of T-Ag on basal apoptotic rate complicates the interpretation of earlier studies of radiation effects that have been carried out using immortalized human fibroblasts.

4. Resistance to checkpoint-suppressing drugs

The clinical usefulness of many anti-cancer drugs is frequently limited by the emergence of drug resistant cell populations. Just as key cell cycle regulators have been conserved throughout eukaryotic evolution, several drug resistance mechanisms that operate in human cells are also found in yeasts. Examination of the ways in which fission yeast cells can become drug-resistant might therefore shed light on novel, analogous mechanisms operating in human cells.

Resistance to caffeine and staurosporine can be conferred in fission yeast either by overexpression of *pad1+* or by mutation of *crm1* (Kumada *et al.*, 1996; Shimanuki *et al.*, 1995). While early studies suggested that this phenotype might relate specifically to checkpoint-inhibitory drugs, it has since been established that the spectrum of agents to which these yeast strains become resistant is quite broad, being restricted principally by the impermeability of the yeast cell wall. Resistance of this sort requires the presence of an otherwise inessential gene, *pap1+*, which encodes an AP-1-like transcription factor. Pap1-responsive genes encode a variety of drug resistance-associated proteins, including thioredoxin, thioredoxin reductase, glutathione synthase, glutathione reductase, catalase and ABC-type transporters. Activation of these genes is normally a transient response to a variety of stresses (Toone and Jones, 1999). Pap1 is activated by the stress-responsive MAP kinase Sty1/Spc1 and is inhibited by Crm1-dependent nuclear export and ubiquitin-dependent proteolysis. Drug resistance results from constitutive activation of this pathway, through activation of Sty1/Spc1, over-expression of Pap1, or inhibition of the mechanisms by which Pap1 is down-regulated.

Many of the components of this fission yeast system have human counterparts, several of which are highly conserved. Studies of this pathway in fission yeast could therefore help to identify equivalent processes that coordinate the activation of diverse drug resistance mechanisms in human cells. Activation of 'master' regulators of this sort could underlie clinical observations of pleiotropic resistance that cannot be explained simply in terms of over-expression of single effector proteins, such as individual ABC transporters.

We have identified a human protein that is closely related to (68% amino acid identity) and a fully functional homologue of the fission yeast *pad1+* gene product. This human protein (Poh1; pad one homologue) confers staurosporine and caffeine resistance in fission yeast, and is capable of inducing resistance to multiple drugs and to ultraviolet light when over-expressed in mammalian cells (Spataro *et al.*, 1997). Pad1/Poh1 is related to the S12/p40 component of the 26S proteasome, the multi-catalytic protease complex responsible for degrading proteins that have been targeted for proteolysis by ubiquitination. We have shown that Poh1 is itself a previously uncharacterized component of the proteasome 19S regulatory cap, and this has since been confirmed in budding and fission yeast by others. Jab1, a further Pad1-related protein that is a component of the COP9/signalosome complex, was recently shown to be required for the nuclear export and ubiquitin-dependent proteolysis of the CDK inhibitor p27 (Tomoda *et al.*, 1999). This finding supports our hypothesis that Pad1/Poh1 is similarly involved in the regulated nuclear export and proteolysis of drug resistance-associated transcription factors.

We and others have also cloned a human homologue of *crm1+* by computer-aided identification of expressed sequence tags and subsequent PCR. The human protein encoded by this gene, hCRM1, was recently shown to be associated with the nucleoporin Nup214/CAN (Fornerod *et al.*, 1997a, 1997b), and may therefore play a key role in determining the nucleo-cytoplasmic distribution of transcription factors relevant to drug resistance.

References

Agarwal, M.L., Agarwal, A., Taylor, W.R. and Stark, G.R. (1995) p53 controls both the G2/M and the G1 cell cycle checkpoints and mediates reversible growth arrest in human fibroblasts. *Proc. Natl Acad. Sci. USA* **92**: 8493–8497.

Barlow, C., Hirotsune, S., Paylor, R., Liyanage, M., Eckhaus, M., Collins, F., Shiloh, Y., Crawley, J.N., Ried, T., Tagle, D. and Wynshaw-Boris, A. (1996) Atm-deficient mice: a paradigm of ataxia telangiectasia. *Cell* **86**: 159–171.

Beamish, H., Williams, R., Chen, P. and Lavin, M.F. (1996) Defect in multiple cell cycle checkpoints in ataxia-telangiectasia postirradiation. *J. Biol. Chem.* **271**: 20486–20493.

Bischoff, J.R., Friedman, P.N., Marshak, D.R., Prives, C. and Beach, D. (1990) Human p53 is phosphorylated by p60-cdc2 and cyclin B-cdc2. *Proc. Natl Acad. Sci. USA* **87**: 4766–4770.

Bundey, S. (1994) Clinical and genetic features of ataxia-telangiectasia. *Int. J. Rad. Biol.* **66**: S23–S29.

Bunz, F., Dutriaux, A., Lengauer, C., Waldman, T., Zhou, S., Brown, J.P., Sedivy, J.M., Kinzler, K.W. and Vogelstein, B. (1998) Requirement for p53 and p21 to sustain G2 arrest after DNA-damage. *Science* **282**: 1497–1501.

Dar, M.E., Winters, T.A. and Jorgensen, T.J. (1997) Identification of defective illegitimate recombinational repair of oxidatively-induced DNA double-strand breaks in ataxia-telangiectasia cells. *Mutation Res.* **384**: 169–179.

Duchaud, E., Ridet, A., Stoppa Lyonnet, D., Janin, N., Moustacchi, E. and Rosselli, F. (1996) Deregulated apoptosis in ataxia telangiectasia: association with clinical stigmata and radiosensitivity. *Cancer Res.* **56**: 1400–1404.

Enoch, T. and Norbury, C. (1995) Cellular responses to DNA damage: cell-cycle checkpoints, apoptosis and the roles of p53 and ATM. *Trends Biochem. Sci.* **20**: 426–30.

Fan, S., Smith, M.L., Rivet, D.J., Duba, D., Zhan, Q., Kohn, K.W., Fornace, A.J., and O'Connor, P.M. (1995) Disruption of p53 function sensitizes breast cancer MCF-7 cells to cisplatin and pentoxifylline. *Cancer Res.* **55**: 1649–1654.

Foray, N., Badie, C., Arlett, C.F. and Malaise, E.P. (1997a) Comments on the paper: The ATM gene and the radiobiology of ataxia-telangiectasia. *Int. J. Radi. Biol.* **71**: 449–450.

Foray, N., Priestley, A., Alsbeih, G., Badie, C., Capulas, E.P., Arlett, C.F. and Malaise, E.P. (1997b) Hypersensitivity of ataxia telangiectasia fibroblasts to ionizing radiation is associated with a repair deficiency of DNA double-strand breaks. *Int. J. Radiol. Biol.* **72**: 271–283.

Ford, M.D., Martin, L. and Lavin, M.F. (1984) The effects of ionizing radiation on cell cycle progression in ataxia telangiectasia. *Mutation Res.* **125**: 115–122.

Fornerod, M., Ohno, M., Yoshida, M. and Mattaj, I.W. (1997a) CRM1 is an export receptor for leucine-rich nuclear export signals. *Cell* **90**: 1051–1060.

Fornerod, M., van Deursen, J., van Baal, S., Reynolds, A., Davis, D., Murti, K.G., Fransen, J. and Grosveld, G. (1997b) The human homologue of yeast CRM1 is in a dynamic subcomplex with CAN/Nup214 and a novel nuclear pore component Nup88. *EMBO J.* **16**: 807–816.

Greenwell, P.W., Kronmal, S.L., Porter, S.E., Gassenhuber, J., Obermaier, B. and Petes, T.D. (1995) TEL1, a gene involved in controlling telomere length in *S. cerevisiae*, is homologous to the human ataxia telangiectasia gene. *Cell* **82**: 823–829.

Hartley, K.O., Gell, D., Smith, G.C., Zhang, H., Divecha, N., Connelly, M.A., Admon, A., Lees-Miller, S.P., Anderson, C.W. and Jackson, S.P. (1995) DNA-dependent protein kinase catalytic subunit: a relative of phosphatidylinositol 3-kinase and the ataxia telangiectasia gene product. *Cell* **82**: 849–856.

Hartwell, L., and Weinert, T. (1989) Checkpoints: controls that ensure the order of cell cycle events. *Science* **246**: 629–634.

Hecker, D., Page, G., Lohrum, M., Weiland, S. and Scheidtmann, K.H. (1996) Complex regulation of the DNA-binding activity of p53 by phosphorylation: differential effects of individual phosphorylation sites on the interaction with different binding motifs. *Oncogene* **12**: 953–961.

Houldsworth, J., and Lavin, M.F. (1980) Effect of ionizing radiation on DNA synthesis in ataxia telangiectasia cells. *Nucl. Acids Res.* **8**: 3709–3720.

Kastan, M.B., Zhan, Q., el Deiry, W.S., Carrier, F., Jacks, T., Walsh, W.V., Plunkett, B.S., Vogelstein, B. and Fornace, A.J., Jr. (1992) A mammalian cell cycle checkpoint pathway utilizing p53 and GADD45 is defective in ataxia-telangiectasia. *Cell* **71**: 587–597.

Kato, R. and Ogawa, H. (1994) An essential gene, ESR1, is required for mitotic cell growth, DNA repair and meiotic recombination in *Saccharomyces cerevisiae. Nucl. Acids Res.* **22**: 3104–3112.

Kuerbitz, S.J., Plunkett, B.S., Walsh, W.V. and Kastan, M.B. (1992) Wild-type p53 is a cell cycle checkpoint determinant following irradiation. *Proc. Natl Acad. Sci. USA* **89**: 7491–7495.

Kumada, K., Yanagida, M. and Toda, T. (1996) Caffeine-resistance in fission yeast is caused by mutations in a single essential gene, crm1+. *Mol. Gen. Genet.* **250**: 59–68.

LaBaer, J., Garrett, M.D., Stevenson, L.F., Slingerland, J.M., Sandhu, C., Chou, H.S., Fattaey, A. and Harlow, E. (1997) New functional activities for the p21 family of CDK inhibitors. *Genes Devel.* **11**: 847–862.

Lavin, M.F. and Shiloh, Y. (1996) Ataxia-telangiectasia: a multifaceted genetic disorder associated with defective signal transduction. *Curr. Opin. Immunol.* **8**: 459–464.

Lu, X. and Lane, D.P. (1993) Differential induction of transcriptionally active p53 following UV or ionizing radiation: defects in chromosome instability syndromes? *Cell* **75**: 765–778.

Meyn, M.S. (1993) High spontaneous intrachromosomal recombination rates in ataxia-telangiectasia. *Science* **260**: 1327–1330.

Meyn, M.S. (1995) Ataxia-telangiectasia and cellular responses to DNA damage. *Cancer Res.* **55**: 5991–6001.

Meyn, M.S., Strasfeld, L. and Allen, C. (1994) Testing the role of p53 in the expression of genetic instability and apoptosis in ataxia-telangiectasia. *Int. J. Radiat. Biol.* **66**: S141–S149.

Morrow, D.M., Tagle, D.A., Shiloh, Y., Collins, F.S. and Hieter, P. (1995) TEL1, an *S. cerevisiae* homolog of the human gene mutated in ataxia telangiectasia, is functionally related to the yeast checkpoint gene MEC1. *Cell* **82**: 831–840.

Muller, E. and Scheidtmann, K.H. (1995) Purification and characterization of a protein kinase which is activated by SV40 large T-antigen and phosphorylates the tumor suppressor protein p53. *Oncogene* **10**: 1175–1185.

Nagasawa, H. and Little, J.B. (1983) Comparison of kinetics of X-ray-induced cell killing in normal, ataxia telangiectasia and hereditary retinoblastoma fibroblasts. *Mutation Res.* **109**: 297–308.

Norbury, C.J. and Nurse, P. (1992) Animal cell cycles and their control. *Annu. Rev. Biochem.* **61**: 441–470.

Nurse, P. (1990) Universal control mechanism regulating onset of M-phase. *Nature* **344**: 503–508.

Ongkeko, W., Ferguson, D.J., Harris, A.L. and Norbury, C. (1995) Inactivation of Cdc2 increases the level of apoptosis induced by DNA damage. *J. Cell Sci.* **108**: 2897–2904.

Powell, S.N., DeFrank, J.S., Connell, P., Eogan, M., Preffer, F., Dombkowski, D., Tang, W. and Friend, S. (1995) Differential sensitivity of p53(–) and p53(+) cells to caffeine-induced radiosensitization and override of G2 delay. *Cancer Res.* **55**: 1643–1648.

Rudolph, N.S. and Latt, S.A. (1989) Flow cytometric analysis of X-ray sensitivity in ataxia telangiectasia. *Mutation Res.* **211**: 31–41.

Russell, K.J., Wiens, L.W., Demers, G.W., Galloway, D.A., Plon, S.E. and Groudine, M. (1995) Abrogation of the G2 checkpoint results in differential radiosensitization of G1 checkpoint-deficient and G1 checkpoint-competent cells. *Cancer Res.* **55**: 1639–1642.

Savitsky, K., Bar Shira, A., Gilad, S., Rotman, G., Ziv, Y., Vanagaite, L., Tagle, D.A., Smith, S., Uziel, T., Sfez, S. *et al.* (1995a) A single ataxia telangiectasia gene with a product similar to PI-3 kinase. *Science* **268**: 1749–1753.

Savitsky, K., Sfez, S., Tagle, D.A., Ziv, Y., Sartiel, A., Collins, F.S., Shiloh, Y. and Rotman, G. (1995b) The complete sequence of the coding region of the ATM gene reveals similarity to cell cycle regulators in different species. *Hum. Mol. Genet.* **4**: 2025–2032.

Seaton, B.L., Yucel, J., Sunnerhagen, P. and Subramani, S. (1992) Isolation and characterization of the *Schizosaccharomyces pombe* rad3 gene, involved in the DNA damage and DNA synthesis checkpoints. *Gene* **119**: 83–89.

Shimanuki, M., Saka, Y., Yanagida, M. and Toda, T. (1995) A novel essential fission yeast gene pad1+ positively regulates pap1(+)-dependent transcription and is implicated in the maintenance of chromosome structure. *J. Cell Sci.* **108**: 569–579.

Spataro, V., Toda, T., Craig, R., Seeger, M., Dubiel, W., Harris, A.L. and Norbury, C. (1997) Resistance to diverse drugs and ultraviolet light conferred by overexpression of a novel human 26 S proteasome subunit. *J. Biol. Chem.* **272**: 30470–30475.

Tomoda, K., Kubota, Y. and Kato, J. (1999) Degradation of the cyclin-dependent kinase inhibitor p27^{Kip1} is instigated by Jab1. *Nature* **398**: 160–165.

Toone, W.M. and Jones, N. (1999) AP-1 transcription factors in yeast. *Curr. Opin. Genet. Dev.* **9**: 55–61.

Waldman, T., Kinzler, K.W. and Vogelstein, B. (1995) p21 is necessary for the p53-mediated G1 arrest in human cancer cells. *Cancer Res.* **55**: 5187–5190.

Wang, Q., Fan, S., Eastman, A., Worland, P.J., Sausville, E.A. and O'Connor, P.M. (1996) UCN-01: a potent abrogator of G2 checkpoint function in cancer cells with disrupted p53. *J. Natl Cancer Inst.* **88**: 956–965.

Wang, S.-W., Norbury, C., Harris, A.L. and Toda, T. (1999) Caffeine can override the S-M checkpoint in fission yeast. *J. Cell Sci.* **112**: 929–937.

Wang, Y. and Prives, C. (1995) Increased and altered DNA binding of human p53 by S and G2/M but not G1 cyclin-dependent kinases. *Nature* **376**: 88–91.

Weinert, T. and Hartwell, L. (1988) The RAD9 gene controls the cell cycle response to DNA damage in *Saccharomyces cerevisiae*. *Science* **241**: 317–322.

Westphal, C.H., Rowan, S., Schmaltz, C., Elson, A., Fisher, D. E. and Leder, P. (1997a) atm and p53 cooperate in apoptosis and suppression of tumorigenesis, but not in resistance to acute radiation toxicity. *Nature Genet.* **16**: 397–401.

Westphal, C.H., Schmaltz, C., Rowan, S., Elson, A., Fisher, D.E. and Leder, P. (1997b) Genetic interactions between atm and p53 influence cellular proliferation and irradiation-induced cell cycle checkpoints. *Cancer Res.* **57**: 1664–1667.

Winters, Z.E., Ongkeko, W.M., Harris, A.L. and Norbury, C.J. (1998) P53 regulates cdc2 independently of inhibitory phosphorylation to reinforce radiation-induced G2 arrest in human-cells. *Oncogene* **17**: 673–684.

Xu, Y. and Baltimore, D. (1996) Dual roles of ATM in the cellular response to radiation and in cell growth control. *Genes Devel.* **10**: 2401–2410.

Yamato, K., Yamamoto, M., Hirano, Y. and Tsuchida, N. (1995) A human temperature-sensitive p53 mutant p53Val-138: modulation of the cell cycle, viability and expression of p53-responsive genes. *Oncogene* **11**: 1–6.

Yao, S.L., Akhtar, A.J., McKenna, K.A., Bedi, G.C., Sidransky, D., Mabry, M., Ravi, R., Collector, M.I., Jones, R.J., Sharkis, S.J., Fuchs, E.J. and Bedi, A. (1996) Selective radiosensitization of p53-deficient cells by caffeine-mediated activation of p34cdc2 kinase. *Nature Med.* **2**: 1140–1143.

17

Tissue transglutaminase in cell death

M. Griffin and E. Verderio

1. Introduction

Transglutaminases (EC 2.3.2.13) are a family of enzymes that catalyse the post-translational modification of proteins either through the formation of intermolecular $\varepsilon(\gamma$-glutamyl) lysine crosslinks and/or the covalent incorporation of diamines and polyamines into the γ-carboximide of peptide bound glutamine residues. Proteins modified by covalent crosslinks using $\varepsilon(\gamma$-glutamyl) lysine bounds are highly stable and show an increased resistance to enzymic, chemical and mechanical disruption.

In mammals the five transglutaminases so far characterized are genetically distinct, have a reactive cysteine at their active site which shares a common amino acid sequence (Y–G–Q–C–W–V), are dependent on Ca^{2+} for their activity and are found both in the intracellular and extracellular environment. Well characterized examples include the plasma transglutaminase (factor XIII) involved in the cross-linking of the fibrin clot during haemostasis and the keratinocyte and epidermal transglutaminases involved in the formation of the cornified envelope during terminal differentiation in the epidermis of skin (for review see Lorand and Conrad, 1984; Greenberg et al., 1991; Aeschlimann and Paulsson, 1994). A further member of the transglutaminase family, which is widely distributed in mammalian tissues and which is the major subject of this chapter is the tissue transglutaminase (Type II, tTgase).

Unlike the plasma, keratinocyte and epidermal transglutaminases, tTgase does not require activation by proteolysis. It is also distinct from the other mammalian transglutaminases in that it can bind and hydrolyse GTP and ATP. Given that tTgase is predominantly a cytosolic enzyme that does not require proteolysis for its activation but can be activated at Ca^{2+} concentrations around 10^{-6} M (Hand et al., 1985), this suggests that its intracellular activity must be regulated by additional means. Under normal physiological conditions it is thought that Ca^{2+} activation of the enzyme is tightly regulated through the binding of GTP and GDP (Lai et al., 1998; Smethurst and Griffin, 1996) which inhibits enzyme activity until Ca^{2+} concentrations reach 100 μm or greater. Although the enzyme binds and hydrolyses ATP at a site thought to be distinct from the GTP binding site, ATP has no effect on the Ca^{2+}-activation of the enzyme. This tight regulation of enzyme activity suggests that tTgase is only active

Programmed Cell Death in Animals and Plants, edited by J.A. Bryant, S.G. Hughes and J.M. Garland.
© 2000 BIOS Scientific Publishers Ltd, Oxford.

when externalized from cells either by controlled secretion (Gaudry *et al.*, 1999; Verderio *et al.*, 1998, 1999) or by cell rupture/leakage (Upchurch *et al.*, 1991) or when the intracellular nucleotide levels are depleted and/or when intracellular Ca^{2+} levels approach those of the extracellular environment. As a consequence of this complex and multifunctional nature, tTgase has been implicated in a number of cellular functions both inside and outside of the cell (Aeschlimann and Paulsson, 1994) one of which is concerned with cell death.

2. The role of tissue transglutaminase in apoptosis

The first report describing the potential involvement of tTgase in apoptosis came from Fesus and colleagues (Fesus *et al.*, 1987) who noted that following induction of liver hyperplasia in rats the level of tTG protein and activity had increased dramatically over several days and was found to correlate to the time when maximum cellular regression occurred.

It was later shown (Fesus *et al.*, 1989) that the isolation of these hepatocytes in reducing agents plus chaotropic reagents resulted in spherical structures enriched in ε(γ-glutamyl) lysine cross-links, which were classified as apoptotic bodies. The hypothesis was that as the cell was dying, tTgase was over-expressed and became activated leading to extensive cross-linking of proteins within the dying cells (Knight *et al.*, 1991). Cross-linking of intracellular components by tTgase was thought to be important in stabilizing the apoptotic cells before their clearance by phagocytosis and was seen as a mechanism to prevent leakage of intracellular, inflammatory substances to the extracellular environment. Since an important feature distinguishing apoptosis from necrosis is the maintenance of cellular integrity, without release of either cytoplasmic or nuclear content, it seemed to be plausible to postulate a role of tTgase in programmed cell death.

Prior to the acceptance of apoptosis as a distinct form of cell death, involvement of tTgase in cell death had been reported in the erythrocyte. The cross-linking of membrane and associated cytoskeletal proteins in the ageing red cell was proposed as a mechanism for membrane stiffening, leading to decreased red cell deformability and removal from the circulation (Lorand and Conrad, 1984). The cross-linked apoptotic bodies isolated by Fesus and other workers (*Figure 1*) were also found to be comparable to the detergent-insoluble cornified envelopes produced by the keratinocyte transglutaminase (now also thought to involve the epidermal Tgase) during the terminal differentiation events occurring in the epidermis (reviewed in Aeschlimann and Paulsson, 1994). Given the evidence drawn from these closely related events, tTgase was proposed as a credible downstream effector in the later stages of apoptosis, since at that time the cellular functions of the enzyme were still only beginning to be understood.

However, a close examination of these early reports indicates that the occurrence of apoptosis and tTgase expression do not always completely overlap. In the early work of Fesus *et al.* (1987) although an increase in tTgase gene expression/protein concentration was reported, close examination of the hepatocytes indicates that only a low percentage of cells stain positive for tTgase, while it was previously established that during the highest point of cellular regression, 50% of the cells were dying by apoptosis (Columbano *et al.*, 1985). In a later report Knight *et al.* (1993), using lymphokine-activated killer cells as a mediator of cell death for the erythroleukemic

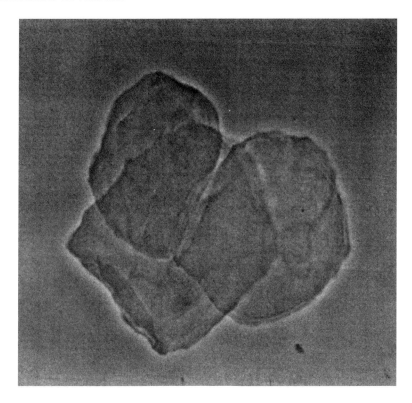

Figure 1. Light micrograph of detergent insoluble apoptotic bodies/envelopes isolated from a hamster fibrosarcoma cell line. The envelopes are between 30 and 40 μm in diameter.

cell line K562, estimated that only 15% of dying cells produced detergent insoluble envelope-like structures. Szende *et al.* (1991) reported that mouse mammary tumours treated with methotrexate had increased levels of tTgase compared to the control; however, some tTGase-positive cells showed no morphological phenotypes of apop-tosis. Since these initial observations, an association of tTgase with apoptosis has been more widely reported in a number of *in vitro* and *in vivo* systems which include neuroblastoma cells (Melino *et al.*, 1994; Piredda *et al.*, 1999), tracheal epithelium (Zhang *et al.*, 1995) and myeloid leukaemia cells (Nagy *et al.*, 1995).

More recently the simultaneous detection of tTgase in epithelial cells showing frag-mented chromatin during post-lactational involution seemed to clearly establish a link between induction of tTgase expression and apoptosis in a physiologically relevant *in vivo* apoptotic system (Nemes *et al.*, 1996). Moreover, measurement of the ε(γ-glutamyl) lysine cross-link in this tissue indicated a two-fold increase over that of controls. In a further *in vivo* model involving involution of the thymus elicited by treatment with either anti CD3 antibody, dexamethasone or by irradiation, induction of tTgase expression was observed in dying cells, which was accompanied by increased serum levels of the free ε(γ-glutamyl) lysine isodipeptide. However in the same report it was shown that induction of tTgase was not required when apoptosis was induced by *fas* receptor stimulation.

Work from many laboratories has shown that tTgase expression is usually induced in the earlier stages of apoptosis, suggesting that tTgase may play a further role in addition to its cross-linking function in the initiation of the apoptotic programme (Fesus and Thomazy, 1988; Nagy *et al.*, 1997; Piacentini *et al*, 1994). In the studies so far undertaken to establish a role for tTgase in apoptosis *in vivo*, the expression of the enzyme has been generally correlated to both the level of nuclear fragmentation and to morphological changes which accompany a cell undergoing apoptosis. Indeed apoptosis has been historically characterized biochemically by endogenous endonuclease activation, which results in the formation of DNA fragments that appear as a distinct ladder on electrophoretic DNA gels and are detectable by *in situ* end labelling (Martin and Green, 1994). However, DNA fragmentation when measured *in situ* does not exclusively diagnose apoptosis as it is also found in well advanced necrosis.

Recent advances indicate that the central component of the specialized apoptotic machinery is a proteolytic system involving a family of proteases called caspases (see Chapter 2). These enzymes participate in a cascade that culminates in the change of a set of proteins resulting in disassembly of the cell (reviewed in Thornberry and Lazebnik, 1998 and in Chapter 2). Although it is not fully understood how caspases contribute to the apoptotic process, it is clear that a subset of caspases is responsible for the cell changes typical of apoptosis. In the light of the essential role of caspases in apoptosis, assessment of caspase-mediated cell death appears to be crucial and perhaps less ambiguous than the estimation of nuclear fragmentation, to establish whether a cell is dying from apoptosis and whether tTgase is an essential effector element in this process. Initial work by Piredda *et al.* (1999) indicates that the cell death occurring in human neuroblastoma cells overexpressing tTgase is only partially inhibited by the general caspase inhibitor Z-VAD. Moreover, two independent reports (Johnson *et al.*, 1998; Melino *et al.*,1994) suggest that tTgase-induced cell death is not susceptible to the apoptosis suppressor bcl-2, which is a known key regulator of apoptotic pathways involving activation of caspase 9 (see Chapter 4). These pathways typically involve death induced via cytotoxic agents, dexamethasone, agents resulting in DNA damage and others whereby mitochondrial dysfunction precedes activation of caspase 9 (reviewed by Adams and Cory, 1998; see also Chapter 6). These initial findings, within the limits of the model cell systems used, tend to be contradictory and seem to suggest that tTgase might not be pivotal to the apoptotic death pathway.

3. Molecular bases of tissue transglutaminase regulation during cell death: role of the enzyme in embryogenesis

There is wide evidence for upregulation of the tTgase gene during cell death, however until recently little was known of the molecular mechanisms regulating its expression. Evidence for the regulation of tTgase expression by hormones and cytokines (retinoic acid, TGF-β1, IL6) (Chiocca *et al.*, 1988; Suto *et al.*, 1993a) has been supported at the molecular level by the isolation of the regulatory regions of the tTGase gene (Lu *et al.*, 1995; Suto *et al.*, 1993b) and identification of a specific tripartite retinoid response element (Nagy *et al.*, 1996) and a $TGF\beta_1/BMP_4$ response element (Ritter and Davies, 1998). Moreover, a recent study by Lu and Davies (1997) suggests that selective demethylation of the tTGase promoter may control tissue-specific expression of the tTgase gene.

Perhaps some of the most convincing lines of evidence indicating that tTgase has a role in physiological programmed cell death has resulted from the information gained examining the activity of its promoter during morphogenesis and development. A 3.8 kb fragment of the mouse tTG promoter was coupled to the reporter gene β-galactosidase and the resulting construct used to make transgenic mice lineages (Nagy *et al.*, 1997). Evidence of selective expression of the transgene in the embryonic limbs in regions of morphogenic apoptosis (interdigital web, anterior necrotic zone) were clearly indicated (*Plate 1*). Expression of the transgene could be detected in cells showing a clear apoptotic morphology but also in morphologically normal cells, in the interdigital mesenchymal cells of a 13.5-day-old embryo. This finding may suggest that there may be an overall increase in tTgase in areas of differentiation and/or a generalized increase in the tTgase transgene in those regions where cells are committed to apoptosis. It may also suggest and confirm early evidence which indicated that the tTgase transgene is activated early in the apoptotic programme. In general, the 3.8 kb of DNA flanking the 5′ end of the mouse tTgase gene appears to include all the regulatory elements responsible for tTgase-tissue specificity, retinoid regulation and apoptosis-induced. Retinoids have been shown to induce apoptosis after cellular differentiation (Martin *et al.*, 1990). Classically, retinoids have long been recognized to have major effects on cell proliferation, differentiation and more recently cell death. Therefore it is plausible that vitamin A and its derivatives which are required for many fundamental life processes are key regulators of tTgase transcription during apoptosis *in vivo* (Nagy *et al.*, 1998).

In a very recent study (Dupe *et al.*, 1999) involving knockout mice lacking the RAR-γ gene and one or both alleles of RAR-β gene it was shown that tTgase promoter activity and expression of stromelysin-3 were down-regulated. These mice display a severe and fully penetrant interdigital webbing caused by the persistence of the fetal interdigital mesenchyme. Since both tTgase and stromelysin-3 contain retinoic acid response elements in their promoter regions it was concluded that retinoic acid might increase the amount of cell death in the interdigital necrotic zones through direct modulation of tTgase expression and contribute to the process of tissue remodelling through the up-regulation of stromelysin-3 expression.

In a further recent study, Thomazy and Davies (1999) examined the level of tTgase expression in the developing limb of the chicken embryo. In agreement with the findings obtained in the transgenic lineages, this work confirmed that tTgase accumulates in clusters of morphologically apoptotic cells in the interdigital web resulting in the separation of the fingers. However, in keeping with data obtained from transgene experiments the enzyme showed heterogeneous expression. At this stage it therefore remains to be elucidated whether the induced tTgase in morphologically apoptotic cells is expressed in a cellular context favourable to its catalytic activity as a Ca^{2+}-dependent cross-linker, or as a GTPase (Nakaoka *et al*, 1994) by virtue of its unique GTP binding and hydrolysis site or as a component which is exported to the extracellular matrix where its cross-linking activity is important for the tissue remodelling process (Aeschlimann *et al*., 1996; Johnson *et al*., 1997, 1999).

4. Role of calcium in apoptosis and in the activation of tissue transglutaminase

Another unsolved key issue on the involvement of tTgase in apoptosis concerns the mechanism whereby different apoptotic signalling pathways can activate the Ca^{2+}-dependent catalytic activity of tTgase, to form Tgase cross-linked polymers which stabilize the apoptotic cells. The role of tissue transglutaminase as a protein cross-linker is strictly dependent on elevations in intracellular calcium and/or a concomitant reduction in GTP/GDP concentration. The involvement of calcium in cell death and apoptosis is well established and the nature of the Ca^{2+}-mediated reactions has been partly elucidated. For example, calcium activates enzymes such as nitric oxide synthase (bNOS) which can elicit NO toxicity, endonucleases responsible for DNA degradation, phospholipase A2 with a consequent rise in lipoxygenase and cycloxygenase metabolites and cysteine proteases (e.g. calpain) (reviewed in McConkey and Orrenius, 1996; Nicotera and Orrenius, 1998). However, there is an overall shortage of experimental data, based on either *in vivo* or *in vitro* models, assessing a quantitative correlation between apoptosis, calcium and tissue transglutaminase activity. Few studies include measurements of intracellular calcium concentration in relation to apoptotic cell death.

Work from McConkey and colleagues shows that in thymocytes, DNA fragmentation and apoptosis induced by CD3/T-cell receptor complex are dependent on an early increase in cytosolic Ca^{2+} concentration due to influx of extracellular Ca^{2+}. Measurements of Ca^{2+} levels in response to CD3-induced programmed cell death show a sustained increase for up to 2 hours in cytosolic $[Ca^{2+}]$ from control values of approximately 100 nM to 250 nM (McConkey *et al.*, 1989). This early rise in intracellular Ca^{2+} might explain the proposed role for tTgase in this apoptotic event (Szondy *et al.*, 1997). Treatment of androgen-independent prostatic cancer cells with thapsigargin (500 nM), which selectively inhibits the sarcoplasmic reticulum and endoplasmic reticulum Ca^{2+}-dependent ATPase pumps (SERCA) results in a three- to four-fold elevation of intracellular free Ca^{2+}. This Ca^{2+} mobilization is sustained by a secondary influx of extracellular Ca^{2+} and results in cell rounding followed by loss of proliferation, fragmentation of the genomic DNA, loss of plasma membrane integrity and cell disassembly into apoptotic bodies (Furura *et al.*, 1994). Thapsigargin is known to induce apoptosis in various cell lines including human hepatoma cells (Tsukamoto and Kaneko, 1993) and mouse lymphoma cells (Lam *et al.*, 1994). Rovere *et al.* (1996) have recently shown that appreciable intracellular Ca^{2+} release is also induced by cross-linking of CD-95 receptor of human T lymphocytes. Therefore, Ca^{2+} seems to play a pivotal role in many examples of apoptotic cell death, although published data seem to indicate that the elevation of calcium measured during programmed cell death may not be sufficient to activate the transglutaminase cross-linking activity efficiently.

Smethurst and Griffin (1996), examining the regulation of tTgase in endothelial cells by electropermeabilization and clamped Ca^{2+} concentrations, estimated that the value of free intracellular calcium necessary for tTgase activation is around 100 μM Ca^{2+} in the presence of low levels of nucleotides. At 10 nM or even 10 μM Ca^{2+} and at physiological levels of cytosolic ATP and GTP, tTgase remains inactive as a protein cross-linker. Subsequent work from Greenberg's group showed that 10 μM Ca^{2+} and 100 nM GTP keep tTgase activity latent (Lai *et al.*, 1998). However, Zhang *et al.* (1998), by investigating the endogenous regulation of tTGase in a human neuroblastoma cell line treated with retinoic acid to promote overexpression of tTgase, measured a modest increase in tTgase activity (approximately 0.5-fold) following Ca^{2+} elevation (up to 200 nM) by thapsigargin (20 μM) treatment. They also demonstrated that the Ca^{2+}-mobilizing drug maitotoxin (5 mM) elevates tTgase activity appreciably, following an increase in intracellular Ca^{2+} concentration of 10 times (~700 nM). However, no correlation between increased $[Ca^{2+}]$, tTgase activity and apoptosis was made in this study. Moreover they noted that if intracellular GTP levels were depleted using the selective drug tiazoflurin, tTgase activity was reduced. Since the binding of tTgase to GTP is known to protect it against proteolysis they raised the possibility that in the presence of elevated Ca^{2+} but decreased GTP levels, tTgase was being degraded by Ca^{2+}-activated proteases, possibly calpain (Zhang et al., 1998). Interestingly, cell viability and the morphology of neuroblastoma cells over-expressing tTgase were not commented upon in this report, despite the fact that earlier reports had shown that the same human neuroblastoma cells become susceptible to apoptosis following increased expression of tTGase (Melino et al., 1994).

The extracellular space, which is characterized by high Ca^{2+} concentration and low levels of nucleotides, has been shown to be favourable to the activation of the enzyme (Aeschlimann et al, 1995; Johnson et al.,1999; Verderio et al., 1998). However, reports so far suggest that tTgase does not seem to play a significant role in extracellular matrix cross-linking during apoptosis, since tTgase appears to be mainly restricted to the intracellular space (Nemes et al., 1996). The question therefore arises as to how tTgase could be activated given that the intracellular [Ca^{2+}] measured during apoptotic cell death appears to be below the required level for full activation? One possibility is that tTgase becomes activated to a lesser extent by sustained small rises of Ca^{2+}, allowing the gradual accumulation of cross-linked products (Lorand, 1996). However it is known that ATP levels decrease slowly in the apoptotic cell since it is thought that energy is required both for maintenance of membrane integrity and for downstream events in the apoptotic cascade (Bossy-Wetzel et al., 1998; Eguchi et al., 1997). A drastic reduction in ATP levels is only thought to occur when the cell is about to enter into necrosis. Presumably under normal physiological conditions phagocytosis of the apoptotic cell would have occurred prior to this. Since ATP levels are maintained it can only be considered that GTP levels are also maintained in the apoptotic cell since there is sparse information in the literature to suggest otherwise.

Given this scenario it might be expected that tTgase remains inactive as a protein cross-linker unless the dying cell proceeds into the early stages of necrosis whereupon massive Ca^{2+} influx from the extracellular space occurs. Activation of tTgase at this late stage could act as a 'fail safe' mechanism to prevent the release of cell components if phagocytosis of the dying cell does not occur. This may explain why cell death involving activation of tTgase is not affected by over-expression of bcl-2, since one of the downstream effects of this protein is to prevent mitochondrial damage resulting in the release of intracellular Ca^{2+} stores but not that resulting from massive influx of extracellular Ca^{2+}.

It cannot be ruled out, however, that accumulation of tTgase within a cell pre-destined to death can also aid in the controlled release of intracellular Ca^{2+} stores either through its GTP-binding activity (Nakaoka et al., 1994) or by helping to reduce dwindling stores of ATP through its ATPase activity (Lai et al., 1998). Such a scenario could involve activation of PLCδ1 (Feng et al., 1996) leading to synthesis of inositol,1,4,5-trisphosphate and mobilization of intracellular Ca^{2+}. However to support this hypothesis apoptotic signals that trigger the GTP binding activity of tTgase need to be identified. The existence of intracellular cofactor(s) that can modify the susceptibility of tTGase to Ca^{2+} has also been recently debated. Work from Greenberg's group has suggested that lipid cofactors could modulate the activation of tTgase by reducing the calcium requirement to obtain cross-linking activity (Lai et al., 1997) This report is interesting since a number of lipid changes are known to occur in the membrane of cells during the apoptotic cascade (Kinloch et al., 1999).

5. Tissue transglutaminase-induced cell death: a novel death pathway

Although a link between tTgase and apoptosis is still not fully established, it cannot be disputed that tTgase is specifically expressed or accumulates in relation to cell death. In particular, it is implicated in situations involving tissue remodelling, such as in the interdigital web during embryogenesis (Nagy et al., 1997), during wound healing (Upchurch et al., 1991) and during tissue fibrosis (Griffin et al., 1979; Johnson et al., 1997; Mirza et al., 1997). There is also increasing evidence to indicate that apoptosis and necrosis, in their classical definition, are the extreme ends of a range of possible modes of death characterized by distinguished morphological and biochemical features (Nicotera and Orrenius, 1998). A wide range of responses to cellular damage do not seem to meet the stereotyped criteria of apoptosis or necrosis, suggesting that there might be alternative modes of cell death, not yet well classified. The duration and extent of increased intracellular calcium concentration, which often accompanies cell death and results in cytotoxicity through activation of a wide range of possible downstream reactions, may partly explain the variety of death pathways and the range of cell morphologies that result. tTgase in its cross-linking mode strictly depends on Ca^{2+} for its activity and is likely to act as an effector molecule in many forms of cell death where Ca^{2+} plays a central role.

Work from our laboratory over the past 2 years has demonstrated that considerable over-expression of tissue transglutaminase in Swiss 3T3 fibroblasts, where the enzyme is transcriptionally activated by a tetracycline-inducible promoter (Verderio et al., 1998) does not trigger cell death. This finding is in disagreement with previous results which relied on tissue transglutaminase-transfected cells where the enzyme is under the control of a constitutive promoter (Gentile et al., 1992; Melino et al., 1994). This new study provides the first data whereby transfected cell lines have been isolated under conditions where tTgase expression is kept repressed and maintained at endogenous levels, thus excluding the isolation of clones naturally 'tolerant' or susceptible to high levels of gene expression. It is also known that a number of cells such as endothelial cells constitutively express high levels of tTgase without any increase in their propensity to undergo apoptosis.

In an endothelial cell model system developed in our laboratory, where the naturally high level of tTgase was markedly reduced by transfection of antisense tTgase cDNA (Jones et al., 1997), cells did not appear to be less susceptible to apoptosis when induced by the general kinase inhibitor staurosporine when compared to the tTgase-rich parental endothelial cells (R. Jones, unpublished data). No nuclear fragmentation was observed in the monolayer of adherent cells in the Swiss 3T3 fibroblasts model system (Verderio et al., 1998) upon increased expression of tissue transglutaminase. However, a greater number of fragmented nuclei was found in the detached cells present in the culture supernatant when the level of the enzyme was increased, despite the finding that the total number of dead cells in the culture supernatant was comparable to that in the corresponding cell line maintained at low endogenous level of transglutaminase expression. This suggests that the transglutaminase protein is facilitating rather than driving the cell death process. This finding was supported by the observation that when the calcium ionophore ionomycin was used to trigger apoptosis of the Swiss 3T3 fibroblasts by calcium mobilization (El Alaoui et al., 1997), cells induced to over-express tTgase displayed an enhanced DNA fragmentation.

The duration and extent of calcium mobilization may determine the mode of cell death, i.e. whether the cell dies by apoptosis or by necrosis/cell lysis. By increasing the intensity of calcium influx using higher concentrations of calcium ionophore, cell lysis was extensively induced in the Swiss 3T3 fibroblasts with low endogenous levels of tTgase. Interestingly, the same treatment of fibroblasts induced to express increased levels of the enzyme, caused a morphologically different type of cell death, as cells displayed changes which were not typical of either apoptosis or necrosis. Many cells remained adherent instead of rounding up and detaching, contained relatively intact cytoplasmic structures and non-fragmented nuclei (*Figures 2A, C and E*).

Visualization of tTgase activity *in situ* indicated that the enzyme was particularly active in the nucleus, in the cytoplasm and at the periphery of the cell at cell–matrix contacts, suggesting that activation of tTgase by sustained calcium influx results in the cross-linking of a large array of proteins (*Figure 2A*). Indeed a wide range of cellular substrates for tTgase have been identified (Knight *et al.*, 1993; Nemes *et al.*, 1997). The observed nuclear activity is supported by the recent finding of a distinct nuclear local-ization of tissue transglutaminase (Lesort *et al.*, 1998) and earlier *in vitro* studies which identified histones as substrates for the enzyme (Ballestar *et al.*, 1996). The cross-linking of calcium-damaged cells by activation of tTgase would contribute to the structural stabilization of the dying cells, whereby cross-linking of both nuclear and cytoplasmic components could prevent their leakage, thus maintaining the integrity of the tissue within which the dying cells are found.

Support for the involvement of tTgase in a form of cell death which does not fit the textbook descriptions of classical apoptotic or necrotic cell death comes from recent work on the molecular mechanisms of renal fibrosis (Johnson *et al.*, 1997). Renal scarring is characterized by loss of renal function and death of tubular cells. In an *in vivo* model of renal fibrosis, tubular cells have been shown to over-express tTgase and to contain excessive levels of $\varepsilon(\gamma$-glutamyl$)$ lysine cross-links, which are likely to influence tubular function and viability. However, these cross-linked tubular cells did not show biochemical and morphological signs of classical apoptosis and appeared to be 'stabilized', although injured, within the tissue (*Plates 2B, D, F and G*). More recent studies (Johnson *et al.*, 1999) have shown that tTgase antigen and activity is also present in the interstitial space of the tubular cells and is particularly concentrated at areas aligning the basement membrane of these cells.

Further evidence for tTgase inducing an alternative form of cell death, characterized by cell stabilization following injury, comes from more recent work by Johnson *et al.* (1998). Treatment of a fibrosarcoma cell line with glucocorticoids, which are strong inducers of tTgase expression, led to an increased level of detergent-insoluble cross-linked 'apoptotic' bodies (see *Figure 1*). Initially these apoptotic bodies were thought to be formed as a result of an increased rate of endogenous apoptosis, but subsequent experiments indicated these bodies to be formed as a consequence of trypsinization damage during cell passage which was significantly increased in cells showing over expression of tTgase. Cross-linking of cellular proteins was not accompanied by either nuclear fragmentation or activation of the ICE protease CPP32 and it was not reduced by over-expression of the apoptosis suppressor protein bcl-2.

In conclusion, from recent and ongoing work our group has proposed an alter-native, transglutaminase-induced form of cell death. It is consequent to sustained intracellular calcium elevation and may represent an immediate, stabilizing response to tissue injury or cell damage such as that resulting from exposure to proteases,

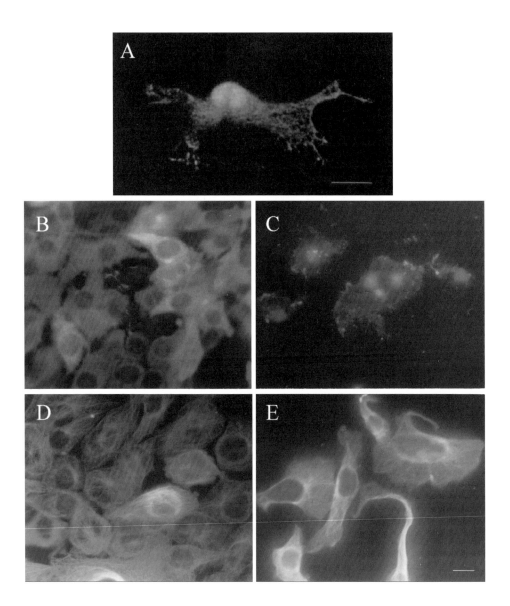

Figure 2. *Adherent 3T3 cells over-expressing tissue transglutaminase after treatment with 20 μM ionomycin (panel A) in the presence of the tissue transglutaminase competitive primary amine substrate fluorescine cadaverine. The fluorescence shows the areas denoting tissue transglutaminase crosslinking. Panels B and C show cells immunostained for tissue transglutaminase and detected using FITC-labelled secondary antibody. B represents control cells while C represents cells treated with ionomycin. Panels D and E show cells immunostained for α-tubulin using a FITC-labelled secondary antibody. D represents control cells while E represents cells treated with ionomycin. The bar represents 5 μm. Adapted form Verderio et al. (1998).*

physical wounding or anoxia. Interestingly, this ability of tTgase to kill cells independently of the apoptotic process has recently been observed in hypertrophic chondrocytes. In this case, the enzyme involved was indicated to be the active *a* subunit of the plasma transglutaminase (Nurminskaya *et al.*, 1998) which becomes over-expressed and proteolytically activated in the hypertrophic cells. Nuclei of dying cells showed no typical apoptotic staining of the nuclei by end labelling (TUNEL) but membrane integrity was lost leading to leakage of some cytoplasmic components.

Ongoing work suggests that tTgase is an important component of a cell/tissue defence system following wounding, thus expanding the original ideas of Patterson and colleagues (Upchurch *et al.*, 1991). For example, when Swiss 3T3 cell monolayers are wounded, a time-dependent elevation of tTgase activity can be visualized *in situ* by inclusion of the fluorescent primary amine substrate fluorescine cadaverine. The increased activity begins at the edge of the wound and propagates outward and is detectable within a few minutes (E. Verderio, B. Nicholas and M. Griffin, unpublished data). Proof that tissue-type transglutaminase is the transglutaminase enzyme responsible for this cross-linking following mechanical injury comes from the observed elevation in cross-linking activity at the edge of the wound, in cells that are over-expressing the enzyme, in the Swiss 3T3 fibroblast model system.

Interestingly, wound-induced intracellular free calcium concentration waves have been recently described in a model of similarly wounded alveolar type II cells (Hinman *et al.*, 1997). Following damage, cells produce a calcium wave that is initiated at the site of injury and propagated throughout the monolayer. Therefore, one role for the calcium wave could be the activation of tissue transglutaminase in order to coordinate the healing process in response to tissue damage. The expression of the enzyme consequent to cell damage may create highly cross-linked polymer networks, thereby contributing rapidly to the structural stabilization of the dying cell and surrounding matrix proteins. The inflammatory response would thus be reduced and helping to preserve the structural integrity of the damaged tissues (Nemes *et al.*, 1996). Interestingly, fas-mediated cell death of thymocytes, which is not accompanied by tTgase induction, appears more dramatic and disorganized in morphology than dexamethasone-mediated apoptosis where tTgase is induced (Nemes *et al.*, 1996).

6. Conclusion

In conclusion, the independent efforts from many different laboratories over the past decade support the idea that tTgase is generally found to be up-regulated in cells destined to undergo cell death. It has also been demonstrated that over-expression of tTgase in cells leads to increased export of the enzyme to the cell surface and into the extracellular matrix (Johnson *et al.*, 1999; Verderio *et al.*, 1998, 1999). Activation of the enzyme in these different cellular compartments would therefore result in the cross-linking of both intracellular and extracellular proteins, leading to stabilization of the dying cell and surrounding matrix, thus maintaining both cellular and tissue integrity. A further extracellular function of the enzyme has also been linked to the activation of TGFβ$_1$ (Nunes *et al.*, 1997), whereby the cross-linking of the latent TGFβ$_1$ binding protein to matrix proteins such as fibronectin (Verderio *et al.*, 1999) and fibrillin (Raghunath *et al.*, 1998) acts as a mechanism for both the storage and activation of this fibrogenic cytokine. Such a multifunctional role for the tTgase could explain its proposed importance in a number of diverse physiological and pathological processes, but more specifically those which involve or are followed by a tissue remodelling event including that of embryogenesis, wound healing, atherosclerotic plaque formation and the fibrotic processes occurring in lung, liver and kidney (Nagy *et al.*, 1998: Bowness *et al.*, 1988, 1994; Griffin *et al.*, 1979; Mirza *et al.*, 1997; Johnson *et al.*, 1997, 1999).

References

Adams, J.M. and Cory, S. (1998). The Bcl-2 protein family: arbiters of cell survival. *Science* **281**: 1322–1326.

Aeschlimann, D. and Paulsson, M. (1994) Transglutaminases: protein crosslinking enzymes in tissues and body fluids. *Thromb. Haemostasis* **71**: 402–415.

Aeschlimann, D., Kaupp, O. and Paulsson, M. (1995) Transglutaminase catalyzed matrix cross-linking in differentiating cartilage: identification of osteonectin as a major glutaminyl substrate. *J. Cell Biol.* **129**: 881–892.

Aeschlimann, D., Mosher, M.D. and Paulsson, M. (1996) Tissue transglutaminase and Factor XIII in cartilage and bone remodeling. *Seminars Thrombosis Hemostasis* **22**: 437–443.

Ashkenazi, A. and Dixit, V. (1998) Death receptors: signaling and modulation. *Science* **281**: 1305–1308.

Ballestar, E., Abad, C. and Franco, L. (1996) Corehistones are glutaminyl substrates for tissue transglutaminase. *J. Biol. Chem.* **271**: 18817–18824.

Bossy-Wetzel, E., Newmeyer, D.D. and Green, D.R. (1998) Mitochondrial cytochrome c release in apoptosis occurs upstream of DEVD-specific caspase activation and independently of mitochondrial transmembrane depolarization. *EMBO J.* **17**. 37–49.

Bowness, J.M., Tarr, A.H. and Wong, T. (1988) Increased tissue transglutaminase activity during skin wound healing in rats. *Biochim. Biophys. Acta* **967**: 234–240.

Bowness, J.M., Venditti, M., Tarr, A.H. and Taylor, J.R. (1994) Increase in epsilon (gamma-glutamyl) lysine crosslinks I atherosclerotic aortas. *Atherosclerosis* **11**: 247–253.

Chiocca, E.A., Davies, P.J.A. and Stein, J.P. (1988) The molecular basis of retinoic acid action. *J. Biol. Chem.* **263**: 11584–11589.

Columbano, A., Ledda-Columbano, G.M., Coni, G., Faa, C., Liguori, G., Santacruz, G. and Pani, P. (1985) Occurrence of cell death (apoptosis) during evolution of liver hyperplasia. *Lab. Invest.* **52**: 670.

Dupe, V., Ghyselinck, N.B., Thomazy, V., Nagy, L., Davies, P.J.A., Chambon, P. and Mark, M. (1999) Essential roles of retinoic acid signaling in interdigital apoptosis and control of BMP-7 expression in mouse autopods. *Devel. Biol.* **203**: 30–43.

Eguchi, Y., Shimizu, S. and Tsymimoto, Y. (1997) Intracellular ATP levels determine cell death fate by apoptosis or necrosis. *Cancer Res.* **57**: 1835–1840.

El Alaoui, S., Lawry, J. and Griffin, M. (1997) The cell cycle and induction of apoptosis in a hamster fibrosarcoma ce line treated with anti-cancer drugs: its importance to solid tumour chemotherapy. *J. Neuro-Oncol.* **31**: 195–207.

Feng, J.F., Rhee, S.G. and Im, M.J. (1996) Evidence that phospholipase delta 1 is the effector in the G(h) (transglutaminase II)-mediated signaling. *J. Biol. Chem.* **271**: 16451–16454.

Fesus, L. and Thomazy, V. (1988) Searching for the function of tissue transglutaminase: its possible involvement in the biochemical pathway of programmed cell death. *Adv. Exp. Med. Biol.* **231**: 119–134.

Fesus, L., Thomazy, V. and Falus, A. (1987) Induction and activation of tissue transglutaminase during programmed cell death. *FEBS Lett.* **224**: 104–108.

Fesus, L., Thomazy, V., Autuori, F., Ceru, M. P., Tarcsa, E. and Piacentini, M. (1989) Apoptotic hepatocytes become insoluble in detergents and chaotrophic agents as a result of transglutaminase action. *FEBS Lett.* **245**: 150–154.

Furura, Y., Lundmo, P., Short, A.D., Gill, D.L. and Isaacs, J. (1994) The role of calcium, pH and cell proliferation in the programmed (apoptotic) death of androgen-independent prostatic cancer cells induced by thapsigargin. *Cancer Res.* **54**: 6167–6175.

Gaudry, C., Verderio, E., Aeschlimann, D., Smith, C., Cox, A. and Griffin, M. (1999) Cell surface localization of tissue transglutaminase is dependent on fibronectin binding site in its N-terminal β-sandwich domain. *J. Biol. Chem.* **274**: 30707–30714.

Gentile, V., Thomazy, V., Piacentini, M., Fesus, L. and Davies, P.J.A. (1992) Expression of tissue transglutaminase in Balb-C 3T3 fibroblasts: effects on cellular morphology and adhesion. *J. Biol. Chem.* **199**: 463–474.

Greenberg, C.S, Birckbichler, P.J. and Rice, R.H. (1991) Transglutaminases: multifunctional cross-linking enzymes that stabilize tissues. *FASEB J.* **5**: 3071–3077.

Griffin, M., Smith, L.L. and Wynne, J. (1979) Changes in transglutaminase activity in an experimental model of pulmonary fibrosis induced by paraquat. *Br. J. Exp. Path.* **60**: 653–661.

Hand, D., Bungay, P.J., Elliott, B.M. and Griffin, M. (1985) Activation of transglutaminase at calcium levels consistent with a role for this enzyme as a calcium receptor protein. *Biosci. Rep.* **5**: 1079–1086.

Hinman, L.E., Beilman, G.J., Groehler, K.E. and Sammak, P.J. (1997) Wound-induced calcium waves in alveolar type II cells. *Am. J. Physiol.* **273**: (Lung Cell. Mol. Physiol. 17) L1242–L1248.

Johnson, T.S., Griffin, M., Thomas, G.L., Skill, J., Cox, A., Yang, B., Nicholas, B., Birckbichler, P.J., Muchaneta-Kubara, C. and El Nahas, A.M. (1997) The role of transglutaminase in the rat subtotal nephrectomy model of renal fibrosis. *J. Clin. Inv.* **99**: 2950–2960.

Johnson, T.S., Scholfield, C.I., Parry, J. and Griffin, M. (1998) Induction of tissue transglutaminase by dexamethasone: its correlation to receptor number and transglutaminase-mediated cell death in a series of malignant hamster fibrosarcomas. *Biochem. J.* **331**: 105–112.

Johnson, T.S., Skill, N.J., El Nahas, A.M., Oldroyd, S.D., Thomas, G.L., Douthwaite, J.A., Haylor, J.L. and Griffin, M. (1999) Transglutaminase transcription and antigen translocation in experimental renal scarring. *J. Am. Soc. Neph.* **10** (10): 2146–2157.

Jones, R., Nicholas, B., Mian, S., Davies, P.J.A. and Griffin, M. (1997) Reduced expression of tissue transglutaminase in a human endothelial cell line leads to changes in cell spreading, cell adhesion and reduced polymerization of fibronectin. *J. Cell. Sci.* **110**: 2461–2472.

Kinloch, R.A., Treherne, J.M., Furness, L.M. and Hayimohamedreza, I. (1999) The pharmacolgy of apoptosis. *Trends Pharmacol. Sci.* **20**: 35–42.

Knight, C.R., Rees, R.C. and Griffin, M. (1991) Apoptosis: a potential role for cytosolic transglutaminase and its importance in tumour progression. *Biochim. Biophys. Acta.* **1096**: 312–318.

Knight, C.R., Rees, R.C., Platts, A., Johnson, T. and Griffin, M. (1993) Interleukin-2 activated human effector lymphocytes mediate cytotoxicity by inducing apoptosis in human leukaemia and solid tumour target cells. *Immunol.* **79**: 535–541.

Lai, T-S., Bielawska, A., Peoples, K.A., Hannun, Y.A. and Greenberg, C.S. (1997) Sphingosylphosphocholine reduces the calcium ion requirement for activating tissue transglutaminase. *J. Biol. Chem.* **26**: 16295–16300.

Lai, T-S., Slaughter, T.F., Peoples, K.A., Hettasch, J.M. and Greenberg, C.S. (1998). Regulation of human tissue transglutaminase function by magnesium-nucleotide complexes. *J. Biol. Chem.* **273**: 1776–1781.

Lam, M., Dubyak, G., Chen, L., Nunez, G., Miesfeld, R.L. and Distelhorst, C.W. (1994) Evidence that bcl-2 represses apoptosis by regulating endoplasmic reticulum associated Ca fluxes. *Proc. Natl Acad. Sci. USA* **91**: 6569.

Lesort, M., Attanavanich, K., Zhang, J. and Johnson, G.V.W. (1998) Distinct nuclear local-ization and activity of tissue transglutaminase. *J. Biol. Chem.* **273**: 11991–11994.

Lorand, L. (1996) Neurodegenerative diseases and transglutaminase. *Proc. Natl Acad. Sci. USA* **93**: 14310–14313.

Lorand, L. and Conrad, S.M. (1984) Transglutaminases. *Mol. Cell. Biochem.* **58**: 9–35.

Lu, S. and Davies, P.J.A. (1997) Regulation of the expression of the tissue transglutaminase gene by DNA methylation. *Proc. Natl Acad. Sci. USA* **94**: 4692–4697.

Lu, S., Saydak, M., Gentile, V., Steins, J.P. and Davies, P.J.A. (1995) Isolation and characteri-zation of the human tissue transglutaminase gene promoter. *J. Biol. Chem.* **270**: 9748–9756.

McConkey, D.J. and Orrenius, S. (1996) The role of calcium in the regulation of apoptosis. *J. Leukocyte Biol.* **59**: 775–783.

McConkey, D.J., Hartzell, P., Amador-Perez, J.F., Orrenius, S. and Jondal, M. (1989) Calcium-dependent killing of immature thymocytes by stimulation via the CD3/T cell receptor complex. *J. Immunol.* **143**: 1801–1806.

Martin, S.J. and Green, D.R. eds. (1994). *Techniques in Apoptosis – A Users Guide*, Portland Press, London.

Martin, S.J., Bradley, J.G. and Cotter, T.G. (1990) HL-60 cells induced to differentiate towards neutrophils subsequently die via apoptosis. *Clin. Exp. Immunol.* **79**: 448–453.

Melino, G. and Piacentini, M. (1998) 'Tissue' transglutaminase in cell death: a downstream or a multifunctional upstream effector? *FEBS Lett.* **430**: 59–63.

Melino, G., Annichiarico-Petruzzelli, M., Pireda, L., Candi, E., Gentile, V., Davies, P.J. and Piacentini, M. (1994) Tissue transglutaminase and apoptosis: sense and antisense trans-fection studies with human neuroblastoma cells. *Mol. Cell. Biol.* **14**: 6584–6596.

Mirza, A., Liu, S.L., Frizell, E., Zhu, J., Maddukuri, S., Martinez, J., Davies, P., Schwarting, R., Norton, P. and Zern, M.A. (1997) A role for tissue transglutaminase in hepatic injury and fibrogenesis, and its regulation by NF-Kβ. *Am. J. Physiol.* **272**: 6281–6288.

Nagy, L., Thomazy, V.A., Shipley, G.L., Fesus, L., Lamph, W., Heyman, R.A., Chandraratna, R.A. and Davies, P.J. (1995) Activation of retinoid X receptors induces apoptosis in HL-60 cell line. *Mol. Cell Biol.* **15**: 3540–3551.

Nagy, L., Saydak, M., Shipley, N., Lu, S., Basilion, J.P., Yan, Z.H., Syka, P., Chandraratna, R.A.S., Stein, J.P., Heyman, R.A. and Davies, P.J.A. (1996) Identification and characteri-zation of a versatile retinoid response element (retinoic acid receptor response element-retinoid X receptor response element) in the mouse tissue transglutaminase gene promoter. *J. Biol. Chem.* **271**: 4355–4365.

Nagy, L., Thomazy, V.A., Saydak, M.M., Stein, J. and Davies, P.J.A. (1997) The promoter of the mouse tissue transglutaminase gene directs tissue-specific, retinoid-regulated and apoptosis-linked expression. *Cell Death Different.* **4**: 534–547.

Nagy, L., Thomazy, V.A., Heyman, R.A. and Davies, P.J.A. (1998) Retinoid-induced apoptosis in normal and neoplastic tissue. *Cell Death Different.* **5**: 11–19.

Nakaoka, H., Perez, D.M., Baek, K.J., Das, T., Husain, A., Misono, K., Im, M.J. and Graham, R.M. (1994) Gh: a GTP-binding protein with transglutaminase activity and receptor signalling function. *Science* **264**: 1593–1596.

Nemes, Z., Friis, R.R., Aeschlimann, D., Saurer, S., Paulsson, M. and Fesus, L. (1996) Expression and activation of tissue transglutaminase in apoptotic cells of involuting rodent mammary tissue. *Eur. J. Cell Biol.* **70**: 125–133.

Nemes, Z., Adony, R., Balazs, M., Boross, P. and Fesus, L. (1997) Identification of cytoplasmic actin as an abundant glutaminyl substrate for tissue transglutaminase in HL60 and VG37 cells undergoing apoptosis. *J. Biol. Chem.* **272**: 20577–20583.

Nicotera, P. and Orrenius, S. (1998) The role of calcium in apoptosis. *Cell Calcium* **23**: 173–180.

Nunes, I., Gleizes, P.E., Metz, C.N. and Rifkin, D.B. (1997) Latent transforming growth factor-β binding protein domains involved in activation and transglutaminase dependent crosslinking of latent transforming growth factor-β. *J. Cell Biol.* **136**: 1151–1163.

Nurminskaya, M., Magee, C., Nurminsky, D. and Linsenmayer, T.F. (1998) Plasma transglutaminase in hypertrophic chondrocytes: expression and cell-specific intracellular activation produce cell death and externalization. *J. Cell Biol*. 142: 1135–1144.

Piacentini, M., Davies, P.J.A. and Fesus, L. (1994) The molecular basis of apoptosis in disease. In: *Apoptosis II* (eds L.D. Tomei and F.O. Cope) *Curr. Comm. Cell. Mol. Biol*. 8: Cold Spring Harbor Laboratory Press, Cold Spring Harbor, NY, pp. 143–164.

Piredda, L., Amendola, A., Colizzi, V., Davies, P.J.A., Farrac, M.G., Fraziano, M., Gentile, V., Uray, I., Piacentini, M. and Fesus, L. (1997) Lack of 'tissue' transglutaminase protein cross-linking leads to leakage of macromolecules from dying cells: relationship to development of autoimmunity in MRLIpr/Ipr mice. *Cell Death Different*. 4: 463–472.

Piredda, L., Farrace, M.G., Lo Bello, M., Malorni, W., Melino, G., Petruzzelli, R. and Piacentini, M. (1999) Identification of 'tissue' transglutaminase binding proteins in neuronal cells committed to apoptosis. *FASEB J*. 13: 355–364.

Raghunath, M., Unsold, C., Kubitscheck, V., Bruckner-Tudeman, L., Peters, R. and Meuli, M. (1998) The cutaneous microfibrillar apparatus contains latent transforming growth factor-β binding protein-1 (LTBP-1) and is a repository for latent TGF-β$_1$. *J. Invest. Dermatol*. 111: 559–564.

Ritter, S. and Davies, P.J.A. (1998) Identification of a transforming growth factor-β1/bone morphogenic protein 4 (TGFβ1/BMP4) response element within the mouse tissue transglutaminase gene promoter. *J. Biol. Chem*. 273: 12798–12806.

Rovere, P., Clementi, E., Ferrarini, M., Heltai, S., Sciorati, C., Sabbadini, MG, Rugarli, C. and Manfredi, AA. (1996) CD95 engagement releases calcium from intracellular stores of long term activated, apoptosis-prone γδ T cells. *J. Immunol*. 156: 4631–4637.

Schittny, J.C., Paulsson, M., Vallan, C., Burri, P.H., Kedei, N. and Aeschlimann, D. (1997) Protein crosslinking mediated by tissue transglutaminase correlates with the maturation of extracellular matrices during lung development. *Am. J. Cell. Molec. Biol*. 17: 334–343.

Smethurst, P.A. and Griffin, M. (1996) Measurement of tissue transglutaminase activity in a permeabilized cell system: its regulation by Ca and nucleotides. *Biochem. J*. 313: 803–808.

Suto, N., Ikura, K. and Sasaki, R. (1993a) Expression induced by interleukin-6 of tissue transglutaminase in human hepatoblastoma HepG2 cells. *J. Biol. Chem*. 268: 7476–7473.

Suto, N., Ikura, K., Shinagawa, R. and Sasaki, R. (1993b) Identification of promoter region of guinea pig liver transglutaminase gene. *Biochim. Biophys. Acta*. 1172: 319–322.

Szende, B., Schally, A.V. and Lapis, K. (1991) Immunocytochemical demonstration of tissue transglutaminase indicative of programmed cell death in hormone sensitive mammary tumours. *Acta Morphol. Hung*. 39: 53–58.

Szondy, Z., Molnar, P., Nemes, Z., Boyiadzis, M., Kedi, N., Toth, R. and Fesus, L. (1997) Differential expression of tissue transglutaminase during *in vivo* apoptosis of thymocytes induced via distinctic signalling pathways. *FEBS Lett*. 404: 307–313.

Thomazy, V.A. and Davies, P. (1999) Expression of tissue transglutaminase in the developing chicken limb is associated both with apoptosis and endochondral ossification. *Cell Death Different*. 6: 146–154.

Thornberry, N.A. and Lazebnik, Y. (1998) Caspases: enemies within. *Science* 281: 1312–1316.

Tsukamoto, A. and Kaneko, Y. (1993) Thapsigargin, a Ca ATPase inhibitor depletes the intracellular Ca^{2+} pool and induces apoptosis in human hepatoma cells. *Cell Biol. Int*. 17: 969.

Upchurch, H.F., Conway, E., Patterson, M.K. Jr. and Maxwell, M.D. (1991). Localisation of cellular band tissue transglutaminase on the extra cellular matrix offer wounding characteristics of the matrix bound enzyme. *J. Cell. Physiol*. 149: 375–383.

Verderio, E., Nicholas, B., Gross, S. and Griffin, M. (1998) Regulated expression of tissue transglutaminase in Swiss 3GT3 fibroblasts: effects on the processing of fibronectin, cell attachment and cell death. *Exp. Cell Res*. 239: 119–138.

Verderio, E., Gaudry, C., Gross, S., Smith, C., Downes, S. and Griffin, M. (1999) Regulation of cell-surface tissue transglutaminase: effects on matrix storage of latent transforming growth factor-β binding protein-1. *J. Histochem. Cytochem*. 47: 1417–1432.

Zhang, L.X., Millis, K.J., Dawson, M.I., Collins, S.J. and Jetten, A.M. (1995) Evidence for the involvement of retinoic acid receptor RAR alpha-dependent signaling pathway in the induction of tissue transglutaminase and apoptosis by retinoids. *J. Biol. Chem.* **270**: 6022–6029.

Zhang, J., Lesort, M., Guttmann, R.P. and Johnson, G.V.W. (1998) Modulation of the *in situ* activity of tissue transglutaminase by calcium and GTP. *J. Biol. Chem.* **273**: 2288–2295.

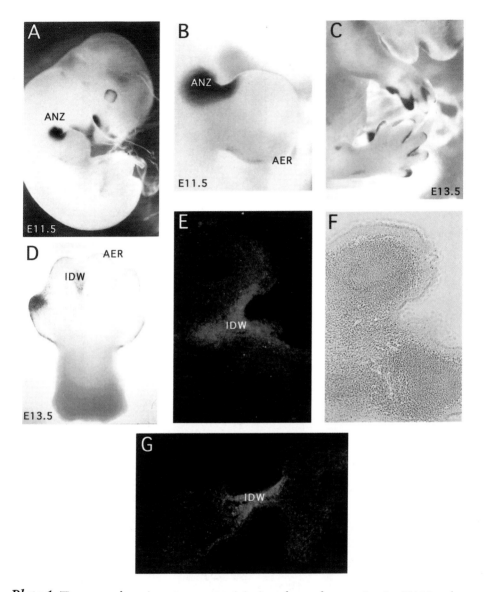

Plate 1. *Tissue transglutaminase promoter activity in embryos of transgenic mice (E11.5 and E13.5 line #26 mice) carrying the 3.8 kb fragment of the mouse tissue transglutaminase promoter fused to the reporter gene β-galactosidase. **A.** β-Galactosidase confined to interior border of forelimb, apical ectodermal ridge, facial processes and cranial mesenchyme at E11.5. **B.** Close-up forelimb of E11.5 embryo, showing expression of the transgene in the anterior necrotic zone (ANZ), apical ectodermal ridge (AER). **C.** Transgene expression at E13.5 embryo is localized to the distal parts of the limbs where it is concentrated over the tips of the fingers and the interdigital areas. **D.** Close up of hindlimb of a E13.5 embryo, expression of the transgene specifically detects areas of programmed cell death in the interdigital webs (IDW) and along the edges of the footplate, the remnants of AER at the tips of the fingers also express the transgene. **E.** and **G.** Detection by dark field microscopy of β-galactosidase expression shows reaction product uniformly distributed in the interdigital mesenchyme (IDW) of E13.5 limbs (E, section in the antero-posterior plane; G, transverse section (dorso-ventral plane). **F.** Same field as **E** in transmitted light to demonstrate the topography. Reprinted from Nagy et al. (1997) Cell Death Different., 4: 534–547 with permission.*

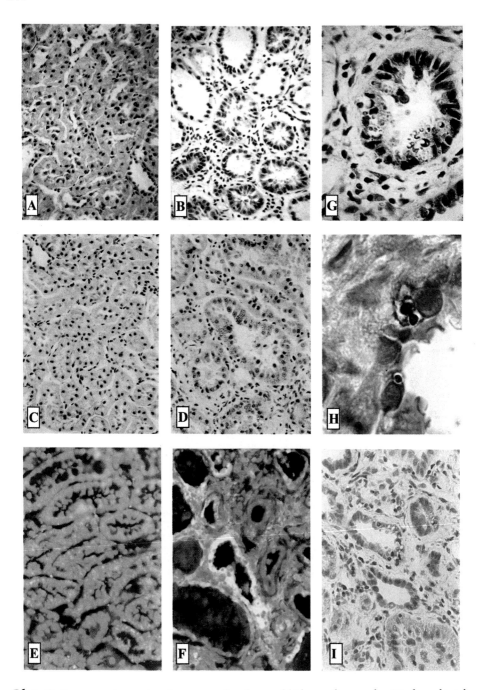

Plate 2. Representative sections from normal and scarred kidneys of rats at day 90 after subtotal nephrectomy. A, B and G are stained for tissue transglutaminase, C and D are stained for ε(γ-glutamyl) lysine cross-link using immunohistochemistry. E and F are stained for ε(γ-glutamyl) lysine using immunofluorescence. A, C and E are sections from control kidneys, while B, D and F are from scarred kidneys. H and I are sections from a scarred kidney that has been in situ end-labelled with a digoxigenin-labelled nucleotide (apoptosis). Reprinted from Johnson et al. (1997) J. Clin. Invest. **99**(12): 2950–2960 with permission.

18

Pathways used by adenovirus E1B 19K to inhibit apoptosis

Kurt Degenhardt, Denise Perez and Eileen White

1. Introduction

Adenovirus relies on its ability to deceive the host cell into providing nearly all the macromolecules necessary for its own replication; thus, the interplay between host and virus leads to an impressive array of actions and reactions. For example, during infection the virus triggers the host cell into cell cycle progression so that the virus may usurp the cellular replication machinery to replicate viral DNA. Upon detection of the de-regulated cell cycle, the host cell initiates apoptosis as a potent host defence against viral infection. It is easy to imagine that the death of an infected cell would act to thwart viral replication and further viral infection. However, the virus has developed anti-apoptotic proteins to counteract host cell apoptosis. These viral proteins rely on the ability to interact with and disable cellular apoptotic machinery (White and Gooding, 1994; White *et al.*, 1994; White, 1998). Thus, the study of the viral anti-apoptotic mechanisms also uncovers insights into cellular apoptotic controls.

Cellular apoptosis in response to viral infection can be initiated and regulated by both intracellular triggers (p53) as well as extracellular signals (TNF-α or Fas ligand; White and Gooding, 1994; White *et al.*,1994; White, 1998). First, adenovirus directly induces p53 activation (Debbas and White, 1993; Lowe and Ruley, 1993; Sabbatini *et al.*, 1995). For example, the adenovirus E1A gene product functions to up-regulate cell division creating an environment conducive to viral replication. Left unabated, the cellular response to E1A-mediated cellular proliferation is the induction of apoptosis through p53-dependent mechanisms (Debbas and White, 1993; White and Gooding, 1994; White *et al.*, 1994; White, 1998). Secondly, the TNF-R1 pathway, which uses similar intracellular components as the Fas receptor to initiate death, is involved in host immune surveillance against virus infection. Virally infected cells are cleared as the immune system initiates apoptosis of infected cells through TNF-R1 receptor mediated death pathway (for review see Nagata, 1997). Adenovirus, in order to

Programmed Cell Death in Animals and Plants, edited by J.A. Bryant, S.G. Hughes and J.M. Garland.
© 2000 BIOS Scientific Publishers Ltd, Oxford.

succeed in replication, has adapted mechanisms to disable these apoptotic pathways through the expression of anti-apoptotic genes.

Viruses have developed anti-apoptotic activities; for example, the poxvirus crm A protein (Ray et al., 1992; Gagliardini et al., 1994; Komiyama et al., 1994) or the baculovirus p35 protein (Clem et al.,1991; Bump et al., 1995) directly inhibit a class of proteases, caspases, that are involved in the execution of multiple death pathways. Arguably, the most common anti-apoptotic activity identified thus far is the utilization of Bcl-2 like proteins. These have been identified in adenovirus, African swine fever virus, and several herpesviruses (for review see White, 1996).

The adenovirus E1B 19K gene product is a Bcl-2 homologue. Not only does E1B 19K share homology with anti-apoptotic members of the Bcl-2 family in important functional domains (Chiou et al., 1994; Han et al., 1996a), but host cells infected by mutant adenovirus lacking E1B 19K, undergo accelerated cell death with impaired viral replication (White et al., 1984; White and Stillman, 1987; Chiou and White, 1998). Therefore, it seemed likely that E1B 19K may disable the apoptotic response by interacting with the host cell regulatory machinery. It became clear that the study of E1B 19K could lead to insights into the cellular mechanisms of apoptosis regulation.

Viruses in general, and adenovirus in particular, have small genomes but still need to perform a complexity of tasks in order to replicate successfully. To accomplish this many viral proteins have multiple functions. E1B 19K is no exception and seems to block apoptosis by several mechanisms. Cells expressing E1B 19K survived with activated p53 but undergo growth arrest. E1B 19K blocks p53-induced Bax-mediated death, but not p21/Waf1-dependent cell cycle arrest (Debbas and White, 1993; Sabbatini et al., 1997). E1B 19K must block apoptosis via another mechanism other than at Bax since E1B 19K blocks Fas-induced apoptosis (Gooding et al., 1991; Hashimoto et al., 1991; White et al., 1992) and Bax does not play a pivotal role in this death signalling cascade. In some situations E1B 19K has also been shown to alter transcriptional activity but the mechanism largely remains unclear (Debbas and White, 1993; Schmitz et al., 1996; Kasof et al., 1998). Taken together, it seems that this small 19 kDa protein provides multiple functional activities for adenovirus.

2. E1B 19K blocks p53-mediated apoptosis

The initial studies on E1B 19K were designed to investigate why both E1A and E1B 19K were necessary for cellular transformation. Genetic dissection of adenovirus early genes showed that, when expressed in mammalian cells, E1A alone can induce cellular proliferation but not transformation (Houweling et al., 1980). In transformation assays, the expression of E1A leads to the formation of cell colonies that appeared transformed but would regress and eventually die (White et al., 1992). This occurs because E1A increases proliferation by de-regulating the cell cycle through the Rb (retinoblastoma protein) pathway. However, this also results in apoptosis because of the activation of p53 (White, 1994), since expression of a dominant negative p53 with E1A prevents apoptosis (Debbas and White, 1993). The co-expression of E1B 19K prevents the regression that leads to the demise of E1A expressing colonies and results in very efficiently transformed cells (Rao et al., 1992; White et al., 1992). Thus, E1A expression induces p53-mediated apoptosis that is blocked by E1B 19K.

2.1 *E1B 19K binds to Bax and prevents apoptosis*

To investigate the mechanism by which E1B 19K inhibits E1A-induced apoptosis a two-hybrid screen using E1B 19K as the bait was performed. A conserved fragment of Bax, the BH3 domain, bound to E1B 19K. Bax was identified as an E1B 19K binding protein and this interaction was verified both *in vitro* and *in vivo* (Han *et al.*, 1996a). The identification of Bax as an E1B 19K binding protein is interesting since Bax expression is regulated by p53 (Miyashita *et al.*, 1994; Miyashita and Reed, 1995). This suggests a mechanism for the inhibition of p53-mediated apoptosis by E1B 19K (Han *et al.*, 1996a). That is, E1A would activate p53, which, in turn upregulates Bax expression which is blocked by E1B 19K (*see Figure 1*).

E1B 19K does not have an obvious catalytic activity so how does the interaction with Bax abrogate apoptosis? One hypothesis is that Bax alone induces apoptosis while Bax bound to E1B 19K, Bcl-2 or another anti-apoptotic family does not induce apoptosis (Oltvai *et al.*, 1993; Yin *et al.*, 1994; Han *et al.*, 1996a; Zha *et al.*, 1996; Han *et al.*, 1998a). This model suggests that the function of the Bax is dependent on its binding partner.

A second model suggests that Bax location regulates Bax function. Bax located at the mitochondrial membrane induces alterations in membrane potential and efflux of

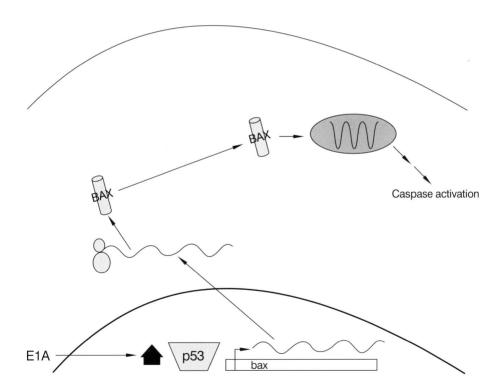

Figure 1. *A schematic representation of E1A-induced apoptosis. Cellular expression of E1A results in the accumulation of p53. Increased levels of p53 promote transcription of many genes including Bax. Bax protein translocates to where it induces the release of cytochrome c, an activator of APAF-1 and ultimately caspase 9. Caspase activation results in a cascade of events that execute the apoptotic programme.*

cytochrome C into the cytosol. This is an important event in apoptosis since cytochrome C in the cytosol interacts with Apaf-1 and this complex activates caspase-9 (Li *et al.*, 1997; Zou *et al.*,1997; see also Chapters 1, 2 and 7). Active caspase 9 is then capable of initiating a caspase cascade that ultimately ends in the death of the cell (Li *et al.*,1997). To avert the activation of the caspase cascade, E1B 19K must prevent Bax-induced cytochrome c release. It seems that E1B 19K can accomplish this through a direct interaction with Bax (Han *et al.*, 1996a; Han *et al.*,1998a). E1B 19K, although not actually a mitochondrial protein, binds to Bax. Indirect immunofluorescence shows that, in the presence of E1B 19K, the majority of Bax will not localize to the mitochondria. Furthermore, the small amount of Bax which does get into the mitochondria also co-localizes with E1B 19K (Han *et al.*, 1996a). A BH3 mutant of Bax was generated which efficiently initiates apoptosis but can only interact with Bcl-2 and not with E1B19K (Han *et al.*, 1998a). Apoptosis initiated by this mutant Bax is blocked by Bcl-2 but not E1B 19K. Taken together these results suggest that E1B 19K inhibits Bax by binding to it and sequestering Bax to a localization where it is non-functional (Han *et al.*, 1996a; Han *et al.*, 1998a). However, some Bax will enter the mitochondria in the presence of E1B 19K. This Bax does not induce apoptosis, suggesting that E1B 19K also inhibits Bax in the mitochondria. Thus, E1B 19K both alters both the localization and function of Bax, through a direct interaction. That is, to inhibit apoptosis, it is essential for E1B 19K to bind to Bax; this prevents Bax from altering the mitochondrial membrane potential so that cytochrome c remains contained within the mitochondria, efficiently inhibiting apoptosis (*see Figure 3*).

3. *The Fas and TNF-R1 pathways are blocked by E1B 19K*

Fas and TNF-R1 are immunologically important cell surface death receptors which have a common mechanism to mediate apoptosis (Nagata, 1997). Upon ligand/receptor interaction, a trimeric receptor complex is formed (Nagata, 1997). The formation of this trimeric complex recruits FADD/caspase 8 (FLICE), either directly, as with Fas (Kischkel *et al.*, 1995), or through TRADD, as in the case of the TNF-R1 pathway (Hsu *et al.*, 1995; Hsu *et al.*, 1996). The association of death receptors with FADD and caspase 8 results in the activation of caspase 8 (Muzio *et al.*, 1996). Active caspase 8 sets into motion the activation of effector caspases and the subsequent death of the cell (*Figure 2*; Nagata, 1997).

Supersensitivity to TNF-α-mediated apoptosis is conferred by expression of the adenoviral protein E1A (Chen *et al.*, 1987; Duerksen-Hughes *et al.*, 1989; Ames *et al.*, 1990). This finding suggests that E1A and TNF-α function on converging pathways or act synergistically to promote cell death. Since E1B 19K blocks apoptosis induced by E1A, the effects of E1B 19K on death receptor-mediated cell death were assessed. In cells that are susceptible to death receptor-mediated apoptosis, the expression of E1B 19K blocks apoptosis triggered by both TNF-R1 and Fas (Gooding *et al.*, 1991; Hashimoto *et al.*, 1991; White *et al.*, 1992).

Since the TNF-R1 and Fas pathways are both blocked by E1B 19K and they share many common components, then it is reasonable to propose that E1B 19K may block both pathways at the same point. To investigate this inhibition of receptor mediated apoptosis, components of the TNF-R1 and Fas pathways were examined to identify the point of E1B 19K inhibition. E1B 19K expression during viral infection did not effect the TNF-α receptors at the cell surface (White *et al.*, 1992).

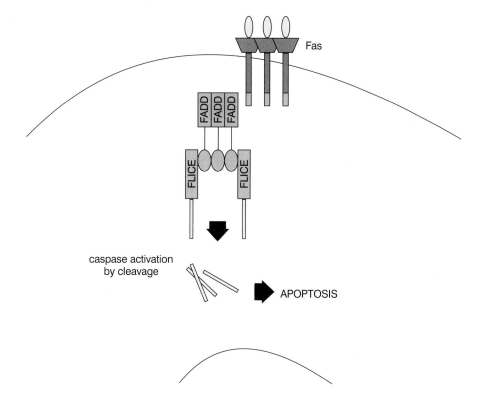

Figure 2. *A schematic representation of Fas-mediated apoptosis. Fas/Fas ligand interaction promotes receptor trimerization. Trimerized receptors recruit heterodimers of FADD-caspase 8 (FLICE) to form a death inducing signalling complex (DISC). DISC formation facilitates caspase activation by proteolytic cleavage of the caspase pro-domain. The activated caspase then initiates apoptosis.*

Transient expression of caspase 8 in HeLa cells results in the induction of apoptosis. Although the co-expression of caspase 8 with CrmA, a caspase inhibitor, blocked apoptosis, the co-expression of E1B 19K was not able to block caspase 8-mediated apoptosis (Perez and White, 1998). These results suggest that E1B 19K acts upstream of caspase 8, blocking the activation of caspase-8 but not the activity of activated caspase 8.

Both the TNF-R1 and Fas pathways use FADD in the activation of caspase 8, suggesting that FADD may be a target for E1B 19K. Transient expression of FADD in HeLa cells induces apoptosis. The viability of FADD expressing cells is significantly increased with the co-expression of E1B 19K (Perez and White, 1998). This observation indicates that E1B 19K blocks receptor-mediated apoptosis at FADD.

3.1 *E1B 19K – FADD and prevents caspase 8 activation*

A clue to E1B 19K inhibition of apoptosis at FADD was derived from the mechanism by which E1B 19K blocks apoptosis at Bax. E1B 19K directly binds to and sequesters Bax, thereby preventing Bax from entering the mitochondria (Han *et al.*, 1996a; Han *et al.*,

1998a). Since E1B 19K binds to and inhibits this death inducing molecule it was reasonable to predict that E1B 19K may be acting similarly with FADD. If this is the case, then E1B 19K must be located in the same cellular compartment as FADD. Indirect immunofluorescence of E1B 19K and FADD showed that the proteins are in the same region of the cell; co-immunoprecipitation experiments of exogenously expressed proteins demonstrate that they are in a complex *in vivo* (Perez and White, 1998).

How does the association of E1B 19K with FADD prevent caspase 8 activation? FADD expressed alone forms filaments in the cell when over-expressed; these filaments are disrupted in cells that also over-express E1B 19K (Perez and White, 1998). The filament form of FADD may result in caspase 8 being in functional proximity to auto(trans)activate other caspase 8 molecules in the complex. E1B 19K prevents this FADD complex from forming (*Figure 3*) and thus prevents caspase 8 from being in functional proximity to auto(trans)activate. Examination of the activation state of caspase 8 shows that it is activated in the presence of FADD, but not if E1B 19K is disrupting FADD multimerization (Perez and White, 1998). Analogously, E1B 19K may be preventing the recruitment of FADD and caspase 8 to the DISC (*see Figure 3*).

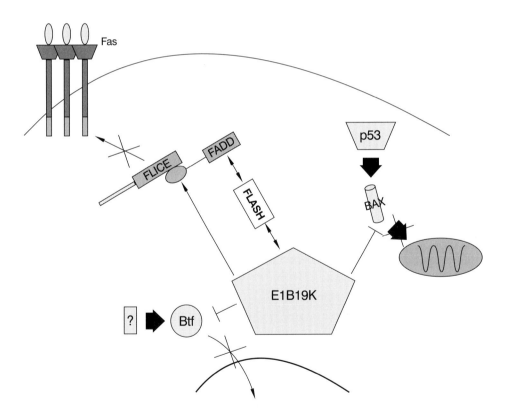

Figure 3. *Inhibition of apoptosis by E1B 19K. Through a series of protein–protein interactions, E1B 19K can block Fas-induced, p53-mediated, and Btf elicited apoptosis. Fas-induced apoptosis is blocked by sequestering the FADD–caspase 8 (FLICE) heterodimer, either directly or through FLASH, from the activated receptor. E1B 19K inhibits p53 mediated cell death by preventing Bax from entering the mitochondria. E1B 19K blocks Btf elicited apoptosis by inhibiting Btf from entering the nucleus. X over arrows represent translocations inhibited by E1B 19K.*

4. E1B 19K inhibits the caspase adapter molecules CED-4 and Apaf-1

Caspases with large pro-domains (caspase 8, caspase 9 and CED-3) interact with adapter proteins that are requisite for caspase activation. As mentioned earlier, the adapter protein FADD binds to both caspase 8 and activated Fas receptor forming a molecular bridge between the two proteins (Kischkel *et al.*, 1995). Analogously, Apaf-1 acts as an adapter for caspase 9 in response to Bax-facilitated release of cytochrome c (Li *et al.*, 1997; Zou *et al.*, 1997). Similarly, the *C. elegans* CED-4 protein, which is a structural and functional homologue of Apaf-1 (Chapter 1 and Zou *et al.*, 1997), acts as an adapter to regulate the activation of the caspase CED-3 (Chapter 1 and Seshagiri and Miller, 1997). These three divergent pathways employ an analogous mechanism for caspase activation.

E1B 19K suppresses the ability of FADD, CED-4 (in the presence of caspase 8) and Apaf-1 to induce apoptosis when exogenously expressed in mammalian cells (Han *et al.*, 1998b; Perez and White, 1998). Studies in *C. elegans* show that the protein CED-9, a Bcl-2 family member, blocks CED-3 activation by binding to and inhibiting CED-4 (Spector *et al.*, 1997; Wu *et al.*, 1997; see also Chapter 1), similar to E1B 19K inhibition of FADD (Perez and White, 1998). E1B 19K was examined for its ability to interact with CED-4. Not surprisingly, E1B 19K bound to CED-4 both *in vivo* and in the yeast two-hybrid assay (Han *et al.*, 1998b). Deletion analysis of CED-4 in the yeast two-hybrid assay identified two regions that bound to E1B 19K, namely the CARD domain and amino acids 87–327 (Han *et al.*, 1998b). The amino acids 87–327 are in a region of CED-4 that is homologous to Apaf-1 (Han *et al.*, 1998b). This region of CED-4 is known as the DED (death effector domain) whereas the region in Apaf-1 is called a CARD (caspase recruitment domain). The DED region is a conserved domain present in several caspase adapter molecules, including FADD (Chinnaiyan *et al.*, 1995; Boldin *et al.*, 1996). The CARD domain is structurally similar to the DED and is also found in molecules that associate with caspases (Aravind *et al.*, 1999). Because E1B 19K inhibits other caspase adapter molecules such as FADD, our laboratory launched studies which showed that Apaf-1 activation of caspase 9 is also prevented by E1B 19K (Y. Shen and E. White, unpublished observations). Furthermore, initial studies suggest that E1B 19K binds to the DED/CARD of Apaf-1 and FADD in addition to the same domain in CED-4. These studies taken together seem to indicate a general mechanism, that is, inhibition of caspase activation occurs through an interaction of E1B 19K with the conserved DED/CARD regions of caspase adapter molecules.

5. E1B 19K interacts with Btf, a novel transcriptional repressor

Our studies on E1B 19K identified a novel gene, *Btf*, that encodes a transcriptional repressor which binds to E1B 19K (Kasof *et al.*, 1998). The Btf protein contains a putative bZIP motif as well as a region homologous to the Myb DNA-binding domain. Both of these motifs are found in many DNA-binding proteins (Kasof *et al.*, 1998). In fact, Btf bound to DNA and repressed transactivation in transcriptional activation assays, showing that Btf was indeed a transcriptional repressor (Kasof *et al.*, 1998). The nuclear localization of Btf is also consistent with Btf functioning as a transcriptional repressor.

Functional analysis revealed that Btf expression results in cell death, presumably through the repression of an anti-apoptotic mediator (Kasof *et al.*, 1998). Btf-induced apoptosis is blocked by co-expression of E1B 19K (Kasof *et al.*, 1998). The abrogation of Btf-mediated apoptosis by E1B 19K occurs as a result of functional sequestration of Btf, as is the case for Bax, FADD, and CED-4. In the presence of E1B 19K, Btf remains in an extranuclear localization as opposed to its nuclear localization in the absence of E1B 19K (Kasof *et al.*, 1998). Thus, E1B 19K seems to sequester Btf, preventing its localization to the nucleus, thus inhibiting transcriptional repression and the initiation of apoptosis (*Figure 3*).

The interaction of Btf is not limited to E1B 19K since Bcl-2 also interacts with Btf. Similarly, Bcl-2 inhibits Btf-mediated apoptosis (Kasof *et al.*, 1998). This suggests a conserved mechanism, i.e. the mammalian Bcl-2 family proteins (homologues of E1B 19K) may regulate transcription by sequestering proteins such as Btf in the cytoplasm.

6. Future directions

Our understanding of the cellular ramifications of E1B ·19K expression is far from complete. Although it is clear that E1B 19K is a potent inhibitor of apoptosis, the mechanisms of several E1B 19K-target proteins are not understood. For example, the two-hybrid screen that identified Bax also identified NBK/Bik as an E1B 19K-binding protein (Boyd *et al.*, 1995; Han *et al.*, 1996b). NBK/Bik induces apoptosis when over-expressed and this apoptosis is abrogated by the co-expression of E1B 19K (Boyd *et al.*, 1995; Han *et al.*, 1996b). However, the mechanism of NBK/Bik-mediated apoptosis has been elusive as have the cellular pathways regulating NBK/Bik activation. Similar questions remain unanswered for Btf and both these areas are currently under investigation.

However, through studies on E1B 19K a general mechanism is emerging: this small viral protein interacts with and disables cellular components essential for the induction of apoptosis. E1B 19K binds to modular domains present in death machinery molecules conserved though evolution, such as the BH3 domain of Bax and the DED/CARD domain of FADD, CED-4, and Apaf-1. By binding to conserved domains E1B 19K can block a variety of pathways at multiple points increasing the efficiency with which it inhibits apoptosis. E1B 19K is an elegant example of viral conservation, a multi-functional viral protein that performs the tasks of an entire class of mammalian Bcl-2-like proteins.

References

Ames, R.S., Holskin, B., Mitcho, M., Shalloway, D. and Chen, M.J. (1990) Induction of sensitivity to the cytotoxic action of tumor necrosis factor alpha by adenovirus E1A is independent of transformation and transcriptional activation. *J. Virol.* **64**: 4115–4122.

Aravind, L., Dixit, V.M. and Koonin, E.V. (1999) The domains of death: evolution of the apoptosis machinery. *Trends Biochem. Sci.* **24**: 47–53.

Boldin, M.P., Goncharov, T.M., Goltsev, Y.V. and Wallach, D. (1996) Involvement of MACH, a novel MORT1/FADD-interacting protease, in Fas/APO-1- and TNF receptor-induced cell death. *Cell* **85**: 803–815.

Boyd, J.M., Gallo, G.J., Elangovan, B., Houghton, A.B., Malstrom, A.B., Avery, B.J., Ebb, R.G., Subramanian, T., Chittendon, T., Lutz, R.S. and Chinnadurai, G. (1995) Bik1, a novel death-inducing protein shares a distinct sequence motif with Bcl-2 family proteins and interacts with viral and cellular survival-promoting proteins. *Oncogene* **11**: 1921–1928.

Bump, N.J., Hackett, M., Hugunin, M., Seshagiri, S., Brady, K., Chen, B., Ferenz, C., Franklin, S., Ghayur, T., Li, P., Licari, P., Mankovich, J., Shi, L.F., Greenberg, A.H., Miller, L.K. and Wong, W.W. (1995) Inhibition of ICE family proteases by baculovirus anti-apoptotic protein p35. *Science* **269**: 1885–1888.

Chen, M.-J., Holskin, B., Strickler, J., Gorniak, J., Clark, M.A., Johnson, P.J., Mitcho, M. and Shalloway, D. (1987) Induction by E1A oncogene expression of cellular susceptibility to lysis by TNF. *Nature* **330**: 581–583.

Chinnaiyan, A.M., O'Rourke, K., Tewari, M. and Dixit, V.M. (1995). FADD, a novel death domain-containing protein, interacts with the death domain of Fas and initiates apoptosis. *Cell* **81**: 505–512.

Chiou, S.-K. and White, E. (1998) Inhibition of ICE-like proteases inhibits apoptosis and increases virus production during adenovirus infection. *Virology* **244**: 108–118.

Chiou, S.-K., Tseng, C.C. and White, E. (1994) Functional complementation of the adenovirus E1B 19K protein with Bcl-2 in the inhibition of apoptosis in infected cells. *J. Virol.* **68**: 6553–6566.

Clem, R.J., Fechheimer, M. and Miller, L.K. (1991) Prevention of apoptosis by a Baculovirus gene during infection of insect cells. *Science* **254**: 1388–1390.

Debbas, M. and White, E. (1993). Wild-type p53 mediates apoptosis by E1A which is inhibited by E1B. *Genes Dev.* **7**: 546–554.

Duerksen-Hughes, P., Wold, W.S.M. and Gooding, L.R. (1989) Adenovirus E1A renders infected cells sensitive to cytolysis by tumor necrosis factor. *J. Immunol.* **143**: 4193–4200.

Gagliardini, V., Fernandez, P.A., Lee, R.K.K., Drexler, H.C.A., Rotello, R.J., Fishman, M.C. and Yuan, J. (1994) Prevention of vertebrate neuronal death by the *crmA* gene. *Science* **263**: 826–828.

Gooding, L.R., Aquino, L., Duerksen-Hughes, P.J., Day, D., Horton, T.M., Yei, P. and Wold, W.S.M. (1991) The E1B-19K protein of group C adenoviruses prevents cytolysis by tumor necrosis factor of human cells but not mouse cells. *J. Virol.* **65**: 3083–3094.

Han, J., Sabbatini, P., Perez, D., Rao, L., Modha, D. and White, E. (1996a) The E1B 19K protein blocks apoptosis by interacting with and inhibiting the p53-inducible and death-promoting Bax protein. *Genes Dev.* **10**: 461–477.

Han, J., Sabbatini, P. and White, E. (1996b) Induction of apoptosis by human Nbk/Bik, a BH3-containing E1B 19K-interacting protein. *Mol. Cell. Biol.* **16**: 5857–5864.

Han, J., Modha, D. and White, E. (1998a) Interaction of E1B 19K with Bax is required to block Bax-induced loss of mitochondrial membrane potential and apoptosis. *Oncogene* **17**: 2993–3005.

Han, J., Wallen, H.D., Nunez, G. and White, E. (1998b) E1B 19,000 molecular-weight protein interacts with and inhibits CED-4-dependent, FLICE-mediated apoptosis. *Mol. Cell. Biol.* **18**: 6052–6062.

Hashimoto, S., Ishii, A. and Yonehara, S. (1991) The E1B oncogene of adenovirus confers cellular resistance to cytotoxicity of tumor necrosis factor and monoclonal anti-Fas antibody. *Int. Immunol.* **3**: 343–351.

Houweling, A., van den Elsen, P.J. and van der Eb, A.J. (1980) Partial transformation of primary rat cells by the leftmost 4.5% fragment of adenovirus 5 DNA. *Virology* **105**: 537–550.

Hsu, H., Xiong, J. and Goeddel, D.V. (1995) The TNF receptor 1-associated protein TRADD signals cell death and NFκB activation. *Cell* **81**: 495–504.

Hsu, H., Shu, H.-B., Pan, M.G. and Goeddel, D.V. (1996) TRADD-TRAF2 and TRADD-FADD interaction define two distinct TNF receptor 1 signal transduction pathways. *Cell* **84**: 299–308.

Kasof, G.M., Goyal, L. and White, E. (1998) Btf: A novel death-promoting transcriptional repressor that interacts with Bcl-2 related proteins. *Mol. Cell. Biol.* **19**: 4390–4404.

Kischkel, F.C., Hellbardt, S., Behrmann, I., Germer, M., Pawlita, M., Krammer, P.H. and Peter, M.E. (1995) Cytotoxicity-dependent APO-1 (Fas/CD95)-associated proteins form a death-inducing signaling complex (DISC) with the receptor. *EMBO J.* **14**: 5579–5588.

Komiyama, T., Ray, C.A., Pickup, D.J., Howard, A.D., Thornberry, N.A., Peterson, E.P. and Salvesen, G. (1994) Inhibition of interleukin-1b converting enzyme by the cowpox virus serpin CrmA. *J. Biol. Chem.* **269**: 19331–19337.

Li, P., Nijhawan, D., Budihardjo, I., Srinivasula, S.M., Ahmad, M., Alnemri, E.S. and Wang, X.D. (1997) Cytochrome c and dATP-dependent formation of Apaf-1/Caspase-9 complex initiates an apoptotic protease cascade. *Cell* **91**: 479–489.

Lowe, S. and Ruley, H.E. (1993). Stabilization of the p53 tumor suppressor is induced by adenovirus-5 E1A and accompanies apoptosis. *Genes Dev.* **7**: 535–545.

Miyashita, T. and Reed, J.C. (1995) Tumor suppressor p53 is a direct transcriptional activator of the human *bax* gene. *Cell* **80**: 293–299.

Miyashita, T., Krajewski, S., Krajewska, M., Wang, H.G., Lin, H.K., Liebermann, D.A., Hoffman, B. and Reed, J.C. (1994) Tumor suppressor p53 is a regulator of *bcl-2* and *bax* gene expression *in vitro* and *in vivo*. *Oncogene* **9**: 1799–1805.

Muzio, M., Chinnaiyan, A.M., Kischkel, F.C., O'Rourke, K., Sheukenko, A., Ni, J., Scaffidi, C., Bretz, J.D., Zhang, M., Gentz, R., Mann, M., Krammer, P.H., Peter, M.E. and Dixit, V.M. (1996) FLICE, A novel FADD-homologous ICE/CED-3-like protease, is recruited to the CD95 (Fas/APO-1) death-inducing signaling complex. *Cell* **85**: 817–827.

Nagata, S. (1997) Apoptosis by death factor. *Cell* **88**: 355–365.

Oltvai, Z.N., Millman, C.L. and Korsmeyer, S. (1993) Bcl-2 heterodimerizes in vivo with a conserved homolog, Bax, that accelerates programmed cell death. *Cell* **74**: 609–619.

Perez, D. and White, E. (1998) E1B 19K inhibits Fas-mediated apoptosis through FADD-dependent sequestration of FLICE. *J. Cell Biol.* **141**: 1255–1266.

Rao, L., Debbas, M., Sabbatini, P., Hockenbery, D., Korsmeyer, S. and White, E. (1992) The adenovirus E1A proteins induce apoptosis which is inhibited by the E1B 19K and Bcl-2 proteins. *Proc. Natl Acad. Sci. USA* **89**: 7742–7746.

Ray, C.A., Black, R.A., Kronheim, S.R., Greenstreet, T.A., Sleath, P.R., Salvesen, G.S. and Pickup, D.J. (1992) Viral inhibition of inflammation: cowpox virus encodes an inhibitor of the interleukin-1b converting enzyme. *Cell* **69**: 597–604.

Sabbatini, P., Lin, J., Levine, A.J. and White, E. (1995) Essential role for p53-mediated transcription in E1A-induced apoptosis. *Genes Dev.* **9**: 2184–2192.

Sabbatini, P., Han, J.H., Chiou, S.K., Nicholson, D.W. and White, E. (1997) Interleukin 1b converting enzyme-like proteases are essential for p53-mediated transcriptionally dependent apoptosis. *Cell Growth Diff.* **8**: 643–653.

Schmitz, M.L., Indorf, A., Limbourg F.R., Stadtler, H., Traencker, E.B. and Baeurle, P. (1996) The dual effect of adenovirus type 5 E1A 13S protein on NF-kappaB activation is antagonized by E1B 19K. *Mol. Cell. Biol.* **16**: 4052–4063.

Sedlak, T.W., Oltvai, Z.N. Yang, E., Wang, K., Boise, L.H., Thompson, C.B. and Korsmeyer, S.J., 1995. Multiple Bcl-2 family members demonstrate selective dimerizations with Bax. Proc. Natl. Acad. Sci. USA 92 (17): 7834 – 7838.

Seshagiri, S. and Miller, L.K. (1997) *Caenorhabditis elegans* CED-4 stimulates CED-3 processing and CED-3-induced apoptosis. *Curr. Biol.* **7**: 455–460.

Spector, M.S., Desnoyers, S., Hoeppner, D.T. and Hengartner, M.O. (1997) Interaction between the *C. elegans* cell-death regulators CED-9 and CED-4. *Nature* **385**: 653–656.

White, E. (1994) Function of the adenovirus E1B oncogene in infected and transformed cells. *Semin. Virol.* **5**: 341–348.

White, E. (1996) Life, death, and the pursuit of apoptosis. *Genes Dev.* **10**: 1–15.

White, E. (1998) Regulation of apoptosis by adenovirus E1A and E1B oncogenes. *Semin. Virol.* **8**: 505–513.

White, E. and Gooding, L.R. (1994) Regulation of apoptosis by human adenoviruses. In: *Apoptosis: The Molecular Basis for Cell Death II* (ed. F. Cope). Cold Spring Harbor Laboratory Press, Cold Spring Harbor, NY, USA: pp. 111–141.

White, E. and Stillman, B. (1987) Expression of the adenovirus E1B mutant phenotypes is dependent on the host cell and on synthesis of E1A proteins. *J. Virol.* **61**: 426–435.

White, E., Grodzicker, T. and Stillman, B.W. (1984) Mutations in the gene encoding the adenovirus E1B 19K tumor antigen cause degradation of chromosomal DNA. *J. Virol.* **52**: 410–419.

White, E., Sabbatini, P., Debbas, M., Wold, W.S.M., Kussher, D.I. and Gooding, L.R. (1992) The 19-kilodalton adenovirus E1B transforming protein inhibits programmed cell death and prevents cytolysis by tumor necrosis factor alpha. *Mol. Cell. Biol.* **12**: 2570–2580.

White, E., Rao, L., Chiou, S.-K., Tseng, C.-C., Sabbatini, P., Gonzalez, M. and Verwaerde, P. (1994) Regulation of apoptosis by the transforming gene products of adenovirus. In: *Apoptosis* (eds E. Mihich and R. T. Shimke). Plenum Press, New York, USA, pp. 47–62.

Wu, D., Wallen, H.D. and Nunez, G. (1997) Interaction and regulation of subcellular localization of CED-4 by CED-9. *Science* **275**: 1126–1128.

Yin, X.-M., Oltvai, Z. and Korsmeyer, S. (1994) BH1 and BH2 domains of Bcl-2 are required for inhibition of apoptosis and heterodimerization with Bax. *Nature* **369**: 321–323.

Zha, H., Aimé-Sempé, C., Sato, T. and Reed, J.C. (1996) Proapoptotic protein Bax heterodimerizes with Bcl-2 and homodimerizes with Bax via a novel domain (BH3) distinct from BH1 and BH2. *J. Biol. Chem.* **21**: 7440–7444.

Zou, H., Henzel, W.J., Liu, X.S., Lutschg, A. and Wang, X.D. (1997) Apaf-1, a human protein homologous to *C. elegans* CED-4, participates in cytochrome c-dependent activation of caspase-3. *Cell* **90**: 405–413.

19

Tumour cell death

Ifor D. Bowen and Farhana Amin

1. Introduction

Cells die either by accident or 'design'. Accidental cell death may be caused by disease or external forces and leads to what pathologists call necrosis. Death by design, often as part of a genetically regulated developmental programme has been called physiological or programmed cell death (Lockshin, 1969, 1971; Lockshin and Williams, 1964). In this context cell death appears to be part of a strategy for survival of the organism. The paradox lies in the fact that programmed cell death, more often than not is essential for the continued life and development of the organism.

It has been realized for some time that cell death forms an essential part of normal embryological development, morphogenesis and metamorphosis (Glucksmann, 1951; Looss, 1889; Saunders, 1966) as an essential counterpoint to mitosis. Cell death in this context is employed to help shape tissues, organs and organisms and can most spectacularly be seen in examples such as the transformation of caterpillar to butterfly, the loss of the tadpole tail and the carving of the pentadactyl limb through interdigital cell death.

Programmed cell death also plays an essential role maintaining the population balance of cells in fully grown organisms. Such organisms are usually in dynamic equilibrium. Cells are born and cells die every minute. The overall shape and size of a tissue is normally maintained by this balance or homeostasis. A good example is the loss of cornified cells from the surface layers of the skin, which is essential for the production and maintenance of a protective skin layer. Cells are born through the process of mitosis at the basal layer of the skin and specialize in the production of keratin as they move up through the intermediate layers finally differentiating to death as flattened dead keratinocytes at the surface. There is a similar renewal and loss of cells in the intestinal villi (Potten, 1992). Another example would be the maintenance of the size and shape of the liver where it is estimated that a cell is born and another must die every minute.

Homeostatic programmed cell death is an important phenomenon and led to the pioneering adoption of a new term called 'apoptosis', functionally defined as an equal and opposite force to mitosis (Kerr et al., 1972; Wyllie, 1980). It was surmised that a disturbance in this subtle balance between mitosis and apoptosis could lead to the formation of a tumour where cell proliferation outstripped cell death. Apoptosis was

Programmed Cell Death in Animals and Plants, edited by J.A. Bryant, S.G. Hughes and J.M. Garland.
© 2000 BIOS Scientific Publishers Ltd, Oxford.

viewed as a positive type of programmed cell death with its own characteristic morphology occurring not only during early development but continuously throughout life. The word 'apoptosis' itself was derived from the classical Greek word meaning a 'falling away', with its useful connotation of suicide, as opposed to the Greek word 'necrosis' to 'make dead', with its pathological connotation of murder (Duke et al., 1996).

The morphological and biochemical basis of apoptosis have been well documented (Wyllie, 1980, 1981, 1987). Since death is ultimately the end-point of apoptosis and necrosis, it is not always easy to distinguish between them, although more and more so-called diagnostic tools are being developed. We may effectively be dealing with the two defined extremes of one continuous spectrum of deaths (Wyllie et al., 1984). Characteristically, however, apoptosis leads to cell condensation and shrinkage whilst necrosis leads to cellular swelling. Apoptotic cells lose water whilst necrotic cells gain water. Membrane pumps continue to function during apoptosis and apoptotic cells exclude vital dyes. Membrane pumps invariably fail early on in necrosis leading to influx of water and sodium along with rapid calcium overload. Energy production is maintained during apoptosis and there may be a synthetic surge leading to the appearance of new mRNA and protein species. In apoptosis there is a dilation of the nuclear envelope accompanied by significant nuclear blebbing, the blebs often being filled with marginating chromatin. DNA is fragmented under the influence of endo-nuclease activation, the enzyme cleaving the DNA at the inter-nucleosomal linker regions. If run on an agarose gel, such regular DNA fragments produce a 'DNA ladder'. Programmed cell death is usually dependent on genetic induction, although the inductive cascade can be accessed via activation of pre-existing transcription factors (Bellamy et al., 1995). Good evidence exists for the genetic basis of apoptosis in the nematode C. elegans (Yuan and Horovitz, 1990; see also Chapter 1) although some of the symptoms of apoptosis may vary in different species and indeed, can also be induced without engaging a genetic cascade.

Necrosis displays characteristic symptoms of cellular swelling, gross calcium overload, phospholipase A activation and irreversible mitochondrial swelling leading an autolytic destruction of the cell. It is not genetically controlled and, provided the induction is via a lethal stimulus, there is an immediate loss of energy production, water balance, volume control and protein synthesis. In the late stages the lysosomes of necrotic cells burst and digestive enzymes are released which further lyse the cell as it breaks up. The severely damaged cell scatters its contents to a degree that often elicits an inflammatory response. Apoptotic cells on the other hand exhibit a florid but discrete fragmentation into physiologically intact spherical apoptotic bodies, which are immediately engulfed by neighbouring cells or macrophages without eliciting an inflammatory response.

It is possible for apoptotic cells to secondarily exhibit the properties of necrosis if they are exposed to a non-physiological environment. Whilst apoptosis is normally induced by physiological stimuli it has been found that chronic noxious stimulation of cells can under certain circumstances induce apoptosis (Cotter et al., 1990). It has been shown that cells continually monitor internal levels of damage, especially DNA damage, and may decide in favour of apoptosis if such damage endangers the integrity and fidelity of future generations of cells (Wertz and Hanley, 1996; see also Chapter 17). Elevation of levels of a key protein, p53, may follow such damage in normal cells and plays a role in the subsequent induction of apoptosis.

There is some confusion in the literature between the terms 'programmed cell death' and 'apoptosis' (Bowen, 1993). The term programmed cell death coined in a developmental context by Lockshin (1969) has priority and, on *a priori* grounds, should have precedence in terms of generic usage. The situation is complicated by the fact that apoptosis as originally envisaged (Kerr *et al.*, 1972) has a broadly morphological and kinetic definition, whilst programmed cell death has a developmental and at least implied genetic basis. Apoptosis can be genetically driven as seen in the work of Ellis *et al.* (1991) on the nematode *C. elegans* (Chapter 1), but some of the symptoms of apoptosis can be induced without engaging in a genetic cascade (Duke *et al.*, 1988). Thus, not all forms of apoptosis are genetically induced. Similarly, not all kinds of genetically programmed cell death are apoptotic in terms of its original definition (Bowen *et al.*, 1993, 1996; Lockshin and Zakeri, 1996; Szende *et al.*, 1995). Clarke (1990) has detailed three basic morphological types of programmed cell death (*Figure 1*).

Non-apoptotic programmed cell death, includes autophagic or vacuolar cell death common in invertebrate tissues (Bowen and Bowen, 1990; Bowen and Ryder, 1974; Bowen *et al.*, 1982; Jones and Bowen, 1980; Lockshin, 1969), and described by Clarke (1990) as Type 2 cell death (Type 1 being regarded as apoptotic), and another category akin to atrophy (Levi-Montalcini and Aloe, 1981) described by Clarke as Type 3 cell death. Both Types 2 and 3 as well as cell differentiation appear to be genetically programmed and usually occur during development. Although these categories are largely based on morphological differences such as the presence or absence of chromatin margination and blebbing, such traits are accompanied by differences in DNA fragmentation and protein synthesis. Lockshin and Zakeri (1996) have called apoptosis 'Type 1 cell death', and non-apoptotic programmed cell death 'Type 2 cell death'. Clarke's Type 3 cell death, referred to as atrophy since it is seen to occur in the case of growth factor withdrawal, is given particular attention in this chapter since it can be induced in tumour cells by a range of protein kinase inhibitors (Amin and Bowen, 1997; Szende *et al.*, 1995). Some evidence is emerging in support of a genetic basis for non-apoptotic programmed cell death (Bowen *et al.*, 1993, 1996; Martin and Johnson, 1991; Martin *et al.*, 1988).

It is logical to regard cellular or cyto-differentiation as differentiation to death since most differentiated cells are post-mitotic and never regain the ability to reproduce and divide. Thus, to a large extent, differentiation, which is known to be based on differential gene activity, should be considered as leading to a special kind of programmed cell death. All differentiating cells move along specialized channels to death.

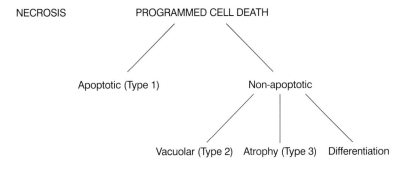

Figure 1. A classification of cell death.

Wangenheim (1987) concludes that differentiation into somatic cell lines makes cell death inevitable, in contrast with the potentially immortal undifferentiated stem or germ cell line. Some cells in certain organisms survive until the organism dies and others, continually produced in self-renewing tissues, differentiate to perform specific functions and then die. An understanding of the factors that control differentiation, could obviously be of value in tumour therapy.

2. Cell death in tumours

One of the reasons why cell death is currently regarded as such an important topic is that its control can be crucial in regulating tumour therapy (see Chapter 20). Tumour cells have a high turnover rate and generally show a high rate of cell loss. Much of this cell loss is due to exfoliation and/or necrosis; however, a critical fraction may be due to apoptosis and the induction of apoptosis through the activation or inhibition of particular genes could alter the kinetics of tumour growth.

2.1 Tumour structure

In a very simple sense, tumours may either be solid or corded (*Figure 2*). Solid tumours grow outwards forming a solid ball of cells, whilst corded tumours grow like liver cords around central blood vessels. Thus, in the case of corded tumours the younger cells grow around these blood vessels, whilst the older cells are pushed peripherally to the outer regions of the tumour. In contrast, within a solid tumour the younger proliferative cells move outwards leaving the older cells in a more or less central position. In such a tumour, the younger tumour cells of the outer zone engender their own vascular blood supply through the production of a chemical signal called angiogenesis factor,

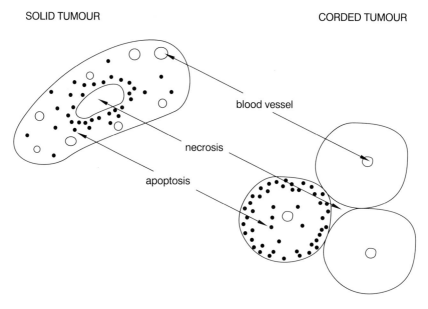

Figure 2. Tumour structure.

which makes the endothelial cells lining the blood capillaries divide and branch, forming a new capillary bed and supplying the tumour cells with sufficient nutrients and oxygen. Vascularization thus follows the younger tumour cells as the population expands outwards. The older tumour cells produce less angiogenesis factor and the older capillary bed in the centre is more easily compromised (Denekamp *et al.*, 1982). Because the blood flow is poor in the centre of such a growing solid tumour, the constituent cells with their high metabolic demands are often starved of glucose and oxygen leading eventually to a necrotic cell death. The centre of growing solid tumours are therefore often 'naturally' necrotic. Moreover, any agents that can selectively compromise or interrupt the vascular supply could have great therapeutic value in that they may extend this necrosis. The cell knock-down achieved by such agents would be considerable, and would arguably be much greater than that achievable through manipulation of apoptosis, which usually occurs at low levels in most solid tumours (Sarraf and Bowen, 1988).

2.2 *Peripheral tumour cell death*

It is important to note that, whilst necrotic death is a major component of cell death in tumours, varying levels of programmed cell deletion, including apoptosis, occur naturally in the proliferative zones of untreated tumours (Sarraf and Bowen, 1986). Interestingly, the incidence of apoptosis always seems to be higher in the hypoxic areas next to the necrosis, where the cells are presumably stressed.

The causes of peripheral cell death in solid tumours are not yet fully understood. Single cell deletions could be due to a withdrawal of the tumour cells to G_o of the cell cycle, followed by subsequent differentiation to death. Some tumour cells do appear to redifferentiate in this manner. A considerable level of tumour cell-targeted hunter–killer cell activity also seems likely. It is known that macrophages, natural killer (NK) cells, especially lymphokine-activated natural killers (LAK), a range of cytotoxic lymphocytes (CTL) and specific tumour-infiltrating lymphocytes (TIL) appear able in various ways to induce a target-mediated apoptosis in tumour cells. The methods of killing include occupying Fas/Apo-1 receptors and also, in the case of CTL and LAK cells, the secretion of granzyme proteases and perforins (see Chapter 5) leading to the activation of endonuclease and apoptosis in individually targeted tumour cells. Although the natural level of apoptosis induced is usually relatively low (between 1% and 3% of the entire population) considerable therapeutic potential would seem to accrue from both cytokine activation and *in vitro* clonal expansion of killer cells which are subsequently returned to the patient (Grobstein *et al.*, 1986; Smyth and Trapani, 1995).

3. Tumour necrosis factor α and tumour cell death

Tumour necrosis factor α (TNF-α) is one important therapeutic cytokine or, more correctly, monokine since it is produced by macrophages (Matthews, 1981). TNF-α induces massive haemorrhagic necrosis in tumours by initiating extensive vascular changes. One major effect is to make the endothelial cells of the blood capillaries more adhesive, promoting platelet adhesion. This leads to blood clotting, stasis, and haemorrhage, which in turn leads to anoxia and tumour cell necrosis. TNF-α appears to work by stimulating cross-linking and adhesion between blood vessel endothelial cells that are induced to produce adhesion molecules ICAM-1 and VCAM-1 and leukocyte adhesion molecule ELAM-1. The therapeutic vascular collapse induced by TNF-α has been reviewed by Mannel *et al.* (1995).

From a clinical point of view, however, TNF-α, although having all the requirements for precipitating massive tumour cell death, has proved disappointing because of its lethal inflammatory effects in humans: systemic TNF-α (sometimes known as cachectin), leads to circulation problems, oedema, endotoxic shock and death. Anchored to the macrophage membrane or targeted on a cell-to-cell basis (Kreigler *et al.*, 1988) the molecule may still hold considerable therapeutic promise. It certainly deserves further investigation, not least for the fact that it can also specifically induce apoptosis in tumour cells.

3.1 *TNF-α and apoptosis*

TNF-α shows specific cytostatic and cytocidal activities against a wide variety of human tumour cell lines. The reception and transduction mechanisms in TNF-α-induced apoptosis can be quite complex. A gene expressing the receptor TNF-R1 has been cloned and it is clear that occupation of this receptor, and indeed the Fas receptor, can, after intermediate transduction via other death-domain-containing proteins, lead to the activation of caspases (see Chapter 2), leading in turn to the activation of endonucleases and of apoptosis. The Fas receptor, also known as CD95/Apo-1, plays an important role in the immune system as a mediator of T-cell death during the removal of autoreactive T cells and for the maintenance of immune system homeostasis. The cytoplasmic regions of TNF-R1 and Fas receptors contain conserved apoptosis or death domains of about 80 amino acid residues, which are essential components of the signalling pathway that triggers apoptosis.

Amin and Bowen (1997) and Amin *et al.* (1999) have shown that TNF-α induces all the symptoms of classical apoptosis in the human breast carcinoma cell line MCF-7. These included early externalization of plasma membrane phosphatidylserine, blebbing or the nuclear and plasma membrane and DNA laddering. In addition, using flow cytometry, a transient increase in wild-type p53 expression was found in the apoptotic cells. Increase in p53 expression has been previously correlated with TNF-α-induced apoptosis in rat glioma cells (Yin *et al.*, 1995).

4. The role of p53 and Bcl-2 in tumour cell apoptosis

The B-cell lymphoma 2 oncogene product (bcl-2) is an integral membrane protein found in mitochondria, the nuclear envelope and endoplasmic reticulum It functions as an inhibitor of apoptosis (see Chapter 4). The function and structure of *bcl-2* is homologous with that of *ced-9*, the gene which inhibits apoptosis in the nematode, *C. elegans* (Hengartner and Horovitz, 1994 and Chapter 1). Over-expression of bcl-2 blocks the apoptotic pathway of a pro-B-lymphocyte precursor cell line and can thus be regarded as a survival factor. The homology between the products CED-9 and Bcl-2 proteins, suggest that some of the molecular mechanisms of apoptosis have been conserved from nematodes to mammals (see Chapter 1). Bcl-2 in fact is only one protein in a large expanding family of related proteins (described more fully in Chapter 4). Functionally the family falls into two groups, one inhibiting apoptosis induced by growth factor deprivation, deregulation of c-myc or genotoxic damage, and the other promoting cell death. It is also clear that the outcome in terms of cell death is dependent on how the members of this complex protein family combine and mix. The proteins can dimerize as homo- or hetero-dimers and the particular composition of a dimer can tip the balance towards or away from cell death. A particular set of family proteins including Bax, Bcl-X, Bak and Bag promote cell death. The combinations possible are complex but one good example relates to combinations between Bcl-2 and Bax (*Figure 3*).

High concentrations of Bax can inhibit the survival functions of Bcl-2. Bax can also form Bax/Bax homodimers, which appear to accelerate rather than inhibit the onset of cell death. Recent developments show that the conformational arrangements of Bcl-2 in the inner mitochondrial membrane may control the leakage of cytochrome c which plays an important role as a co-factor in activating caspase activity and apoptosis (Chapters 6 and 7).

Since up-regulation of Bcl-2 inhibits apoptosis it is not surprising to find that increased bcl-2 expression is often found in a wide range of tumours such as glioma, lymphoma, melanoma, breast and colorectal cancer. Up-regulation of Bcl-2 may, therefore, compromise the process of natural cell death in a tumour thus leading to a poor prognosis. It is thought that prolonged cell survival could provide additional time for the neoplastic cells to acquire more genetic mutations, creating a reservoir of viable tumour cells from which metastasis may arise. Conversely, inhibition of bcl-2 expression, or the up-regulation of *Bax* expression or its equivalent, should maximize natural apoptosis and prove therapeutic. Some cell types such as melanocytes normally have elevated levels of Bcl-2 and, in this case, if the cells become cancerous they are less susceptible to apoptosis, producing an aggressive melanoma. In this instance repression of the *bcl-2* gene or neutralization of the product Bcl-2 could be therapeutic. This topic is also related to the oncogenic viruses that inhibit apoptosis (see Chapter 19). It is of course always advantageous if not essential for viruses to inhibit apoptosis so that they can use the protein synthesis machinery of their live hosts. Thus, the Epstein–Barr virus BHRFI protein, a homologue of Bcl-2, inhibits apoptosis

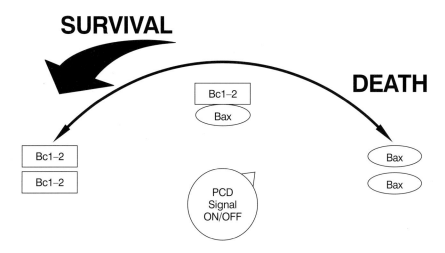

Figure 3. Bcl-2 homodimerization or heterodimerization with Bax protects the cells from apoptosis. Bax homodimers promote apoptosis.

and, indeed, the LMP-1 protein from the same virus induces a zinc finger protein that inhibits TNF cytotoxicity. The adenovirus *E1B* gene (Chapter 18) encodes a protein homologue of Bcl-2 that inhibits p53.

Significantly, stimulation of the *bax* gene may be correlated with wild type *p53* oncogene activity. The *bax* gene promoter region contains p53 binding sites. It has been proposed, therefore, that *p53* up-regulation induces raised Bax protein levels which, in excess, will hetero-dimerize with Bcl-2 resulting in apoptosis.

4.1 *Tumour suppression by p53*

Tumour suppressor phosphoprotein p53 has a molecular weight of 53 000 and is involved among other things with mediating the cellular response to DNA damage and the maintenance of genomic integrity. Damaged cells that are irreparable undergo apoptosis if p53 is expressed by the cell (Chapter 16). In p53-null cells DNA damage may lead to malignancy. Oncogenic mutations are reduced by p53 because of its ability to facilitate DNA repair. This indirect method of tumour suppression is increased by the ability of p53 to promote apoptosis. Loss of *p53* gene function can therefore contribute to tumorigenesis by allowing inappropriate cell survival (*Figure 4*).

4.2 *p53 mutations and tumorigenesis.*

Deletions or mutations of *p53* have been detected in over 50% of human cancers and occur in a wide range of tumour type. Such mutations are often a late event in tumorigenesis and are frequently associated with aggressive cancers. Mutant mice homozygous for *p53* deletions are predisposed to spontaneous and carcinogen-induced tumours. Multiple myeloma and acute lymphoblastic leukaemia often have *p53* mutations in relapse phase tumours. By contrast, *p53* mutations are rare in highly curable tumours.

Apoptosis triggered by p53 can be inhibited by Bcl-2, but not p53-induced G_1 arrest of the cell cycle, demonstrating that the two functions of p53 are independent. There are, apparently, transcriptionally dependent and independent mechanisms by which p53 controls apoptosis, and this sometimes varies with cell type. *Bax* activation and/or *bcl-2* repression are examples of how p53 might shift the balance in favour of apoptosis. Bcl-2/Bax ratio is influenced by over-expression of p53 and both the decrease in Bcl-2 and an increase in Bax proteins and message are associated with p53-induced apoptosis. The extent to which basal levels of p53 influence *bcl-2* is, however, highly tissue-specific.

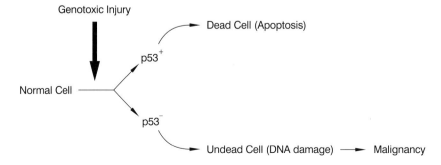

Figure 4. *Damaged cells that cannot be repaired undergo apoptosis if p53 is repressed. In mutant or p53-null cells the DNA damage can lead to malignancy.*

Polyak *et al.* (1997) emphasized the role of hypoxic *p53* mutants in the development of apoptotically resistant tumour cells and metastases. Many tumour cells also appear insensitive to growth factor withdrawal and cannot enter a p53-independent apoptotic path; others are insensitive to immune and Fas-relayed apoptosis. It has therefore been strongly argued that tumours are as much a result of apoptotic failure as of a deregulation of the cell cycle. Therapeutic approaches could thus include the introduction of wild type *p53* gene into cancers as part of an integrated gene therapy programme along with the blocking of growth factor reception or transduction.

5. Non-apoptotic programmed cell death?

To some extent the concept of a non-apoptotic programmed cell death is still controversial, but is gradually gaining acceptance as has been outlined in the Introduction to this chapter. Kinase inhibitors can produce aberrant forms of cell death in tumour cells that do not conform to the definitions of 'classical' apoptosis. Szende *et al.* (1995) have demonstrated such programmed cell death induced by an EGF receptor tyrosine kinase inhibitor. Protein and RNA synthesis inhibitors blocked the apparently non-apoptotic programmed cell death. Similar results were reported by Szegedi *et al.* (1997) using the tyrosine kinase inhibitor AG-213 on human tumour cell lines. The same phenomenon is being studied by Amin (1999) and has been briefly reported by Amin and Bowen (1997). In this instance it has been demonstrated that tyrphostin AG-213 induces a non-apoptotic form of programmed cell death in human MCF-7 breast carcinoma cells (Amin *et al.*, 1999). The cells do not shrink and bleb as in apoptosis but appear to become heavily vacuolated very much like necrotic cells. They do become annexin V positive after treatment, indicating externalization of phosphatidylserine, but the pattern of fluorescence is heavy and mosaic in appearance. The treated cells do not show any DNA fragmentation as demonstrated by the TUNEL technique or Comet assay. The same cells treated with TNF-α produced all the classical symptoms of apoptosis including DNA fragmentation and transient upregulation of p53, whereas no such changes were found in the tyrphostin AG 213-treated cells.

It has recently been shown that under certain circumstances, such as caspase inhibition by z-VAD, cells die in a fashion which does not follow the morphology of apoptotic cell death and they do not show DNA fragmentation (Hickman *et al.*, 1998; Xiang *et al.*, 1996). It appears that specific participation of caspases together with nucleases and protein kinases are required for the full proteolytic events typical of apoptosis (Salvesen and Dixit, 1997). There is also emerging evidence that caspase-independent programmed cell death displays a partly necrotic morphology.

The acquired resistance of tumour cells to apoptosis is a major concern in cancer therapy (Harrison,1995) and further studies of non-apoptotic programmed cell death, and its induction using specific kinase inhibitors, may open up new avenues of research and non-surgical intervention.

References

Amin, F. (1999) Modes of cell death in MCF-7 breast carcinoma cells. *PhD University of Wales.*
Amin, F. and Bowen, I.D. (1997) A comparative study of the different modes of cell death in MCF-7 breast cancer cells. *Cell Prolif.* **30**: 484.

Amin, F., Bowen, I.D., Szegedi, Z., Rudolf, M. and Szende, B. (1999) Apoptotic and non-apoptotic modes of programmed cell death in MCF-7 human breast carcinoma. *Cell Biol. Int.,* in press.

Bellamy, C.O.C., Malcomson, R.D.G., Harrison, D.J. and Wyllie, A H. (1995) Cell death in health and disease: the biology and regulation of apoptosis. *Cancer Biol.* **6**: 3–16.

Bowen, I.D. (1993) Apoptosis or Programmed Cell Death? *Cell Biol. Int.* **17**: 365–380.

Bowen, I.D. and Bowen S.M. (1990) *Programmed Cell Death in Tumours and Tissues.* Chapman and Hall, London and New York.

Bowen, I.D. and Ryder, T.A. (1974) Cell autolysis and deletion in the planarian *Polycelis tenuis* Iijima. *Tissue Cell Res.* **154**: 265–274.

Bowen, I.D. Den Hollander, J.E. and Lewis, G.H.J. (1982) Cell death and acid phosphatase activity in the regenerating planarian *Polycelis tenuis. Differentiation* **21**: 160–170.

Bowen, I.D., Morgan S.M. and Mullarkey, K. (1993) Cell death in the salivary glands of meta-morphosing *Calliphora vomitoria. Cell Biol. Int.* **17**: 13–33.

Bowen, I.D., Mullarkey, K. and Morgan, S.M. (1996) Programmed cell death in the salivary glands of the blow fly *Calliphora vomitoria. Microsc. Res. Tech.* **34**: 202–207.

Clarke, P.G.H. (1990) Developmental cell death: morphological diversity and multiple mecha-nisms. *Anat. Embryol.* **181**: 195–206.

Cotter, T.G., Lennon, S.V. and Martin, S.J. (1990) Apoptosis: programmed cell death. *J. Biomed. Sci.* **2**: 72–80.

Denekamp, J., Hill, S.A. and Hobson, B. (1982) Vascular occlusion and tumour cell death. *Eur. J. Cancer Clin. Oncol.* **19**: 271–275.

Duke, R.C., Sellins, K.S. and Cohen, J.J. (1988) Cytolytic lymphocyte-derived lytic granules do not induce DNA fragmentation in target cells. *J. Immunol.* **141**: 2191–2194.

Duke, R.C., Ojcius D.M. and Young, J.D.-E. (1996) Cell suicide in health and disease. *Sci. Am.* Dec., 48–55.

Ellis, R.E., Yuan, J. and Horovitz, H.R. (1991) Mechanisms and functions of cell death. *Annu. Rev. Cell Biol.* **7**: 663–698.

Grobstein, K.H., Urdal, D.L., Tushinshi, R.J. *et al.* (1986) Induction of macrophage tumori-cidal activity by granulocyte-macrophage colony-stimulating factor. *Science* **232**: 506–508.

Glucksmann, A. (1951) Cell deaths in normal vertebrate ontogeny. *Biol. Rev.* **26**: 59–86.

Harrison, D.J. (1995) Molecular mechanisms of drug resistance in tumors. *J. Pathol.* **175**: 7–12.

Hengartner, M.O. and Horvitz, H.R. (1994) *C. elegans* cell survival gene ced-9 encodes a func-tional homolog of the mammalian proto-oncogene bcl-2. *Cell* **76**: 665–676.

Hickman, J.A., Dive, C., Brady, G., Gumby R., Dubrez, L., Morgan, C. and Griffiths, G. (1998) Commitment to death: molecular events preceding the engagement of apoptosis. *Cell Prolif.* **30**: 445 (abstr).

Jones, G.W. and Bowen, I.D. (1980) The fine structural localisation of acid phosphatase in pore cells of embryonic and newly hatched *Deroceras reticulatum* (Pulmonata: Stylommatophora). *Cell Tissue Res.* **204**: 253–265.

Kerr, J.F.R., Wyllie, A.H. and Currie, A.R. (1972) Apoptosis: a basic biological phenomenon with wide-ranging implications in tissue kinetics. *Br. J. Cancer* **26**: 239–257.

Kreigler, M., Perez, C., DeFaey, K. *et al.* (1988) A novel form of TNF-cachectin is a cell surface cytotoxic transmembrane protein: ramifications for the complex physiology of TNF. *Cell* **53**: 45–53.

Levi-Montacini, R. and Aloe, L. (1981) Mechanisms of action of nerve growth factor in intact and lethaly injured sympathetic nerve cells in neonatal rodents. In *Cell Death in Biology and Pathology* (eds I.D. Bowen and R.A. Lockshin), Chapman and Hall, London and New York, pp. 295–327.

Lockshin, R.A. (1969) Programmed cell death: activation of lysis by a mechanism involving the synthesis of protein. *J. Insect Physiol.* **15**: 1505–1516.

Lockshin, R.A. (1971) Programmed cell death: nature of the nervous signal controlling breakdown of intersegmental muscles. *J. Insect Physiol.* **17**: 149–158.

Lockshin, R.A. and Williams C.M. (1964) Programmed Cell Death II. Endocrine potentiation of the breakdown of the intersegmental muscles of silkmoths. *J. Insect Physiol.* **10**: 643–649.

Lockshin, R.A. and Zakeri, Z. (1996) The biology of cell death and its relationship to ageing. In *Cellular Aging and Cell Death* (eds N.J. Holbrook, G.R. Martin and R.A. Lockshin). Wiley-Liss, New York, pp. 167–180.

Looss, A. (1889) Uber degeneration serchein im Terreich, besonders uber die reduktion des forschlarvenschwanzes und die in verlag disselben auttertenden histolytishen. *Proz. Preisschr. Jablonowsk* Ges **10**.

Mannel, D., Murray, C., Risau, W. and Clauss, M. (1995) Tumour necrosis: factors and principles. *Immunol. Today* **17**: 254–256.

Martin, D.P. and Johnson, J. (1991) Programmed cell death in the peripheral nervous system. In: *Apoptosis: The Molecular Basis of Cell Death* (eds D.L. Tomei and F.O. Cope). (Current communications in cell and molecular biology, Vol. 3.) Cold Spring Harbor Laboratory Press, New York, pp. 247–261.

Martin, D.P., Schmidt, R.E., De Stefano, P.S., Lowry,O.H., Carter, J.G. and Johnson, E.M. Jr. (1988) Inhibitors of protein synthesis prevent neuronal death caused by nerve growth factor deprivation. *J. Cell Biol.* **106**: 829–840.

Matthews, N. (1981) Tumour necrosis factor from the rabbit v. synthesis *in vitro* by mononuclear phagocytes from various tissues of normal and BCG-injected rabbits. *Brit. J. Cancer* **44**: 418–412.

Polyak, K., Xia, Y., Zweier, J.L., Kinzler, K.W. and Vogelstein, B. (1997) A model for p53-induced apoptosis. *Nature* **389**: 300–305.

Potten, C.S. (1992) The significance of spontaneous and induced apoptosis in the gastrointestinal tract. *Cancer Metastasis Rev.* **11**: 179–195.

Salvesen, G.S. and Dixit, V.M. (1997). Caspases: intracellular signalling by proteolysis. *Cell* **91**: 443–446.

Sarraf, C.E. and Bowen, I.D. (1988) Kinetic studies on a nuerine sarcoma and an analysis of apoptosis. *Br. J. Cancer* **54**: 989–998.

Sarraf, C.E. and Bowen, I.D. (1988) Proportions of mitotic and apoptotic cells in a range of untreated and treated experimental tumours. *Cell Tissue Kinet.* **21**: 45–49.

Saunders, J.W. (1966) Death in embryonic systems. *Science* **154**: 604–612.

Server, A.C and Mobley, W.C. (1991) Neuronal cell death and the role of apoptosis. In: *Apoptosis: The Molecular Basis of Cell Death* (eds L.D. Tomei and F.O. Cope). (Current communications in cell and molecular biology, Vol. 3). Cold Spring Harbor Laboratory Press, pp. 263–278.

Smyth, J. and Trapani, J.A. (1995) Granzyme: exogenous proteinases that induce target cell apoptosis. *Immunol. Today* **16**: 202–204.

Szegedi, Z., Stetak, A., Nagy, J. *et al*. (1997) Characterization of cell death caused by tyrphostin AG-213 on human cell lines. *Cell Prolif.* **29**: 6.

Szende, B., Keri, Gy., Szegedi, Z.S., Benedecsky, I., Csikos, A., Orfi, L. and Gazit, A. (1995) Tyrphostin induces non-apoptotic programmed cell death in colon tumour cells. *Cell Biol. Int.* **19**: 903–911.

Wangenheim, K.-H. (1987) Cell death through differentiation: potential immortality of somatic cells: a failure in control of differentiation. In: *Perspectives on Mammalian Cell Death* 6 (ed. C.S. Potten). Oxford University Press, Oxford, pp. 129–159.

Wertz, I.E. and Hanley, M.R. (1996) Diverse molecular provocation of programmed cell death. *Trends Biochem. Sci.* **21**: 359–364.

Wyllie, A.H. (1980) Glucocorticoid-induced thymocyte apoptosis is associated with endogenous endonuclease activation. *Nature* **284**: 555–556.

Wyllie, A.H. (1981) Cell death: a new classification separating apoptosis from necrosis. In: *Cell Death in Biology and Pathology* Chapter 1. (eds I.D. Bowen and R.A. Lockshin). Chapman and Hall, London and New York, pp. 9–34.

Wyllie, A.H. (1987) Apoptosis: Cell death in tissue regulation. *J. Pathol.* **153**: 313–316.

Wyllie, A.H., Duvall, E. and Blow, J.J. (1984) Intracellular mechanisms in cell death in normal and pathological tissues. In: *Cell Ageing and Cell Death* (eds I. Davies and D.C. Sigee). Cambridge University Press, Cambridge, pp. 269–284.

Yin, D., Kondo, S., Barnett, G.H., Morimura, T. and Takeuchi, J. (1995). Tumor necrosis factor-alpha induces p53 dependent apoptosis in rat-glioma cells. *Neurosurgery* **27**: 762–763.

Yuan, J. and Horovitz, H.R. (1990) The *Caenorhabditis elegans* genes ced-3 and ced-4 act autonomously to cause programmed cell death. *Dev. Biol.* **138**: 33–41.

Xiang, Q., Chao, D.T. and Korsmeyer, S.J. (1996) Bax induced cell death may not require interleukin 1-beta-converting enzyme-like proteases. *Proc. Natl Acad. Sci. USA* **93**: 14559–14563.

20

Controlling apoptosis: implications for carcinogenesis?

Alan R. Clarke

1. Introduction

The induction of cell death has been closely associated with the response to cellular stress, and in particular the response to DNA damage. With the advent of gene targeting it has been possible to demonstrate genetic dependencies for this process. Such dependency was established for the tumour suppressor gene *p53* several years ago using *p53* mutant murine strains, and supports similar data obtained from human cell lines. These observations have given rise to one hypothesis that may explain the role of *p53* as a tumour suppressor; namely that *p53* normally functions to delete cells bearing DNA damage. When this pathway is compromised, either by loss of function or dominant negative mutations, cells normally destined to die persist. These are presumed to carry increased levels of DNA damage and therefore an increased mutation rate and predisposition to malignancy. This hypothesis predicts a number of different endpoints for cell populations in which the normal apoptotic response is compromised. First, there should be an increase in clonogenic survival following insult. Second, there should be an increase in the mutation burden within surviving cells. Finally, this increase should be associated with an increased rate of malignant development.

To test this hypothesis we have analysed these endpoints in a number of different systems in mice deficient for *p53*. We have also analysed some of these endpoints in mice deficient for the mismatch repair gene *Msh2* as we have shown that, as with *p53*, this gene is essential for the normal induction of apoptosis following exposure to certain types of DNA damage. This observation raised a similar hypotheses to that detailed above for *p53*, namely that *Msh2* deficiency may predispose to malignancy by virtue of failure to engage cell death pathways.

Programmed Cell Death in Animals and Plants, edited by J.A. Bryant, S.G. Hughes and J.M. Garland.
© 2000 BIOS Scientific Publishers Ltd, Oxford.

2. *p53*

Although *p53* was first described as an oncogene, it is now well recognized as a tumour suppressor gene. Much recent work has attributed a plethora of different activities to this protein (e.g. reviewed in Giaccia and Kastan, 1998). These include the induction of apoptosis, cell cycle arrest and a direct exonuclease activity. Many of these functions are mediated through transcriptional activation and repression of a host of target genes. Amongst these genes it is notable that *p53* mediates up-regulation of some genes associated with DNA repair (e.g. Rafferty *et al.*, 1996).

Loss of function of the *p53* tumour suppressor gene has been strongly associated with the development of human malignancies (Hollstein *et al.*, 1991). This is perhaps most convincingly shown by the existence of the human familial predisposition to diverse malignant tumours which comprises the Li-Fraumeni syndrome, characterized by germ line *p53* mutation (Malkin *et al.*, 1990). Inactivation of *p53* function is associated both with genomic instability (e.g. Livingstone *et al.*, 1992) and inability to delete cells by apoptosis following DNA damage.

To investigate *p53* function *in vivo*, several groups independently generated *p53*-deficient murine strains (Clarke *et al.*, 1993; Donehower *et al.*, 1992, Lowe *et al.*, 1993a; Tsukuda *et al.*, 1993). The reported phenotypes of these strains are remarkably similar. The majority of mice null for *p53* develop normally to birth, although a percentage of females develop anencephaly *in utero*, a phenotype that appears to be highly background-dependent and has led to the suggestion that *p53* plays a strong protective role during development (Armstrong *et al.*, 1995; Sah *et al.*, 1995). Adult *p53*-deficient mice rapidly succumb to a variety of tumours, this spectrum being dominated by lymphoma. Mice heterozygous for *p53* develop both lymphoma and sarcoma, although the precise spectrum of tumours is again dependent upon genetic background (Donehower *et al.*, 1992; Harvey *et al.*, 1993a, 1993b; Purdie *et al.*, 1994).

Prior to the development of nullizygous strains, cell culture experiments had established an *in vitro* role for *p53* in mediating the cell death response to DNA damaging agents (e.g. Yonish-Rauach *et al.*, 1991). We wished to address whether this response reflected a true *in vivo* dependency. To achieve this, we analysed the apoptotic response of thymocytes maintained in short-term culture which were derived from wild type and *p53*-deficient mice (Clarke *et al.*, 1997). We found a clear requirement for *p53* to induce death following exposure to ionizing radiation, a partial requirement following exposure to the topoisomerase inhibitor etoposide and a *p53*-independent response following exposure to non-clastogenic agents such as the steroid methylprednisolone and the calcium ionophore ionomycin. These results confirmed a primary role for *p53* in the induction of apoptosis following DNA damage.

Thymocytes represent a somewhat unusual population of cells in that the vast majority of this cell type is destined to die *in vivo* in the course of normal thymic ontogeny, and short-term cultures enter a death pathway irrespective of conditions following 24–48 hours of culture. We therefore wished to extend these observations to other cell types, for which baseline apoptotic responses had been well established. We first analysed the epithelial cells of the small and large intestine, partly because of the strong association between loss of function of *p53* and malignancy within the human intestine (Clarke *et al.*, 1995).

Mice mutant for *p53* were exposed to 5 gray (Gy) γ-irradiation, and apoptosis scored within the intestinal crypts (Clarke *et al.*, 1994). *p53*-deficient mice were found, once again, to be completely resistant to the induction of apoptosis following gamma-irradiation. Further, heterozygotes were found to have a partially diminished response, indicating a clear dose-dependency for *p53*. We extended this analysis to other cell types, namely IL7-dependent pre-b cells (Griffiths *et al.*, 1997) and splenic T cells (Clarke *et al.*, 1997). Both of these cell types again showed a *p53*-dependent death response following clastogenic injury.

This type of response to DNA damage is, however, not consistent across all cell types. A comparison of the responses of fibroblasts (Lowe *et al.*, 1993b,), thymocytes (Clarke *et al.*, 1993; Lowe *et al.*, 1993a), haematopoietic cells (Clarke *et al.*, 1997; Griffiths *et al.*, 1997; Lotem and Sachs, 1993) and intestinal epithelium cells (Clarke *et al.*, 1994; Merritt *et al.*, 1994) shows that not all cell types exhibit *p53*-dependent apoptosis following DNA damage. Indeed, the majority of cell types fail to enter a death programme following ionizing radiation (Clarke *et al.*, 1997; Midgley *et al.*, 1995). Some cells types up-regulate *p53* but fail to enter a death pathway; others, such as hepatocytes, will even fail to stabilize *p53* in response to DNA damage. However, it is also clear that the *p53*-dependent death pathway is intact in many of these cell types, perhaps most clearly demonstrated by embryonic fibroblasts, which acquire a *p53*-dependent apoptotic response on transfection with the adenovirus *e1a* gene. Of the other cell types analysed, it is perhaps significant that the immediate induction of *p53*-dependent apoptosis is observed in cells that normally proliferate rapidly, such as those of the haematopoietic and intestinal systems. This may reflect an intolerance of these populations to DNA damage that might otherwise become fixed and expanded as a mutant sub-clone.

These issues become further complicated where an extended time course is analysed following DNA damage. In the intestine a *p53*-independent wave of apoptosis is evident at delayed time points, peaking at 36 hours following exposure to ionizing irradiation (Clarke *et al.*, 1997). The morphological appearance of these delayed apoptotic events suggests that death is occurring at the G_2 stage of the cell cycle (Merritt *et al.*, 1997). Somewhat surprisingly, we found that the presence of this delayed, *p53*-independent apoptosis was not a general feature of cells which normally engage *p53*-dependent apoptosis, as no prominent delayed wave of death was observed in the mantle zone lymphocytes of the spleen (Clarke *et al.*, 1997). This striking difference may go some way to explain the tumour predisposition observed in *p53* null animals. If apoptosis does play a key role in suppressing neoplasia, then those tissues entirely reliant upon *p53* to perform this function might be predicted to be most tumour prone in a *p53* null background. Thus, *p53* may be effectively redundant in tissues that engage *p53*-independent death, and only crucial where no such delayed wave exists. In support of this hypothesis, *p53* null mice are highly predisposed to T-cell malignancies, precisely that cell type shown not to exhibit *p53*-independent death.

The generation of *p53* mutant mice clearly demonstrates that *p53* deficiency does predispose to malignancy and that *p53* is essential for much of the normal apoptotic response to DNA damage. To address whether a concomitant increase in mutation frequency was occurring in *p53*-deficient cells, we analysed the mutation rate at the *Hprt* locus in wild type and *p53* mutant IL7-dependent B-cell precursors (Griffiths *et al.*, 1997). The *Hprt* locus was chosen because loss of function mutations within this gene can be quantified *in vitro* by the use of appropriate selective media. We observed several key phenomena in these experiments. First, we confirmed the presence of a clear *p53*-dependent apoptotic response, even at relatively low levels of DNA damage

(1 Gy). Second, we found that the absolute failure to engage death pathways, scored 8 hours after exposure, did not translate into an equivalent change in long-term clonogenicity. We did observe a clear increase in clonogenic survival in the *p53* null cells; however, there was a marked loss in survival relative to untreated cells. This again indicates that apoptosis scored at a single time point is not a perfect indicator of long-term survival, invoking the possibility of a *p53*-independent wave of cell death in this cell type and underlining the fact that *p53* deficiency only confers an increase in long-term survival, not absolute protection. Nonetheless, we did observe a clear 10–100 fold difference in clonogenic survival conferred by *p53* deficiency.

Analysis of the surviving clones revealed an *Hprt* mutation rate of 7×10^{-5} in *p53* null cells. Unfortunately, insufficient wild type cells survived to score an equivalent frequency in this genotype. However, the value obtained for *p53* –/– cells is striking in that, by comparison with earlier data (Thacker, 1979), this mutation frequency is within the normal range for the level of radiation exposure used. This leads to the unexpected conclusion that a *p53* null environment does not predispose to an increase in mutation rate, as would be predicted given the multiple roles identified for *p53* in cell cycle arrest and DNA repair. These experiments do, however, show *p53* deficiency to increase mutation burden, simply by virtue of its impact upon clonogenic survival (see *Figure 1*). If these results can be translated into a 'pure' *in vivo* setting, this identifies a critical role for *p53* in preventing the propagation of mutation, not by mediating DNA repair but by the effective deletion of damage bearing cells.

To directly test this hypothesis, we returned to intestinal epithelial cells, in which it is possible to score loss of function mutations at the *Dolichos biflorus* (*Dlb-1*) locus immunohistochemically by failure to bind the *Dolichos biflorus* lectin. We scored mutation frequency in mice mutant for *p53* and heterozygous (a/b) at the *Dlb-1* locus. We found that *p53* status did not affect the spontaneous mutation rate, confirming for an endogenous locus the findings of two independent experiments using an exogenous *lacI* target transgene in addition to an inactivating *p53* mutation (Nishino *et al.*, 1995; Sands *et al.*, 1995). We did however observe a *p53*-dependent difference following exposure to γ-radiation, but only at the highest dose used, 6 Gy. At this dose, the mutation rate at the *Dlb-1* locus was elevated five- to six-fold in *p53* –/– animals compared to that in *p53* wild type animals (Clarke *et al.*, 1997)

The *Dlb-1* study did therefore demonstrate that *p53* deficiency can lead to an increase in the mutation rate, from which it is possible to infer that mutations are more likely within other genes critical for tumorigenesis. However, this increase only occurred at high, non-physiological levels of DNA damage. Possibly this reflects a level of DNA damage for which both *p53*-dependent and *p53*-independent pathways must be intact to efficiently suppress mutation. Taken together, these experiments therefore reveal a complex picture of *p53* reliance, showing that the presence or absence of an apoptotic pathway does not necessarily influence mutation rate, and that the ability of *p53* to act as a 'guardian of the genome' appears strongly cell type-specific (see also Chapter 16).

We next wished to assess the influence of *p53* upon malignant development. To achieve this, we intercrossed the *p53* mutants with mice mutant for the *adenomatous polyposis coli* (*Apc*) gene. In humans, germline heterozygosity predisposes to the development of a large number of adenomas along the length of the colon, a proportion of which can develop into invasive carcinoma. This syndrome is known as familial adenomatous polyposis (FAP). Several different murine strains, characterized

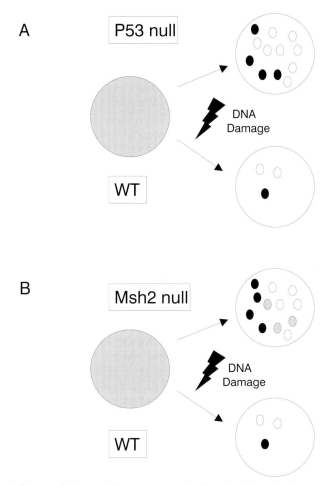

Figure 1. *The influence of clonogenicity upon mutation burden. Mutation frequency, for example at the* Hprt *locus, can often be scored through the simple application of selective media. These diagrams illustrate such experiments. Confluent tissue culture plates (indicated by the filled grey circle) are exposed to DNA damage and subsequently plated in selective medium. Approximately 2 weeks later clones become visible and these can be scored as individual colonies. **A.** Mutation burden dependent upon p53 (illustration relevant to mutation and survival of IL7-dependent pre-B cells). Following exposure to ionizing radiation it is possible to score an increase in clonogenic survival (four-fold in this illustration) in p53 null cells compared to wild type (WT) cells. Of these clones, a proportion are mutant and survive appropriate selection conditions (black circles). Non-mutant clones (white circles) will fail to survive. In this illustration the mutation rate is equal in wild type and p53 mutant cells (33%), but the mutation burden is four-fold higher simply by virtue of increased clonogenic survival. **B.** A similar experiment illustrating mutation burden in wild type (WT) and* Msh2 *null cells (illustration relevant to mutation and survival embryonic stem cells). Following exposure to DNA damage (in this example the methylating agent temozolomide) it is possible to score an increase in clonogenic survival (four-fold in this illustration) in* Msh2 *null cells compared to wild type (WT) cells. Of these clones, a proportion are mutant and survive appropriate selection conditions (black circles). Non-mutant clones (white circles) will fail to survive. In this illustration the number of mutant clones is higher than that predicted from the increase in clonogenic survival (these 'additional' clones are coloured grey), indicating that the mutation rate is higher in* Msh2 *null cells than in wild type cells. Thus, in this example, the increase in mutation burden derives both from an increase in clonogenic survival and a reduced ability to repair DNA damage.*

by germline mutation of *Apc*, have been developed. In all of these models, constitutive inactivation of both *Apc* alleles results in early embryonic lethality. Heterozygotes are characterized by the development of adenomas along the length of both the small and large intestine (Su *et al.*, 1992). As in FAP patients, the development of these lesions is invariably associated with loss of the remaining wild type *Apc* allele.

We wished to address two possibilities. First, that *p53* deficiency may increase the mutation rate at the remaining wild type allele and thereby increase the rate of adenoma formation. Second, that *p53* deficiency would promote neoplastic progression, as in humans loss of *p53* is associated with the later stages of colorectal neoplasia; *p53* mutant mice were therefore intercrossed to *Apc* heterozygotes.

Survival and tumorigenesis were scored in age-matched cohorts of mice which were mutant for *Apc* and *p53* (Clarke *et al.*, 1995). Survival rates on the *Apc* heterozygote background were found to be similar, regardless of *p53* genotype, over the course of the experiment (120 days). Irrespective of *Apc* genotype, *p53*-deficient mice developed lymphoma and sarcoma at the incidence predicted from previous studies (Jacks *et al.*, 1994; Purdie *et al.*, 1994). Surprisingly, analysis of intestinal neoplasia also revealed no differences associated with *p53* deficiency, either in tumour incidence or stage of progression.

However, *Apc* mutant, *p53* null mice additionally developed pancreatic lesions at a very high incidence, with only 17% of these animals with this genotype showing normal pancreatic morphology. A subset of these lesions progressed to pancreatic adenocarcinoma, analysis of which all showed that loss of the remaining wild type *Apc* allele had occurred. These results apparently reveal a very tissue specific role for *p53* in the prevention of neoplasia. In the intestine, *p53* deficiency does not increase the spontaneous rate of loss of the wild type *Apc* allele, as implied by the failure to see an increase in adenoma burden. This is in reasonable agreement with the *Dlb-1* experiments described above, where again no evidence was found for a role for *p53* at spontaneous or low levels of DNA damage. In contrast, absence of *p53* appears to increase the rate of loss of the *Apc* allele in the pancreas, or at least it leads to the rapid progression of *Apc*-deficient lesions.

Taken as a whole, the data obtained from the *p53* null strains presents a complex, and at times contradictory picture of the role of *p53* in carcinogenesis. This can perhaps best be rationalized by treating each cell type in isolation, as the significance of *p53* deficiency appears to markedly differ between cell types.

3. *Msh2*

Structural distortions produced by nucleotides which are either unpaired or paired with non-complementary nucleotides are recognized by proteins encoded by the mismatch repair (MMR) genes. To date, six human MMR genes have been cloned, MSH2, MLH1, PMS1, PMS2, MSH3 and GTPB. Loss of function of these genes in the human has been particularly associated with the inherited cancer-susceptibility syndrome of hereditary non-polyposis colorectal cancer. Mutant murine strains have recently been produced for nearly all of these strains. The phenotype of these shows some similarities with *p53*-deficient mice, as all homozygous null mice are viable but prone to the rapid development of neoplasia.

The Msh2 protein normally recognizes mismatched nucleotides and mediates their removal. Failure of this process is thought to lead directly to an increase in mutation frequency and thereby an increased predisposition to malignancy. However, *Msh2 –/–* cells also show increased survival following exposure to methylating agents *in vitro.* These properties suggest that the Msh2 protein may parallel *p53*, in that it functions both by directly repairing DNA lesions and also by deleting DNA-damage-bearing cells by initiating apoptosis. To directly test this hypothesis, we used ES cells and mice bearing a gene-targeted inactivation of *Msh2* (de Wind *et al.*, 1995). Using these reagents we investigated the influence of *Msh2* status on both and the initiation of apoptosis following DNA damage and mutation frequency (Toft *et al.*, 1999).

We scored apoptosis in the intestine following exposure to the methylating agents MNNG and temozolomide, and also to the cross-linking agent cisplatin as cisplatin adducts may be recognized by *Msh2*. In response to all these drugs, wild type mice showed an apoptotic response which peaked between 8 and 16 hours post-treatment. *Msh2* null mice showed markedly weaker responses to all three drugs, although the degree of *Msh2* dependency varied, being greatest for temozolomide (an approximate five-fold difference) and least for cisplatin (a 5–10% reduction; Toft *et al.*, 1999).

We again then used the *Dlb-1* assay to score the *in vivo* mutation frequency within intestinal epithelium. Using this assay, we found that *Msh2 –/–* mice had a significantly increased mutation frequency both at endogenous levels of DNA damage and following exposure to the methylating agent temozolomide (Toft *et al.*, 1999).

These experiments had successfully established an *in vivo* role for *Msh2* in signalling apoptosis following DNA damage and also in preventing the propagation of mutation. However, it was not clear from these studies if a direct relationship exists between these two phenomena. In an attempt to address this, *in vitro* studies were performed which were designed to identify the roles played by DNA repair and cell death. ES cells, either wild type or null for *Msh2*, were exposed to the methylating agent temozolomide and then clonogenic survival and mutation frequency (at a thymidine kinase transgene) were scored. Results obtained from this approach showed *Msh2* deficiency to confer a 13-fold increase in clonogenicity and an overall 110-fold increase in the number of mutant clones. This implies that both altered cell survival and altered DNA repair influence mutation burden, as a significant component of the *Msh2*-dependent elevation in mutation frequency is directly attributable to the increased survival of *Msh2* null cells (see *Figure 1*; Toft *et al.*, 1999).

4. *Msh2* and *p53*

As *p53* mediates the apoptotic response following a number of different types of injury, we also investigated whether *Msh-2*-dependent death was occurring through a *p53*-dependent pathway. By analysing the response of *p53*-deficient mice we found that MNNG, temozolomide and cisplatin all induced apoptosis in a *p53*-dependent manner, but that there was also a *p53*-independent wave of death which peaked approximately 72 hours after treatment (Toft *et al.*, 1999). We also investigated the apoptotic response of doubly mutant animals. Surprisingly, these failed to show the delayed *p53*-independent apoptosis. This observation suggests that mutations in *p53* and *Msh2* should show cooperativity in the development of neoplasia. Indeed, mice mutant for both genes do show very rapid development of neoplasia, predominantly of the lymphoid system (Cranston *et al.*, 1997; N.J. Toft, unpublished data). Furthermore, the doubly mutant animals show an increased rate of developmental abnormality (Cranston *et al.*, 1997), although this appears to be very background-dependent (Toft *et al.*, 1998).

5. Conclusions

We have used mice deficient for *Msh2* and *p53* in an attempt to address the role played by apoptosis in the prevention of malignancy. Specifically, we have addressed the hypothesis that failure to engage cell death will increase the number of cells with DNA-damage within a given population, that this will lead to an increase in the mutation frequency and thereby lead to an increased predisposition to malignancy. Through a number of different studies we have begun to reveal a very complex pattern of gene dependency, such that answers derived in one tissue type will not necessarily be relevant to other cell types.

It has been possible to characterize several key features. First, the nature of *p53* dependency in the apoptotic response differs between cell types. For example, intestinal cells show both strong *p53*-dependent and -independent apoptotic responses, yet splenic mantle zone cells appear to rely entirely upon *p53* to elicit a death response. In any given tissue the presence or absence of *p53*-independent forms of death may well determine the reliance upon *p53* of that tissue type to prevent neoplasia. Second, *p53* deficiency can be shown to lead to an increase in mutation frequency in some cell types. We have been able to show this phenomenon for IL7-dependent pre-b cells. In this cell type we have also been able to show that this increase probably occurs not through a deficit of DNA repair, but simply through an increase in clonogenic survival. This finding strongly supports the notion that evasion of apop-tosis can play a critical role in carcinogenesis. However, other tissues can be found (such as the intestine) in which *p53* deficiency fails to influence the mutation frequency, even though the principal apoptotic response of that tissue has been ablated. This raises questions about the relevance of *p53*-independent pathways in mutation prevention and it also queries the fundamental purpose of the very efficient *p53*-dependent apoptotic pathway present in the intestine.

We have also found that the role of *p53* in tumour prevention differs markedly between tissue types. Hence, in the intestine, *p53* deficiency apparently does not impact upon neoplastic development or progression. By contrast, it does play a vital role in the prevention of pancreatic malignancy. This latter tissue is not characterized by a *p53*-dependent apoptotic response following exposure to damage, strongly suggesting that this protection is mediated through other functions other than those associated with cell death, such as DNA repair.

Our analysis of mice deficient for the mismatch repair gene *Msh2* has shown that p53 is unique in its ability to mediate both cell death and DNA repair. We have estab-lished that, in addition to its direct role in mediating DNA repair, *Msh2* is required for the normal induction of apoptosis following DNA damage of methylation type. Further, in the absence of Msh2 there is a dramatic increase in susceptibility to mutation. The key question is similar to that posed for p53, namely whether this increase arises as a consequence of failed DNA repair or of failed cell death. *In vitro* experiments designed to answer this question indicate, at least in cultured embryonic stem cells, that both these pathways play a protective role. The challenge now remains to establish the relevance of this finding to different cell types *in vivo*.

We have been able to show that these two pathways (*p53*- and *Msh2*-mediated) are to some extent interdependent, raising the possibility that these two types of mutation, usually considered mutually exclusive events in human cancer, may indeed interact. This possibility is enhanced by the recent observation that hemizygosity of *p53* is sufficient, at least in the mouse, to predispose to malignancy (Venkatachalam *et al.*,

1998). Such a reduction in protein would not normally be identified in screens of human material, making it possible that an interaction is indeed occurring, but that it has not yet been recognized.

In summary, we have used strains mutant for *p53* and *Msh2* to make a series of observations relevant to the hypothesis that neoplastic predisposition is a reflection of the ability to engage apoptosis. We have been able to demonstrate that loss of a given apoptotic pathway can lead to an increase in mutation burden, and that deficiency of genes controlling these pathways predisposes to neoplasia. However, we have also shown that the relationship between these endpoints is far from simple, and that observations made in one tissue type are rarely directly relevant to others. These studies therefore form part of the ongoing challenge to establish, for any given cell type, which pathways are of true relevance to both carcinogenesis and chemoprevention. Only with such detailed knowledge will it become possible to efficiently identify and exploit new targets for therapeutic intervention.

References

Armstrong, J.F., Kaufman, M.H., Harrison, D.J. and Clarke, A.R. (1995) High frequency developmental abnormalities in p53 deficient mice consistent with acquired mutations. *Curr. Biol.* 5: 931–936.

Clarke, A.R., Purdie, C.A., Harrison, D.J., Morris, R.G., Bird, C.C., Hooper, M.L. and Wyllie, A.H. (1993) Thymocyte apoptosis induced by *p53*-dependent and independent pathways. *Nature* 362: 849–851.

Clarke, A.R., Gledhill, S., Hooper, M.L., Bird, C.C. and Wyllie, A.H. (1994) p53 dependence of early apoptotic and proliferative responses within the mouse intestinal epithelium following gamma-irradiation. *Oncogene* 9: 1767–1773.

Clarke, A.R., Cummings, M.C. and Harrison, D.J. (1995) Interaction between murine germline mutations in p53 and Apc predisposes to pancreatic neoplasia but not to increased intestinal malignancy. *Oncogene* 11: 1913–1920.

Clarke, A.R., Howard, L.A., Harrison, D.J. and Winton, D.J. (1997) p53 mutation frequency and apoptosis in the murine small intestine. *Oncogene* 14: 2015–2018.

Cranston, A., Bocker T., Reitmair A., Palazzo J., Wilson T., Mak, T. and Fishel, R. (1997) Female embryonic lethality in mice nullizygous for both *Msh2* and *p53*. *Nature Genet.* 17: 114–118.

de Wind, N., Dekker, M., Berns, A., Radman, M. and te Riele, H. (1995) Inactivation of the mouse *Msh-2* gene results in mismatch repair deficiency, methylation tolerance, hyperrecombination and predisposition to cancer. *Cell* 82: 321–330.

Donehower, L.A., Harvey, M., Slagle, B.L., McArthur, M.J., Montgomery, C.A., Butel, J.S. and Bradley, A. (1992) Mice deficient for p53 are developmentally normal but susceptible to spontaneous tumors. *Nature* 356: 215–221.

Giaccia, A.J. and Kastan, M.B. (1998) The complexity of p53 modulation: emerging patterns from divergent signals. *Genes Dev.* 12: 2973–2983.

Griffiths, S.D., Clarke, A.R., Healy, L.E., Ross, G., Ford, A.M., Hooper, M.L., Wyllie, A.H. and Greaves, M. (1997) Absence of p53 promotes propagation of mutant cells following genotoxic damage. *Oncogene* 14: 523–531.

Harvey, M., Sands, A.T., Weiss, R.S., Hegi, M.E., Wiseman, R.W., Pantazis, P., Giovanella, B.C., Tainsky, M.A., Bradley, A. and Donehower, L.A. (1993a) *In vitro* growth-characteristics of embryo fibroblasts isolated from p53-deficient mice. *Oncogene* 8: 2457–2467.

Harvey, M., McArthur, M.J., Montgomery, C.A., Bradley, A. and Donehower, L.A. (1993b) Genetic background alters the spectrum of tumors that develop in p53-deficient mice. *FASEB J.* 7: 938–943.

Hollstein, M., Sidransky, D., Vogelstein, B. and Harris, C.C. (1991) p53 mutations in human cancers. *Science* **253**: 49–53.

Jacks, T., Remington, L., Williams, B.O., Schmitt, E.M., Halachmi, S., Bronson, R.T. *et al.* (1994) Tumour spectrum analysis in *p53*-mutant mice. *Curr. Biol.* **4**: 1–7.

Livingstone, L.R., White, A., Sprouse, J., Livanos, E., Jacks, T. and Tlsty, T.D. (1992) Altered cell cycle arrest and gene amplification potential accompany loss of wild-type *p53*. *Cell* **70**: 923–935.

Lotem, J. and Sachs, L. (1993) Haematopoietic cells from mice deficient in wild-type *p53* are more resistant to induction of apoptosis by some agents. *Blood* **82**: 1092–1096.

Lowe, S.W., Schmitt, E.S., Smith, S.W., Osborne, B.A. and Jacks, T. (1993a) *p53* is required for radiation-induced apoptosis in mouse thymocytes. *Nature* **362**: 847–849.

Lowe, S.W., Ruley, H.E., Jacks, T. and Housman, D.E. (1993b) *p53*-dependent apoptosis modulates the cytotoxicity of anticancer agents. *Cell* **74**: 957–967.

Malkin, D., Li, F.P., Strong, L.C., Fraumeni, J.F., Nelson, C.E., Kim, D.H., Kassel, J., Gryka, M.A., Bischoff, F.Z., Tainsky, M.A. and Friend, S.H. (1990) Germ line *p53* mutations in a familial syndrome of breast cancer, sarcomas and other neoplasms. *Science* **250**: 1233–1238.

Merritt, A., Potten, C.S., Kemp, C.J., Hickman, J.A., Balmain, A., Lane, D.P. and Hall, P.A. (1994) The role of p53 in spontaneous and radiation induced apoptosis in the gastrointestinal tract of normal and p53 deficient mice. *Cancer Res.* **54**: 614–617.

Merritt, A.J., Allen, T.D., Potten, C.S. and Hickman, J.A. (1997) Apoptosis in small intestinal epithelial from p53-null mice: evidence for a delayed, p53-independent G2/M-associated cell death after gamma-irradiation. *Oncogene* **14**: 2759–2766.

Midgley, C.A., Owens, B., Briscoe, C.V., Thomas, D.B., Lane, D.P. and Hall, P.A. (1995) Coupling between gamma irradiation, p53 induction and the apoptotic response depends upon cell type in vivo. *J. Cell Sci.* **108**: 1843–1841.

Nishino, H., Knoll, A., Buettner, V.L., Frisk, C.S., Maruta, Y., Haavik, J. and Sommer, S.S. *et al.* (1995) p53 wild type and nullizygous Big Blue transgenic mice have similar frequencies and patterns of observed mutation in liver, spleen and brain. *Oncogene* **11**: 263–270.

Purdie, C.A., Harrison, D.J., Peter, A., Dobbie, L., White, S., Howie, S.E.M., Salter, D., Bird, C.C., Wyllie, A.H., Hooper, M.L. and Clarke, A.R. (1994) Tumour incidence, spectrum and ploidy in mice with a large deletion in the p53 gene. *Oncogene* **9**: 1767–1773.

Rafferty, J.A., Clarke, A.R., Sellapan, D., Koref, M.S., Frayling, I.M. and Margison, G.P. (1996) Induction of murine O⁶-alkylguanine-DNA-alkyltransferase in response to ionising radiation is p53 gene dose dependent. *Oncogene* **12**: 693–697.

Sah, V.P., Attardi, L.D., Mulligan, G.J., Williams, B.O., Bronson, R.T. and Jacks, T. (1995) A subset of p53-deficient embryos exhibit exencephaly. *Nature Genet.* **10**: 175–180.

Sands, A.T., Suraokar, M.B., Sanchez, A., Marth, J.E., Donehower, L.A. and Bradley, A. (1995) p53 deficiency does not affect the accumulation of point mutations in a transgene target. *Proc. Natl Acad. Sci. USA* **92**: 8517–8521.

Su, L.K., Kinzler, K.W., Vogelstein, B., Preisinger, A.C., Moser, A.R., Luongo, C., Gould, K.A. and Dove, W.F. (1992) Multiple intestinal neoplasia caused by mutation in the murine homolog of the APC gene. *Science* **256**: 1114–1116.

Thacker, J. (1979) Involvement of repair processes in radiation-induced mutation of cultured mammalian cells. In: *Radiat. Res.* (Proc. 6th Int. Cong. Radiation Res. Tokyo, eds S. Okada *et al.*) pp. 611–620.

Toft, N.J., Arends, M.J., Wyllie, A.H. and Clarke, A.R. (1998) No female embryonic lethality in mice nullizygous for *Msh2* and *p53*. *Nature Genet.* **18**: 17.

Toft, N.J., Winton, D.J., Kelly, J., Howard, L.A., Dekker, M., Te Riele, H., Arends, M.J., Wyllie, A.H., Margison. G.P. and Clarke, A.R. (1999) Msh2 status modulated both apoptosis and mutation frequency in the murine small intestine. *Proc. Natl Acad. Sci. USA* **96**: 3911–3915.

Tsukada, T., Tomooka, Y., Takai, S., Ueda, Y., Nishikawa, S.-I., Yagi, T., Tokunaga, T., Takeda, N., Suda, Y., Abe, S., Matsuo, I., Ikawa, Y. and Aizawa, S. (1993) Enhanced proliferative potential in culture of cells from *p53*-deficient mice. *Oncogene* **8**: 3313–3322.

Venkatachalam, S., Shi, Y.P., Jones, S.N., Vogel, H., Bradley, A., Pinkel, D. and Donehower, L.A. (1998) Retention of wild-type p53 in tumors from p53 heterozygous mice: reduction of p53 dosage can promote cancer formation. *EMBO J.* **17**: 4657–4667.

Yonish-Rouach, E., Resnitzky, D., Lotem, J., Sachs, L., Kimchi, A. and Oren, M. (1991) Wild-type p53 induces apoptosis of myeloid leukaemic cells that is inhibited by interleukin-6. *Nature* **352**: 345–347.

21

Type I interferons inhibit the resolution of chronic inflammation

Dagmar Scheel-Toellner, Arne N. Akbar, Darrell Pilling, Catherine H. Orteu, Christopher D. Buckley, Keqing Wang, Paul R. Webb, Janet M. Lord and Mike Salmon

1. Introduction

Under normal circumstances an immune response develops rapidly following the recognition of a 'non-self' antigen within the body. The antigen, whether it be a virus, bacteria or cancer cell, is then eliminated by the pleiotropic actions of the innate and adaptive immune systems. Resolution of the immune response requires the removal of the majority of immune cells that were expanded and recruited during the active phase of the response. In addition, a small number of lymphocytes are retained to ensure an efficient response if the antigen is encountered again, so-called immune memory (Ahmed and Gray, 1996; Sprent, 1997). Therefore the successful resolution of an immune response is a tightly regulated balancing act between keeping some of the antigen-specific T cells alive to allow a memory response and losing the majority of the activated, effete inflammatory cells.

Over the last few years new insight has been gained into the mechanisms that underlie the organized removal of inflammatory-cell infiltrates that develop in a localized immune reaction. Work from our group and others has shown that these cells are removed by apoptosis and that the inhibition of this process contributes significantly to chronic inflammation.

2. The resolution of immune responses is mediated by apoptosis

Apoptosis, or programmed cell death is a complex procedure by which an effete cell shuts down its normal function, including proliferation and DNA repair mechanisms and packages its constituents so that they can be safely phagocytosed by either a professional phagocyte or a neighbouring cell (Kerr *et al.*, 1972). This mechanism

Programmed Cell Death in Animals and Plants, edited by J.A. Bryant, S.G. Hughes and J.M. Garland.
© 2000 BIOS Scientific Publishers Ltd, Oxford.

allows the body to remove unwanted cells at various stages of development without release of intracellular constituents that might cause breakdown of immunological tolerance or inflammation.

We have shown previously that in viral infection, the initial increase in T-cell numbers seen within a few days, is followed by a wave of apoptosis occurring in the activated T cells (Akbar *et al.*, 1993). This organized involution of the immune response appears to be disregulated in chronic inflammatory diseases (Salmon *et al.*, 1997). In this article we have reviewed the sequence of events involved in the resolution of T-cell-mediated responses in healthy individuals, and then considered the data that inform on the aspects of this process that may be compromised in chronic inflammation.

3. Pathways leading to T-cell apoptosis

There are two major pathways by which T-cell apoptosis is induced: cytokine deprivation induced apoptosis (Duke and Cohen, 1986) and activation-induced apoptosis, usually mediated by Fas/FasL interactions. Apoptosis induced by cytokine deprivation is caused by a lack of survival signals that are transduced by the γ chain of the IL2 receptor. Several cytokine receptors, such as those for IL4, IL7, IL13 and IL15, also associate with the IL2 receptor γ chain following receptor ligation and use it to activate downstream signalling pathways. All of these cytokines are capable of inducing survival in activated T cells (Akbar *et al.*, 1996). Deprivation of survival signals propagated through this pathway leads to loss of the expression of apoptosis preventing molecules of the Bcl-2 family (Akbar *et al.*, 1996; Hawkins and Vaux, 1997). This pathway acts relatively slowly, activated T cells typically begin to die within 24 h of being deprived of survival-mediating cytokines.

Activation-induced apoptosis is triggered through cross-linking of the FAS molecule (CD95, APO1) on the cell membrane by FAS ligand. This signal can be mimicked by cross-linking the FAS receptor by anti-FAS antibodies (Trauth *et al.*, 1989). FAS is a member of the TNF receptor family. Among these a growing number of receptors are found that can induce apoptosis, these are now termed death receptors 1–5 (Peter *et al.*, 1998; Ashkenazi and Dixit, 1998). The signalling pathway leading to FAS-induced apoptosis is currently under intense investigation (Peter *et al.*, 1998). Upon cross-linking of the FAS receptor, a multimolecular complex of the receptor with several intracellular signalling molecules is formed. This complex has been named the DISC (death inducing signalling complex). One of the molecules in the DISC, caspase 8 (FLICE), is cleaved and activated as a result of receptor oligomerization and DISC formation. Caspase 8 is a member of a family of cysteine proteases that form a cascade reminiscent of the cascade in complement activation (Thornberry and Lazebnik, 1998). The caspase cascade can be divided into initiator and effector caspases. The initiator caspases, 8, 9 and 10, are involved in the upstream signals that induce the process of apoptosis. When activated, they cleave and thus activate effector caspases like 3 and 6. The effector caspases have a multitude of targets within the cell and thus directly or indirectly lead to the cellular changes characteristic of apoptosis. FAS-induced apoptosis occurs rapidly; in highly responsive cell lines the morphological changes characteristic for end-stage apoptosis, such as membrane blebbing and nuclear condensation, occur within 3 hours.

4. Resolution of T-cell mediated immune responses in healthy individuals

We have recently analysed the resolution of T-cell-mediated immune responses in the skin of healthy individuals (Orteu *et al.*, 1998). The lesions were induced by cutaneous injection of mycobacterial protein preparations (purified protein derivative). This injection leads to a classical T-cell-mediated immune response termed the Mantoux reaction (Turk, 1980). It typically develops within 3 days and is resolved within 14 days. We followed the expression of molecules involved in the regulation of apoptosis as well as proliferation and apoptosis. While the number of infiltrating T cells peaked at day 7, by day 14 the number of T cells had begun to decline. The maximal expression of FAS ligand coincided with maximal T-cell proliferation, bcl-2 expression and IL2 production, suggesting that it occurred coincident with T-cell activation. T-cell apoptosis occurring at this time-point, is likely to represent FAS-induced apoptosis, since there was ample IL2 and IL15 present to prevent aptosis induced by cytokine deprivation. The levels of T-cell apoptosis were higher at day 14 than on day 7. At this time-point, markedly less FAS ligand and IL2 were detectable and bcl-2 expression had been down-regulated in T cells. This is more likely to actually represent the resolution phase of the immune response to the mycobacterial proteins. The lack of FAS ligand expression, combined with the lack of IL2, IL15 and Bcl-2 expression at this time point, suggests that the increased rate of apoptosis was triggered by cytokine deprivation.

Taken together, the data from these experiments indicate that FAS-mediated apoptosis occurs at the peak of inflammation and may provide a mechanism for preventing over-stimulation of the immune response. The actual resolution of the T-cell-mediated immune response appears to be mediated by cytokine deprivation-induced apoptosis.

5. The resolution of immune responses is inhibited in chronic inflammation

One of the most intriguing features of rheumatoid arthritis is that it represents an inflammatory response that does not resolve. During their lifetime most people will suffer from transient events of synovitis or even arthritis, typically associated with infections, but usually these will resolve spontaneously after a short time. In rheumatoid arthritis, the inflammation becomes chronic and persistent. We have recently shown that the persistence of T cells in the rheumatoid joint can be explained by inhibition of apoptosis (Salmon *et al.*, 1997). We looked extensively for apoptotic T cells in synovial fluid of patients with rheumatoid arthritis, using almost every available method. Although we found apoptotic granulocytes, we did not detect any apoptotic T cells. When we applied the same methods to the synovial fluid of patients with gout, we detected both apoptotic granulocytes and T cells.

The lack of T-cell apoptosis in rheumatoid synovial fluid cannot be explained by an intrinsic resistance of these cells to apoptosis, since when isolated from the synovial microenvironment, they died rapidly (Salmon *et al.*, 1997). The available data suggest that T cells in the rheumatoid synovium are a normal population of highly differentiated cells (Matthews *et al.*, 1993). Moreover, their low level of CD45RB expression suggests that these cells have been through multiple cycles of reactivation (Salmon *et al.*, 1994). This phenotype is associated with a high susceptibility to apoptosis. Cells with this phenotype can be found in the peripheral blood of normal healthy individuals, suggesting that synovial fluid T-cells are a population of highly selected, but otherwise normal cells. The low response of these cells to activation (Kitas *et al.*, 1988) can thus be explained by their differentiation stage, rather than by an intrinsic cellular defect. The question that now remains is, if these cells are highly susceptible to apoptosis, what is keeping them alive in the synovium of patients with rheumatoid arthritis?

5.1 *Stromal cells provide an anti-apoptotic microenvironment for T cells*

Kurnick and co-workers showed that activated T cells can be rescued from apoptosis by co-culture with fibroblasts (Scott *et al.*, 1990). When we cultured synovial T cells in the presence of synovial fibroblasts, or with their conditioned medium, we found that this delayed their apoptosis dramatically (Salmon *et al.*, 1997). In co-cultures of T cells with fibroblasts the T cells up-regulated their expression of Bcl-X_L, a molecule that increases resistance to apoptosis in many cellular systems (Boise *et al.*, 1994). However Bcl-2, a closely related anti-apoptotic molecule was not affected. These data support the involvement of fibroblasts in the rescue of T cells from apoptosis, since we found a similar phenotype of high Bcl-X expression and low Bcl-2 expression in synovial T cells. Our results suggest that the inflammatory T cell infiltrate persists, at least in part, because apoptosis is actively suppressed by the synovial micro-environment. It is very unlikely that T-cell apoptosis is completely blocked. We suggest that during active flares of rheumatoid disease activity, probably associated with random infections, cells are actively recruited to the joint in a largely non-specific manner. During more stable periods of disease, the suppression of T-cell apoptosis will maintain the infiltrating population for extended periods of time.

6. Stromal rescue is mediated by IFN-β

After spending several years on testing a large number of cytokines or adhesion molecules, we identified interferon-β as the factor by which fibroblasts prevent T-cell apoptosis (Pilling *et al.*, 1999). Actually, both IFN-α and IFN-β and their multiple isoforms can inhibit cytokine deprivation induced apoptosis in activated T cells. They share the same receptor and are grouped under the term type I interferon. Fibroblasts only produce IFN-β, but IFN-α produced by macrophages and other cell types can also provide a rescue signal for T cells. Type I interferon has been detected in the rheumatoid synovium by ELISA (Feldmann *et al.*, 1996), by immunohistological staining and by a highly sensitive bioasssay using translocation of STAT1 as a measure of IFN-signalling activity (Pilling *et al.*, 1999). Interestingly, we found far lower levels of Type I interferon in samples from patients with gout and osteoarthritis. As mentioned above, we detected far higher levels of T-cell apoptosis in gout compared to rheumatoid synovial fluid, again suggesting that the *in vivo* inhibition of T-cell apoptosis can be explained by IFN-β.

Other groups have shown that type I interferons such as IFN-α and IFN-β can act as survival factors for T cells, myeloma cells and for other B-cell lines (Kaneko *et al.*, 1997; Milner *et al.*, 1995; Panayiotidis *et al.*, 1994). This appears to be an evolutionary conserved pathway, since Marrack *et al.* (1999) have recently reported rescue of activated T cells from apoptosis by type I interferons in mice type I interferons are able to prevent apoptosis of activated T cells without inducing proliferation (Gombert *et al.*, 1996), this could potentially play an important role in the maintenance of T-cell memory. In the resolution phase of a T-cell-mediated immune response some of the activated T cells need to be kept alive to maintain a memory response upon re-encounter with the antigen. It is possible that the stromal (IFN-β)-mediated rescue of activated T cells from apoptosis is essentially a physiological mechanism for the generation and maintenance of T-cell memory, that under certain circumstances is subverted, leading to the perpetuation of chronic inflammation.

6.1 *IFN-β blocks FAS-induced apoptosis in T cells*

As described above, type I interferons have a profound inhibitory effect on cytokine deprivation induced apoptosis in T cells. As we mentioned above, this is only one of the two major pathways in induction of T-cell apoptosis. When we tested the effect of IFN-β on FAS-mediated apoptosis, we found that in activated T-cell lines and T-cell clones IFN-β was able to reduce FAS-induced apoptosis (Scheel-Toellner *et al.*, 1999). These observations suggest that IFN-β has a general ability to block T-cell apoptosis.

7. Inhibition of spontaneous neutrophil apoptosis *in vitro* and in the rheumatoid joint

The dominant cell populations in the synovial fluid of patients with rheumatoid arthritis are not lymphocytes, but neutrophil granulocytes. In the synovial tissue, though, they are quite rare. Interestingly, those found in the tissue are situated at the pannus/cartilage interface, the area in which most damage occurs (Edwards and Hallett, 1997). Although we found apoptotic neutrophils in the rheumatoid synovial fluid, these occurred at relatively low levels, considering the short half-life of these cells *in vivo*. This led us to consider a role for the inhibition of neutrophil apoptosis by the rheumatoid micro-environment.

Neutrophils are very short-lived cells *in vivo*, they have a half-life of about 8 hours. When isolated from the peripheral blood and taken into culture approximately 60% of the cells spontaneously enter apoptosis within 20 hours. When we tested the effect of IFN-β on neutrophil spontaneous apoptosis we found that their apoptosis was significantly delayed. IFN-β had an even higher effect on neutrophil apoptosis than on T cells (K.Q. Wang, unpublished observations). Importantly, when we tested the ability of IFN-β rescued neutrophils to undergo a respiratory burst after triggering them with the general neutrophil activator FMLP, they were still capable of being activated. We therefore concluded that IFN-β maintains the neutrophils in a functional non-apoptotic state.

For neutrophils isolated from synovial fluid, IFN-β had a less dramatic, but still significant effect. As mentioned above we found higher levels of T-cell apoptosis and lower levels of IFN-β in synovial fluid of gout patients. To determine whether a similar effect would be seen for neutrophil apoptosis, we prepared smears of fresh synovial fluids of patients with gout or RA and stained them to detect the characteristic nuclear changes of apoptotic neutrophils. In line with our results on T-cell apoptosis in rheumatoid patients, we detected significantly less neutrophil apoptosis in the rheumatoid synovial fluids.

We therefore conclude that type I interferons can inhibit the resolution of chronic inflammation by inhibiting the apoptosis of T cells as well as neutrophil granulocytes in the rheumatoid synovium.

8. Apoptosis of stromal cells in rheumatoid arthritis

Several groups have studied apoptosis in the rheumatoid synovium, both as a biological process, as a way to understand the disease but also as a simple therapeutic tool. Firestein *et al.* (1995) have also reported unexpectedly low levels of apoptosis in the synovium, particularly noting the absence of T-cell death. Others have concentrated instead on mechanisms to induce apoptosis in synovial cells therapeutically; including cytotoxic drugs and gene-therapy approaches (Migita *et al.*, 1997; Okamoto *et al.*, 1998). Intriguingly, drugs commonly used in rheumatoid arthritis may operate through induction of apoptosis, including steroids and sulphasalazine (VelaRoch *et al.*, 1998).

Firestein's group has developed the hypothesis that synovial fibroblasts are inhibited from dying by apoptosis by mutations in control genes, such as p53 (Firestein *et al.*, 1997). Such mutations are a prerequisite, but are insufficient for the development of malignancy. Intriguingly, many of the mutations they identified were identical to those found previously in malignant cells. They propose that oxygen free radicals produced as part of the inflammatory process lead to mutations in p53 in fibroblasts, rendering them less susceptible to apoptosis. This model integrates very effectively with the model for chronic inflammation derived from our own studies described earlier, as the proliferating fibroblasts would then maintain survival of inflammatory cells within the lesion via production of IFN-β. Oxidant-induced somatic mutations of p53 in fibroblasts would then provide the point at which a specific acute inflammatory process switches to a chronic phase and loses its specificity. At this point it simply becomes chronic inflammation, within a site determined by the primary reaction. Synovitis will therefore persist in the joints, inflammatory bowel disease in the gut, myositis in the muscles.

9. How does type I interferon inhibit T-cell apoptosis?

The role of type I interferons in the inhibition of the resolution of chronic inflammation makes it an interesting target for therapy. One possible way forward would be to block IFN action using antibodies. This approach is limited, however, since there are multiple isoforms of type I interferons. An alternative approach would be to interrupt the signalling pathways involved in type I interferon-mediated rescue. In recent years we have tried to define intracellular signalling mechanisms that could explain the rescue effect. There are three candidate mechanisms so far; they all interfere with different steps in the pathway to apoptosis, and therefore could complement each other.

9.1 *Type I interferon up-regulates the expression of Bcl-X$_L$*

Up-regulation of Bcl-X$_L$ expression by type 1 interferons (Gombert *et al.*, 1996; Pilling *et al.*, 1999) is probably induced by the activation of STAT 1 and 2 molecules to form homo- and hetero-dimers, that translocate to the nucleus and induce gene transcription (*Figure 1*) (O'Shea, 1997). Bcl-X$_L$ is mostly localized in the mitochondria and inhibits apoptosis through an as yet unexplained mechanism. One of the possibilities is that it inhibits the activation of caspase 9 by APAF1 and cytochrome c, since it has been shown to bind to APAF 1 (Pan *et al.*, 1998).

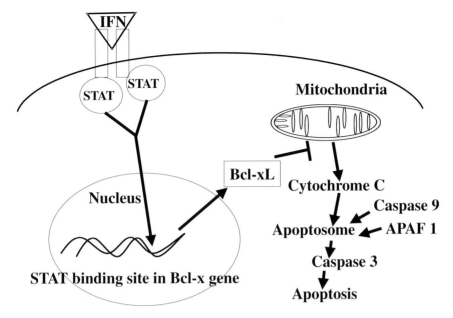

Figure 1. *Type I interferon upregulates Bcl-X$_L$ expression. Cross-linking of the type I interferon receptor by IFN-β leads to the activation and dimerization of STAT molecules. These translocate to the nucleus and bind to STAT binding sites in the promoter region of the Bcl-X gene. The expression of the Bcl-X gene is increased, and the resulting increase of the Bcl-X$_L$ level inhibits the mitochondrial release of cytochrome c. Inhibition of cytochrome c release by Bcl-X leads to a decreased susceptibility to apoptosis.*

Up-regulation of Bcl-X$_L$ is relatively slow, it takes at least 6 hours until an up-regulation of Bcl-X$_L$ protein can be detected (D. Pilling *et al.*, unpublished observation). This mechanism can therefore explain the inhibition of cytokine deprivation-induced apoptosis, but it cannot explain the inhibition of FAS-induced apoptosis, which progresses within hours (Trauth *et al.*, 1989). In addition Bcl-X$_L$ has been shown not to be upregulated in the type I interferon-mediated inhibition of neutrophil spontaneous apoptosis (P. Webb, unpublished observations). Type I interferon-mediated up-regulation of Bcl-X$_L$ can thus mediate a reduced susceptibility to apoptosis, but is unlikely to explain the blockage of the more rapid pathways to apoptosis. We were therefore investigating more rapid pathways that could actually lead to a block of the apoptotic pathway.

9.2 *Type I interferons reverse and suppress the nuclear translocation of protein kinase C-δ*

Protein kinase C (PKC)-δ is one of 11 isoforms of PKC, which are independently regulated and are differentially involved in the regulation of proliferation, differentiation and apoptosis (Deacon *et al.*, 1997). Several lines of evidence suggest that PKC-δ is involved in the regulation of apoptosis. Over-expression of the constitutively active form leads to apoptosis (Ghayur *et al.*, 1996) and activation by its specific activator, bistratene A, mediates cell cycle arrest and apoptosis (Griffith *et al.*, 1996; Watters *et al.*, 1994). PKC-δ translocates to the nuclear membrane early in apoptosis and at a later stage, is cleaved there by caspase 3 (Emoto *et al.*, 1995). As illustrated in *Figure 2*, the

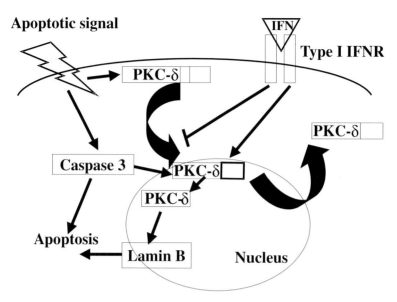

Figure 2. *Type I interferons regulate the subcellular localization of PKC-δ. Induction of apoptosis triggers the translocation of PKC-δ to the nucleus. In the nucleus PKC-δ is cleaved by activated caspase 3. The resulting fragment is constitutively active and phosphorylates Lamin B, this is a prerequisite for the dissolution of the nuclear lamina in the early phases of apoptosis. The translocation of the intact PKC-δ can be inhibited and reversed by type I interferon. This mechanism can contribute to the active inhibition of nuclear changes in apoptosis by type I interferons.*

translocation of PKC-δ to the nuclear membrane can be reversed by type I interferons (Scheel-Toellner *et al.*, 1999). Type I interferons also inhibit the caspase 3-mediated proteolytic activation of PKC-δ. One of the targets for phosphorylation by PKC-δ is lamin B, which is one of the constituents of the nuclear lamina. In apoptosis lamin B is phosphorylated by PKC-δ and subsequently is cleaved by caspase 6 (Earnshaw, 1995; M. Lord *et al.*, unpublished observation). One of the effects of type I IFN could be the inhibition of the dissolution of the nuclear lamina in apoptosis since it removes PKC-δ from the vicinity of its target, lamin B. It also removes PKC-δ from the subcellular location in which it is proteolytically activated by caspase 3.

This signalling pathway could explain the effects of IFN-β on downstream events in the effector phase of apoptosis, but we recently found that activation of caspase 3 is also inhibited by IFN-β. We therefore need to identify a rescue signal that fulfils the requirements of being fast and intersecting with the apoptotic signalling pathway upstream of caspase 3 activation.

9.3 *Do type I interferons inhibit apoptosis by a PI3 kinase-dependent pathway?*

We are currently investigating the involvement of a PI3 kinase dependent pathway in the interferon-mediated rescue of T cells and neutrophils from apoptosis (*Figure 3*). AKT or PKB is a serine/threonine kinase that plays a central role in the regulation of apoptosis (Downward, 1998). It is activated through a PI3 kinase-dependent pathway (Toker and Cantley, 1997). AKT deactivates the pro-apoptotic member of the bcl-2

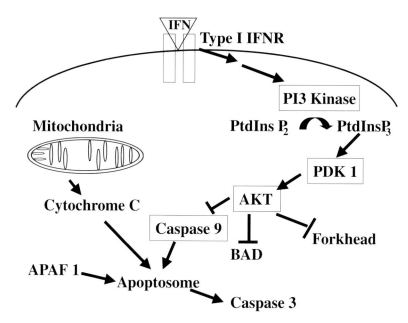

Figure 3. *Type I interferons activate a PI-3 kinase-dependent anti-apoptotic pathway. Triggering of the type I interferon receptor leads to activation of a PI-3 kinase-dependent pathway, leading to activation of AKT 1. This pathway may directly block activation of caspase 3 by deactivating upstream molecules such as caspase 9. This provides a mechanism by which type I interferons can directly block actively induced apoptosis.*

family BAD by phosphorylating it (Datta *et al.*, 1997). AKT has also been shown to phosphorylate transcription factors of the forkhead family (Brunet *et al.*, 1999) which are involved in the regulation of transcription of pro-apoptotic genes, such as FAS-L. Recently, AKT was also implicated in the regulation of caspase activation, as it has been shown to phosphorylate caspase 9 and prevent its activation (Cardone *et al.*, 1998). Caspase 9 is involved in the induction of apoptosis through the mitochondrial pathway (Green and Reed, 1998) Many apoptotic signals, such as FAS receptor cross-linking, irradiation and staurosporine lead to release of cytochrome c from the mito-chondriae. In the cytosol cytochrome c binds to APAF1 and pro-caspase 9. Formation of this complex, known as the apoptosome, leads to proteolytic activation of caspase 9, which then cleaves and activates caspase 3 (Li *et al.*, 1997). Caspase 3 cleaves a large number of target proteins within the cell and thus leads to many of the characteristic features of apoptosis, like DNA cleavage and nuclear condensation.

A recent publication showed that caspase 9 can be regulated by phosphorylation. Phosphorylation of caspase 9 on serine 196 by the serine threonine kinase Akt prevented the proteolytic activation of caspase 9 and its subsequent activation of caspase 3 (Cardone *et al.*, 1998). We are currently investigating the effects of IFN-β on AKT activation.

10. Conclusion

We have discussed mechanisms that could explain the inhibition of the normal reso-lution of the inflammatory response in the rheumatoid synovium. This does not explain how and why an inflammatory infiltrate initially develops in the synovial tissue, but does indicate a clear mechanism leading to its persistence. Patients with multiple sclerosis who are treated with high doses of IFN-β not infrequently develop arthritis as a side-effect (Levesque, 1999). The symmetrical polyarthritis in these patients shared many similarities with rheumatoid arthritis, supporting the importance of type-I interferons in the pathology of rheumatoid arthritis. A clear understanding of the mode of action of this cytokine in rheumatoid arthritis is very likely to lead to the identification of novel therapeutic targets for this common inflammatory disease.

References

Ahmed, R. and Gray, D. (1996) Immunological memory and protective immunity – understanding their relation. *Science* **272**: 54–60.

Akbar, A.N., Borthwick, N.J., Salmon, M., Gombert, W., Bofill, M., Shamsadeen, N., Pilling, D., Pett, S., Grundy, J.E. and Janossy, G. (1993) The significance of low Bcl-2 expression by CD45RO-T-cells in normal individuals and patients with acute viral infections – The role of apoptosis in T-cell memory. *J. Exp. Med.* **178**: 427–438.

Akbar, A.N., Borthwick, N.J., Wickremasinghe, R.G., Panayiotidis, P., Pilling, D., Bofill, M., Krajewski, S., Reed, J.C. and Salmon, M. (1996) Interleukin-2 receptor common gamma-chain signalling cytokines regulate activated T-cell apoptosis in response to growth factor withdrawal-selective induction of anti-apoptotic (Bcl-2, Bcl-x(L)) but not pro-apoptotic BAX, Bcl-x(S)) gene expression. *Eur. J. Immunol* **26**: 294–299.

Ashkenazi, A. and Dixit, V.M. (1998) Death receptors: signaling and modulation. *Science* **281**: 1305–1308.

Boise, L.H., Gonzalesgarcia, M., Postema, C.E., Ding, L.Y., Lindsten, T., Turka, L.A., Mao, X.H., Nunez, G. and Thompson, C.B. (1994) Bcl-x, a bcl-2 related gene that functions as a dominant regulator of apoptotic cell death. *Cell* **74**: 597–608.

Brunet, A., Bonni, A., Zigmond, M.J., Lin, M.Z., Juo, P., Hu, L.S., Anderson, M.J. Arden, K.C., Blenis, J. and Greenberg, M.E. (1999) Akt promotes cell survival by phosphorylating and inhibiting a Forkhead transcription factor. *Cell* **92**: 857–868.

Cardone, M., Roy, N., Stennicke, H.R., Salvesen G.S., Franke, T. F., Stanbridge, E., Frisch, S. and Reed, J.C. (1998) Regulation of cell death protease Caspase-9 by phosphorylation. *Science* **282**: 1318–1320.

Datta, S.R., Dudek, H., Tao, X., Masters, S., Fu, H., Gotoh, Y. and Greenberg, M. (1997) AKT phosphorylation of BAD couples survival signals to the cell-intrinsic death machinery. *Cell* **91**: 231–241.

Deacon, E.M., Pongracz, J., Griffiths, G. and Lord, J.M. (1997) Isoenzymes of protein kinase C: differential involvement in apoptosis and pathogenesis. *J. Clin. Path. Mol. Path.* **50**: 124–131.

Downward, J. (1998) Mechanisms and consequences of activation of protein kinase B/Akt. *Curr. Op. Cell Biol.* **10**: 262–267.

Duke, R.C. and Cohen, J.J. (1986) IL-2 addiction-withdrawal of growth-factor activates a suicide programme in dependent T-cells. *Lymphokine Res.* **5**: 289–299.

Earnshaw, W.C. (1995) Nuclear changes in apoptosis. *Curr. Opin. Cell Biol.* **7**: 337–343.

Edwards, S. and Hallett, M.B. (1997) Seeing the wood for the trees: the forgotten role of neutrophils in rheumatoid arthritis. *Immunol. Today* **18**: 320–324.

Emoto, E., Manome, Y., Meinhardt, G., Kisaki, H., Kharbanda, S., Robertson, M., Ghayur, T., Wong, W.W., Kamen, R., Weichselbaum, R. and Kufe, D. (1995) Proteolytic activation of protein kinase C δ by an ICE-like protease in apoptotic cells. *EMBO J.* **14**: 6148–6156.

Feldmann, M., Brennan, F.M. and Maini, R.N. (1996) Role of cytokines in rheumatoid arthritis. *Annu. Rev. Immunol.* **14**: 397–440.

Firestein, G.S., Yeo, M. and Zvaifler, N.J. (1995) Apoptosis in rheumatoid arthritis synovium. *J. Clin. Invest.* **96**: 1631–1638.

Firestein, G.S., Echeverri, F., Yeo, M., Zvaifler, N.J. and Green, D.R. (1997) Somatic mutations in the p53 tumor suppressor gene in rheumatoid arthritis synovium. *Proc. Natl Acad. Sci. USA* **94**: 10895–10900.

Ghayur, T., Hugunin, M., Talanian, R.V., Ratnofsky, S., Quinlan, C., Emoto, Y., Pandey, P., Datta, R., Huang, Y.Y., Kharbanda, S., Allen, H., Kamen, R., Wong, W. and Kufe, D. (1996) Proteolytic activation of protein kinase C delta by an ICE/CED3-like protease induces characteristics of apoptosis *J. Exp. Med.* **184**: 2399–2404.

Gombert, W., Borthwick, N.J., Wallace, D.L., Hyde, H., Bofill, M., Pilling, D., Beverley, P. C.L., Janossy, G., Salmon, M. and Akbar, A.N. (1996) Fibroblasts prevent apoptosis of IL-2 deprived T-cells without inducing proliferation – a selective effect on Bcl-x(L) expression. *Immunology* **89**: 397–404.

Green, D.R. and Reed, J.C. (1998) Mitochondria and apoptosis. *Science* **281**: 1309–1312.

Griffiths, G., Garrone, B., Deacon, E., Owen, P., Pongracz, J., Mead, G., Bradwell, A., Watters, D. and Lord, J.M. (1996) The polyether bistratene-A activates protein-kinase-C-delta and induces growth arrest in HL-60 cells. *Biochem. Biophys. Res. Comm.* **222**: 802–808.

Hawkins, C.J. and Vaux, D.L. (1997) The role of the Bcl-2 family of apoptosis regulatory proteins in the immune system. *Semin. Immunol.* **9**: 25–33.

Kaneko, S., Suzuki, N., Koizumi, H., Yamamoto, S. and Sakane, T. (1997) Rescue by cytokines of apoptotic death induced by IL-2 deprivation of human-antigen specific T cell clones. *Clin. Exp. Immunol.* **109**: 185–193.

Kerr, J.F.R., Wyllie, A.H., Currie, A.H. (1972) Apoptosis, a basic biological phenomenon with wide ranging implications in tissue kinetics. *Br. J. Cancer.* **26**: 239–257.

Kitas, G.D., Salmon, M., Farr, M., Young, S.P. and Bacon, P.A. (1988) T-cell functional defects, intrinsic or extrinsic. *J. Autoimmun.* **1**: 339–351.

Levesque, M.C., Ward, F.E., Jeffery, D.R. and Weinberg, J.B. (1999), Interferon-β1A-induced polyarthritis in a patient with the HLA-DRB1*0404 allele. *Arthritis Rheum.* **42**: 569–573.

Li, P., Nijhawan, D., Budihardjo, I., Srinivasula, S.M., Ahmad, M., Alnemri, E.S. and Wang, X.D. (1997) Cytochrome c and dATP-dependent formation of Apaf-1/caspase-9 complex initiates an apoptotic protease cascade. *Cell* **91**: 479–489.

Marrack, P., Kappler, J. and Mitchell, T. (1999) Type I interferons keep activated T-cells alive. *J. Exp. Med* **189**: 521–529.

Matthews, N., Emery, P., Pilling, D., Akbar, A.N. and Salmon, M. (1993) CD45RB expression by T-lymphocytes from patients with rheumatoid arthritis. *Arthritis Rheum.* **36**: 603–607.

Migita, K., Eguchi, K., Ichinose, Y., Kawabe, Y., Tsukada, T., Aoyagi, T. and Nagataki, S (1997) Effects of rapamycin on apoptosis of rheumatoid synovial cells. *Clin. Exp. Immunol.* **108**: 199–203.

Milner, A.N., Grand, R.J.A. and Gregory, C.D. (1995) Effects of interferon-alpha on human B-cells repression of apoptosis and prevention of cell growth are independent responses of Burkitt-lymphoma lines. *Int. J. Cancer* **61**: 348–354.

Okamoto, K., Asahara, H., Kobayashi, T. *et al.* (1998) Induction of apoptosis in the rheumatoid synovium by Fas ligand gene transfer. *Gene Therapy* **5**: 331–338.

Orteu, C.H., Poulter, L.W., Rustin, M.H.A., Sabin, C.A., Salmon, M. and Akbar, A.N. (1998) The role of apoptosis in the resolution of T-cell mediated cutaneous inflammation. *J. Immunol.* **161**: 1619–1629.

O'Shea, J.J. (1997) Jaks, STATs cytokine signal transduction, and immunoregulation: are we there yet? *Immunity* **7**: 1–11.

Pan, G.H., O'Rourke, K. and Dixit, V.M. (1998) Caspase 9, Bcl-x and Apaf 1 form a ternary complex. *J. Biol. Chem.* **273**: 5843–5845.

Panayiotidis, P., Ganeshaguru, K., Jabbar, S.A.B. and Hoffbrand, A.V. (1994) Alpha interferon (αIFN) protects B-chronic lymphocytic-leukemia cells from apoptotic cell death in vitro. *Br. J. Haematol.* **86**: 169–173.

Peter, M.E., Scaffidi, C., Medema, J.P., Kischkel, F.C. and Krammer, P.H. (1998) The death receptors. In *Apoptosis, Problems and Diseases* (ed. S. Kumar). Springer, Heidelberg, pp. 24–63.

Pilling, D., Akbar, A.N., Girdlestone, J., Orteu, C.H., Amft, E.N., Borthwick, N.J., Scheel-Toellner, D., Buckley, C.D. and Salmon, M. (1999) Interferon-β mediates stromal cell rescue of T-cells from apoptosis. *Eur. J. Immunol.* **29**: 1041–1050.

Salmon, M., Pilling, D., Borthwick, N.J., Viner, N., Janossy, G., Bacon, P.A. and Akbar, A.N. (1994) The progressive differentiation of primed T-cells is associated with an increasing susceptibility to apoptosis. *Eur. J. Immunol.* **24**: 892–899.

Salmon, M., Scheel-Toellner, D., Huissoon, A.P., Pilling, D., Shamsadeen, N., Hyde, H., D'Angeac, A.D., Bacon, P.A., Emery, P. and Akbar, A.N. (1997) Inhibition of T-cell apoptosis in the rheumatoid synovium. *J. Clin. Invest.* **99**: 439–446.

Scheel-Toellner, D., Akbar, A.N., Pilling, D., Hardie, D., Lombardi, G., Salmon, M. and Lord, J.M. (1999) Inhibition of T-cell apoptosis by IFN-β rapidly reverses nuclear translocation of protein kinase C-δ. *Eur. J. Immunol.* **29**: 2603–2612.

Scott, S., Pandolfi, S. and Kurnick, J.T. (1990) Fibroblasts mediate T-cell survival – a proposed mechanism for retention of primed T-cells. *J. Exp. Med.* **172**: 1873–1876.

Sprent, J. (1997) Immunological memory. *Curr. Opin. Immunol.* **9**: 371–379.

Thornberry, N.A. and Lazebnik, Y. (1998) Caspases: enemies within. *Science* **281**: 1312–1316.

Toker, A. and Cantley, L. (1997) Signalling through the lipid products of phosphoinositide-3-OH kinase *Nature* **387**: 673–676.

Trauth, B.C., Klas, C., Peters, A.M., Matzku, S., Moller, P., Falk, W., Debatin, K.M. and Krammer, P.H. (1989) Monoclonal antibody mediated tumor regression by induction of apoptosis. *Science* **245**: 301–305.

Turk, J.L. (1980) Production of delayed type hypersensitivity and its manifestations. In: *Delayed Hypersensitivity*, 3rd edn. Elsevier/North Holland, Amsterdam, p. 13.

VelaRoch, N., Kong, L.P., Talal, N. and Dang, H. (1998) Sulphasalazine and its metabolites may mediate their therapeutic effects through apoptosis. *Arthritis Rheum.* **41**: 9SS, p. 765.

Watters, D.J., Beamish, H.J., Marshall, K.A., Gardiner, R.A. and Seymour, G.J. (1994) Accumulation of HL60 leukemia cells in G2/M and inhibition of cytokinesis caused by 2 marine compounds, bistratene-A and cyclooxazoline. *Cancer Chemother. Pharmacol.* **33**: 399–409.

22

Death of plant cells and their contribution to rumen function

A.H. Kingston-Smith and M.K. Theodorou

1. An introduction to the rumen ecosystem

Livestock farming has become increasingly dependent upon the feeding of high levels of supplemental protein to overcome the inability of grass as forage to meet the nutrient demands of animals of progressively higher genetic merit. One of the disadvantages of this practice was dramatically demonstrated by the outbreak of bovine spongiform encepalopathy (BSE) in the United Kingdom's herds in the 1980s and 1990s. In the future, in the face of declining product prices, agricultural policy reforms and widespread consumer concern, farmers will need to reduce their reliance on purchased supplements and make greater use of the protein in farm-grown forages. However, moves towards more sustainable systems of livestock production are impeded by the disappointingly low efficiency with which ruminants utilize forage proteins. For example, between 70 and 90% of dietary nitrogen is not incorporated into meat and milk but is excreted from the animal (Dewhurst *et al.*, 1996).

The nutrition of the ruminant animal is dependent upon the large and diverse population of bacteria, protozoa and fungi in the rumen. These micro-organisms degrade freshly ingested plant biomass, providing the host animal with usable forms of carbon and nitrogen (Theodorou and France, 1993). However, soon after ingestion of fresh herbage, the rate of degradation of plant carbohydrate is insufficient to meet the energy demands of the rumen microbial population and therefore, plant proteins are used as an energy source. This is an inefficient process whereby plant proteins are degraded to amino acids and deaminated with the consequential production of copious quantities of ammonia, destined to be excreted from the animal in the form of urea. Ironically, the practice of adding easily degradable protein supplements to the ruminant diet exacerbates the situation even further, resulting in the excretion of more polluting nitrogen.

Significant improvements in the use of forage proteins by ruminants can be achieved by either (i) making energy more readily available in the rumen in the hours following

Programmed Cell Death in Animals and Plants, edited by J.A. Bryant, S.G. Hughes and J.M. Garland.
© 2000 BIOS Scientific Publishers Ltd, Oxford.

ingestion of the plant biomass (e.g. by supplying high-energy concentrates), or (ii) by decreasing the degradation of forage proteins in the rumen. For example, the main reason for the increased nutritive value of spring-fed compared to autumn-fed grass is the increased efficiency of use of amino acids in the former. This is mainly because the content of fructan and other easily degradable soluble sugar in spring grass is about twice that of autumn grass. It is our contention that a more comprehensive understanding of the mechanisms associated with the degradation of foliar proteins in the rumen is needed in order to evolve new feeding strategies for livestock that are less reliant on supplemental feeding and more acceptable, both in terms of nutrition and the environment. This chapter considers the previously neglected role played by freshly ingested plant cells in the rumen and the significance of their contribution to rumen function.

Grazing livestock ingest considerable quantities of fresh herbage (about 100 kg of fresh forage for an average dairy cow), in several meals over the course of a day. The forage is excised into relatively long lengths, depending on sward height (Gibb *et al.*, 1998; Orr *et al.*, 1997) and the actions of tongue and teeth roll it into a ball (or bolus) which is swallowed with minimal chewing. The environment in the rumen is highly anaerobic, with a redox potential of between –300 and –350 mV and the temperature remains relatively static at 38–42°C, due in part to heat generated during fermentation, but mainly to the homeothermic metabolism of the animal. In grazing animals, the majority of plant cells entering the rumen are intact, as is evident in cattle where the rumen is characteristically stratified, with the development of a substantial 'raft' of plant biomass floating on top of the rumen fluid. Buffering capacity in the rumen is provided by the production of copious quantities of saliva containing bicarbonate and phosphate salts, enabling the rumen to be maintained at a pH of 6–7. Mixing of rumen contents occurs by repeated rhythmic contractions of the rumen wall. Following initial chewing, most of the physical breakdown of ingested plant biomass is brought about through rumination, or 'chewing the cud' (re-chewing of regurgitated digesta). Plant biomass is incubated for 2–3 days on entering the rumen, thus providing time for microbial degradation of plant cell walls.

The degradation of plant proteins in the rumen is considered to be a microbial process (*Figure 1*). However, the widespread use of dried and ground forage in ruminant nutrition has ignored the intrinsic properties of intracellular compartmentation and the potential impact of plant enzymes during the initial stages of digestion of fresh forage. Obviously, when dried and ground material are evaluated, compartmentation and enzyme activity are not considered, as they will have been destroyed during processing of the forage. We propose that freshly ingested forage is more likely to acclimate partially to what is perceived as a novel stress (elevated temperature and an anaerobic environment) than to die instantly on excision from the main plant. We consider that the stresses imposed on leaves during anaerobic incubations in the rumen cause a cellular response, similar to, but more rapid than that observed during natural senescence and that as a consequence, plant proteases mediate the degradation of plant proteins.

By far the most abundant protein in plants is ribulose bisphosphate carboxylase-oxygenase (RUBISCO; up to 60% of soluble leaf protein) located within the chloroplast stroma (Dean and Leech, 1982). Thus if the microbial population was responsible for the degradation of RUBISCO, it would need to breach the plant cell-wall, plasma membrane and the chloroplastic double envelope to gain access to the intact protein.

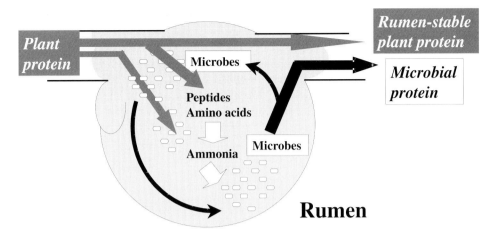

Figure 1. *Schematic representation of the traditional view of proteolysis in the rumen.*

Given that most of the plant cells are intact on entering the rumen, this requires microbial attachment to plant biomass and subsequent fibrolytic activity to be extremely high to account for the observed rates of degradation of protein in the rumen. However, while plant cell wall degrading activity is usually measured in days after ingestion, protein degradation usually occurs within hours (Theodorou and France, 1993). Moreover, although many rumen micro-organisms are proteolytic, measured activities of their extracellular proteases are minimal, and have even been estimated to be only sufficient to account for internal protein turnover (Asoa *et al.*, 1993). Thus we contend that the intrinsic contribution made by the ingested plant material to digestion in the rumen has largely been ignored. We believe that plant proteins are rapidly degraded in the rumen through rumen-induced perturbation of plant metabolism post-grazing, and this has a significant impact on the ultimate efficiency of protein utilization in livestock.

2. Potential responses of plant metabolism to the rumen

Plants are sessile organisms and therefore have evolved a multitude of responses to adverse conditions in order to survive. Environmental conditions often encountered are extremes of temperature, water status or nutrient supply. Plant material is often used as a feed for insects and larger herbivores and much research effort in the past has concentrated on resource allocation in the plant parts remaining after herbivory and so we are well aware of systemic damage responses involving chemical signalling to induce new nucleic acid and protein synthesis. To our knowledge, little if anything is known about metabolism of the excised plant parts prior to and in the early stages of digestion so we are forced to extrapolate from observed metabolic responses resulting from the imposition of stresses in related experimentation.

The initial response of living plant cells entering the rumen is likely to be adaptation to the prevailing conditions. In the rumen this is constant darkness, a temperature of 39°C and anoxia. These stresses have all been shown to affect protein turnover in intact plant tissues. Light is a regulator of activity and expression of enzymes, especially those associated with photosynthesis (Andrews and Ballment, 1983; Leegood, 1990). Heat stress initially induces *de novo* synthesis of specific heat-shock proteins within minutes of the ambient temperature exceeding the intrinsic temperature tolerance of the plant (Howarth and Ougham, 1993). It is thought that the newly formed heat-shock proteins protect other proteins and enzymes that are critical to cell function from incorrect assembly or partial degradation during periods of heat stress. However, prolonged heat stress can also cause membrane damage (Blum and Ebercon, 1980; Mittler and Lam, 1995) which would eliminate the functional integrity of the cell. This would allow vacuolar proteases access to cytoplasmic- and possibly organelle located-proteins. The effect of this is likely to be large-scale proteolysis within the confines of the plant cell wall.

Unlike heat and light stresses, the above-ground parts of most plants are unlikely to encounter anaerobic conditions. The commonest cause of anaerobiosis in plants is flooding, which results in waterlogged soil in which the roots have difficulty extracting oxygen as air pockets between soil particles are now filled with water. Some plants, for example rice, are tolerant of this stress, partly because they have aerenchyma, air channels formed in roots by programmed cell death, which allow air to be transported from above the water to submerged parts, allowing continuation of aerobic metabolism (see Chapter 13). In plants that are not tolerant of anaerobic conditions, lack of oxygen has a time-dependent, dual effect on plant metabolism. Short-term anoxia results in up-regulation of transcription and translation of the 'anaerobic proteins', lactate dehydrogenase and alcohol dehydrogenase (Ricard *et al.*, 1994), but prolonged periods without oxygen result in cell death (Drew, 1997; Perata and Alpi, 1993; Ricard *et al.*, 1994). The major obstacle for survival of plant tissues when deprived of oxygen is the adequate generation of ATP. Restrictions to the energy supply will severely handicap protein and nucleic acid synthesis and may promote cell death simply because of the inability of metabolism to continue functioning with limited reserves. Switching from oxidative phosphorylation to fermentation generates a limited energy supply but the long-term prospects for reliance on this resource is questionable for eukaryotic organisms adapted to an oxygen-enriched environment. It is also possible that the products of fermentation (acetate, ethanol, acetaldehyde and, in some plants, lactate) are toxic to plant metabolism (Perata *et al.*, 1992). Hence, although plant cells may be able to survive within the rumen in the short-term, the imposed stresses cannot be tolerated over a period of many hours. Thus, it is highly likely that the rate of progression of cell death is the most pressing determinant of proteolytic rate within the ingested tissues.

3. Cell death in plants

Cell death in plant cells is important for the viability of the whole organism as it is in animal cells, although the signal transduction pathways appear to be dissimilar (Beers, 1997; Greenberg, 1996). For example, strong evidence in favour of the presence of the caspase family of proteases in plants has not been found (but see Chapter 1), and therefore it is likely that another protein or series of proteins performs this role in plants. Plant cells may die in an apparently non-programmed manner in response to prevailing adverse conditions such as through oxidative damage (Prasad, 1996; Prioul, 1996), or in response to invasion by pathogens (necrotic lesions of the hyper-sensitive response; Mittler and Lam, 1995; Mittler *et al.*, 1997 but see also Chapters 10 and 11).

Damage by pathogens can range from small lesions to the loss of the entire plant. The controlled death of specific plant cells results in the formation of xylem vessels (Chapter 14), dehiscence of pollen, fruit and leaves (Chapter 16), aerenchyma formation under anaerobic conditions (Chapter 13), and dissected leaf formation (Greenberg, 1996; Groover et al., 1997; Jones and Dangl, 1996; McCabe and Pennel, 1996). During these processes the contents of certain files of cells are removed by increased activities of among other things proteases and nucleases (Groover et al., 1997; Niki et al., 1995; Stephenson et al., 1996).

In programmed cell death it is likely that proteins targeted for degradation are identified either by chemical modification such as oxidative tagging, interaction with ubiquitin (a small protein which regulates proteolysis), or a structural rearrangement to place proteolytic target sequences on accessible surfaces (Garbarino and Belknap, 1994; Genschik et al., 1994; Vierstra, 1996). Natural senescence is a programmed cell death process of plant cells. During senescence, proteins and carbohydrates are remobilized from old parts of the plant for transport to younger, actively growing parts (Hillman et al., 1994; Chapters 10 and 11). This is most readily observed as the visible loss of chlorophyll from the older leaves of a plant which is accompanied by degradation of protein and enucleation, mediated by proteases and nucleases, and is driven by consumption of glycolytic substrates from within the senescing leaves (Chapter 10). Senescence can also be induced artificially when leaf blades are excised and placed in the dark for a period of days (Morris et al., 1996).

During induced senescence the processes of chlorophyll breakdown and proteolysis, which occur together during natural senescence, can be observed independently. Both natural and induced senescence involve up-regulation of acidic cysteine proteases which are probably located in the vacuole (Feller, 1986; Matile, 1997). These high activity proteases undoubtedly have a role in the degradation of protein during senescence, but the vacuolar location suggests obligate transportation of proteins from other cell compartments. In senescence, unlike many other forms of cell death, the chloroplast remains relatively intact until the latter stages (past the 'point of no return'; see Chapter 10). In addition, given that the majority of the foliar protein is present as RUBISCO within the chloroplast, it is not surprising that evidence in favour of in situ chloroplastic proteolysis is accumulating (Anastassious and Argyroudi-Akoyunoglou, 1995).

4. Post-ingestion plant metabolism in the rumen; does it resemble cell death?

4.1 The new hypothesis

Theoretically, when compared with ambient conditions encountered by plants in the field, the multiple stresses of the rumen are likely to promote proteolysis which may be linked to cell death processes (Theodorou et al., 1996). Thus it has been possible to construct a new model of proteolysis in the rumen in which the involvement of plant enzymes is not overlooked (Figure 2). In our model, plant material entering the rumen is forced into premature cell death with the associated increase in intrinsic proteolytic activity (if a senescence-like programmed cell death occurs) or opportunity for protease to encounter protein substrates (if physical response to the environmental shock leads to loss of compartmentation). Thus, protein degradation can proceed

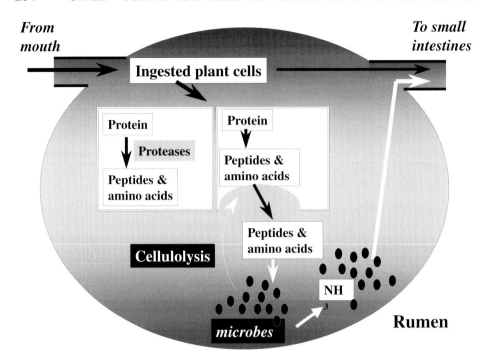

Figure 2. *New hypothesis: the role of plant enzymes in rumen function.*

rapidly, almost from the moment the plant tissue enters the rumen but without the involvement of the rumen microbial population and the necessity for prior cell wall degradation.

Simultaneously with the alterations to intracellular plant metabolism, the microbial population of the rumen can begin to attach to the plant material and break down the cell walls. Hence, plant proteases are responsible for the degradation of soluble plant proteins to smaller molecular weight peptides and amino acids, of a size sufficiently small to be released from the intact cells and available for uptake by the microbial population. In this scheme, the plant enzymes have a fundamental role in the initial stages of proteolysis of plant protein entering the rumen, especially during the first few hours of digestion.

4.2 *The in vitro approach*

Current investigations into the role of plant enzymes in the rumen employ simple, *in vitro* techniques to assess the response of plant tissues to the stresses within the rumen (Beha *et al.*, 1998). A predetermined weight of plant material is incubated in a buffered solution similar to that of the rumen, but in the absence of the rumen microbial population. Recovery of the material at intervals allows assessment of protein content and enzyme activities. Contamination by micro-organisms attached to the leaf surface cannot be eliminated but is minimized by (i) growing the plant material in a 'clean environment' in a glasshouse, or for preference controlled environment cabinets, and (ii) surface sterilization of leaf tissue in ethanol prior to incubation. Given that the difference in the microbial titre between the incubation buffer and an equivalent volume of rumen fluid is at least five orders of magnitude, it is unlikely that the microbial population in the buffer contributes in a significant way to protein degradation in the plant cells. Moreover, rumen-like microbial activity can be routinely assessed by determination of volatile fatty acids (acetate, propionate and butyrate) produced during fermentation; these have never been detected in significant quantities during incubations of plant material in this *in vitro* system.

4.3 *Protein degradation and cell death in vitro*

Protein disappearance from excised leaf blades from ryegrass (*Lolium perenne*) incubated anaerobically at 39°C was dramatic. Losses of up to 70% of the soluble protein pool were observed over 8–12 h (*Figure 3a*). The protein was not recovered in the incubation buffer indicating that protein degradation was being observed and not just a simple process of diffusive release from damaged cells. It was interesting that although both heat and anaerobic stress have been shown to involve a degree of protein synthesis during the early stages of the response, the net protein content of the leaves decreased rapidly during incubation under these combined conditions. The rate of net protein decrease within leaves during anaerobic incubations corresponds with observed rates of loss of nuclear DNA. Use of the DNA-binding fluorochrome, DAPI (4′,6-diamidino-2-phenylindole) showed that the bright nuclei observed in mesophyll cells of healthy pea leaves became noticeably less intense after anaerobic incubation at 39°C for 2 h, and were completely absent after 12 h (A. Kingston-Smith, unpublished observations). These observations correspond with the timescale expected for the onset of anoxia-mediated cell death in non-tolerant species (an acclimation phase of less than 30 minutes followed by death over several hours; Pradet and Raymond, 1983; Roberts *et al.*, 1984a, 1984b).

The initial rates of net protein loss can be used to calculate a kinetic constant $t_{1/2}$ which represents the minimum time required to degrade half of the original protein pool. From these values it is possible to compare the stability of protein from various crops or as a result of incubation conditions. For ryegrass leaf blades under anaerobic, high temperature treatment the half time was in the region of 3 hours (Beha *et al.*, 1998). This compared dramatically with measurements made when leaf blades were incubated aerobically at 25°C, where significant protein loss was not observed. Rapid loss of protein from leaf blades or discs cut from leaves has been observed in two grass and eight leguminous species studied to date when incubated anaerobically and at high temperature. However the kinetics of the protein loss were found to vary between crops, suggesting that there were differences between species in terms of protein stability under these conditions. This in turn could have implications for the availability of peptides and amino acids in the rumen, and hence the rate of growth of the microbial population and ammonia production. The underlying reasons for species-specific variation in protein stability are unclear. One possibility is that there could be species-specific differential induction of protease activity in plant cells entering the rumen, similar to the increased protease activity found in senescent leaves as compared with mature leaves (Chapter 10). However, measurements of electrolyte leakage from leaf discs incubated anaerobically at 39°C showed that membrane damage, including loss of the tonoplast, occurs within 6 hours. This suggests that protein stability in the rumen is probably due to an interaction of biochemical and physical factors.

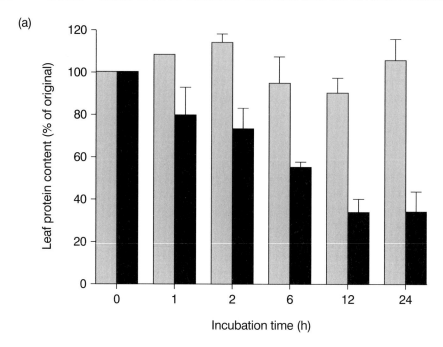

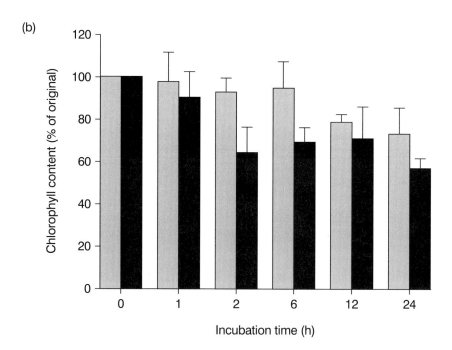

Figure 3. Effect of incubation of ryegrass leaves in buffer maintained at either ambient (25°C aerobic; hatched bars) or 'rumen-like' (39°C anaerobic; solid bars) conditions. Means plus (SE) of three determinations are shown.

5. Future prospects for improved livestock agriculture

In this article we question the assertion that the degradation of proteins in freshly ingested plant material in the rumen is a purely microbial process. Results obtained to date from an *in vitro* system where rumen micro-organisms were excluded refutes the current dogma. In our experiments, fresh plant material has been observed to continue metabolism, albeit altered for many hours post-ingestion. The result of this is rapid loss of foliar protein occurring simultaneously with cell death processes. Some parameters, such as loss of nuclear DNA and increased protease activity correspond with similar events known to occur during natural senescence. Thus it is possible that on entering the rumen plant cells are forced into a premature, and rapid senescence; events such as increased protease activity and damage to DNA occur over days during senescence, but appear to take place within hours in the *in vitro* experimental system used to simulate rumen conditions. However, losses of chlorophyll were not extensive during incubation of grass in anaerobic buffer at 39°C compared with ambient conditions (*Figure 3b*) and the majority of the chlorophyll pool was undegraded after 24 h. Therefore it is unlikely that we are simply observing a more rapid execution of natural senescence (*Figure 4*).

The gathering of information relating to the control of proteolysis within plant tissue post-ingestion is of considerable importance in the development of future forage breeding strategies. By determining the key events relating to the control of plant-mediated proteolysis during ruminal incubation, it may be possible to identify

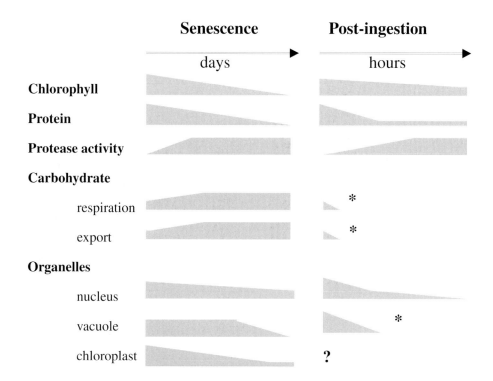

Figure 4. Similarities between senescence and post-ingestion metabolism of fresh forage.
(*) preliminary indications; (?) response unknown at present.

traits that contribute to decrease the rate of ruminal proteolysis. In this way it should be possible to develop a conventional plant-breeding programme to improve the efficiency of the rumen fermentation by selecting for decreased rates of proteolysis and cell death in grasses thus establishing better synchronization of the availability of protein and energy in the rumen. By establishing the above hypothesis, we wish to sanction the use of 'natural production' systems for ruminant livestock and demonstrate that grasses bred for quality can have a significant positive impact on animal performance.

Acknowledgements

This work is currently being undertaken by the authors together with Ms Elke Beha (PhD student) and with the technical assistance of Ms Andrea Bollard. The authors are grateful to BBSRC, MAFF, MDC, MLC and Germinal Holdings Ltd for financial support.

References

Andrews, T.J. and Ballment, B. (1983) The function of the small subunit of ribulose bisphosphate carboxylase–oxygenase. *J. Biol. Chem.* **258**: 7514–7518.

Anastassious, R. and Argyroudi-Akoyunoglou, J.H. (1995) Thylakoid-bound proteolytic activity against LHCII apoprotein in bean. *Photosynth. Res.* **43**: 241–250.

Asoa, N., Ushida, K. and Kojima, Y. (1993) Proteolytic activity of rumen fungi belonging to the genera *Neocallimastix* and *Piromyces*. *Lett. App. Microbiol.* **16**: 247–250.

Beers, E.P. (1997) Programmed cell death during plant growth and development. *Cell Death Different.* **4**: 649–661.

Beha, E.M., Kingston-Smith, A.H. and Theodorou, M.K. (1998) Death of plant cells in the rumen, a hostile, anaerobic, hot (39°C) and dark environment. *J. Exp. Bot.* **49**: P7.68.

Blum, A. and Ebercon, A. (1980) Cell membrane stability as a measure of drought and heat tolerance in wheat. *Crop Sci.* **21**: 43–47.

Dean, C. and Leech, R.M. (1982) Genome expression during normal leaf development. *Plant Physiol.* **69**: 904–910.

Dewhurst, R.J., Mitton, A.M., Offer, N.W. and Thomas, C. (1996) Effects of the composition of grass silages on milk production and nitrogen utilisation by dairy cows. *Animal Sci.* **62**: 25–34.

Drew, M.C. (1997) Oxygen deficiency and root metabolism: Injury and acclimation under hypoxia and anoxia. *Annu. Rev. Plant Physiol. Plant Mol. Biol.* **48**: 223–250.

Feller, U. (1986) Proteolytic enzymes in relation to leaf senescence. In: *Plant Proteolytic Enzymes, Volume II* (ed. M.J. Dalling). CRC Press, Boca Raton, pp. 49–68.

Garbarino, J.E. and Belknap, W.R. (1994) Isolation of a ubiquitin-ribosomal protein gene (ubi3) from potato and expression of its promoter in transgenic plants. *Plant Mol. Biol.* **24**: 119–127.

Genschik, P., Durr, A. and Fleck, J. (1994) Differential expression of several E2-type ubiquitin carrier protein genes at different developmental stages in *Arabidopsis thaliana* and *Nicotiana sylvestris*. *Mol. Gen. Genet.* **244**: 548–556.

Gibb, M.J., Huckle, C.A. and Nuthall, R. (1998) Effect of time of day on grazing behaviour by lactating dairy cows. *Grass Forage Sci.* **53**: 41–46.

Greenberg, J.T. (1996) Programmed cell death: a way of life for plants. *Proc. Natl Acad. Sci. USA* **93**: 12094–12097.

Groover, A., DeWitt, N., Heidel, A. and Jones, A. (1997) Programmed cell death of plant tracheary elements differentiating *in vitro*. *Protoplasma* **196**: 197–211.

Hillman, J.R., Glidewell, S.M. and Deighton, N. (1994) The senescence syndrome in plants: an overview of phytogerontology. *Proc. Roy. Soc. Edinburgh* **102B**: 447–458.

Howarth, C.J. and Ougham, H.J. (1993) Gene expression under temperature stress. *New Phytol.* **125**: 1–26.

Jones, A.M. and Dangl, J.L. (1996) Logjam at the styx: programmed cell death in plants. *Trends Plant Sci.* **1**: 1360–1385

Leegood, R.C. (1990) Enzymes of the Calvin cycle. In: *Methods in Plant Biochemistry, Volume 3.* (ed. P.M. Dey, J.B. Harbornne). Academic Press, London, pp. 15–37.

McCabe, P.F. and Pennell, R.I. (1996) Apoptosis in plant cells in vitro. In: *Techniques in Apoptosis* (eds T.G. Kotter and S.J. Martin). Portland Press, London, pp. 301–326.

Matile, P. (1997) The vacuole and cell senescence. In: *The Plant Vacuole; Advances in Botanical Research, Volume 25* (ed. R.A. Leigh). Academic Press, London, pp. 87–112.

Mittler, R. and Lam, E. (1995) Identification, characterization and purification of a tobacco endonuclease activity induced upon hypersensitive response cell death. *Plant Cell* **7**: 1951–1962.

Mittler, R., Simon, L. and Lam, E. (1997) Pathogen-induced programmed cell death in tobacco. *J. Cell Sci.* **110**: 1333–1344.

Morris, K., Thomas, H. and Rodgers, L. (1996) Endopeptidases during the development and senescence of *Lolium temulentum* leaves. *Phytochem.* **41**: 377–384.

Niki, T., Gladish, D.K., Lu, P. and Rost, T.L. (1995) Cellular changes precede cavity formation in the vascular cylinders of pea roots (*Pisum sativum* L. cv. Alaska). *Int. J. Plant Sci.* **156**: 290–302.

Orr, R.J., Penning, P.D., Harvey, A. and Champion, R.A. (1997) Diurnal patterns of intake rate by sheep grazing monocultures of ryegrass or white clover. *Appl. Animal Behav. Sci.* **52**: 65–77.

Perata, P. and Alpi, A. (1993) Plant responses to anaerobiosis. *Plant Sci.* **93**: 1–17.

Perata, P., Vernieri, P., Armellini, D., Bugnoli, M., Tognoni, F. and Alpi, A. (1992) Immunological detection of acetaldehyde protein adducts in ethanol-treated carrot cells. *Plant Physiol.* **98**: 913–918.

Pradet, A. and Raymond, P. (1983) Adenine nucleotide ratios and adenylate energy charge in energy metabolism. *Annu. Rev. Plant Physiol.* **34**: 199–224.

Prasad, T.K. (1996) Mechanisms of chilling-induced oxidative stress injury and tolerance in developing maize seedlings: changes in antioxidant systems, oxidation of proteins and lipids, and protease activities. *Plant J.* **10**: 1017–1026.

Prioul, J.-L. (1996) Corn. In: *Photoassimilate Distribution in Plants and Crops: Source-sink Relationships* (ed. E. Zamski and A.A. Schaffer). Marcel Dekker, New York, pp. 549–594.

Ricard, B., Couee, I., Raymond, P., Saglio, PH., Saint-Ges, V. and Pradet, A. (1994) Plant metabolism under anoxia. *Plant Physiol. Biochem.* **32**: 1–10.

Roberts, J.K.M., Callis, J., Jardetsky, D., Walbot, V. and Freeling, M. (1984a) Cytoplasmic acidosis as a determinant of flooding intolerance in plants. *Proc. Natl Acad. Sci. USA* **81**: 6029–6033.

Roberts, J.K.M., Callis, J., Wemmer, D., Walbot, V. and Jardetzky, O. (1984b) Mechanism of cytoplasmic pH regulation in hypoxic maize root tips and its role in survival under hypoxia. *Proc. Natl Acad. Sci. USA* **81**: 3379–3383.

Stephenson, P., Collins, B.A., Reid, P.D. and Rubinstein, B. (1996) Localization of ubiquitin to differentiating vascular tissues. *Am. J. Bot* **83**: 140–147.

Theodorou, M.K. and France, J. (1993) Rumen microorganisms and their interactions. In: *Quantitative Aspects of Ruminant Digestion and Metabolism* (eds J.M. Forbes and J. France). CAB International, Oxon, pp. 145–162.

Theodorou, M.K., Merry, R.J. and Thomas, H. (1996) Is proteolysis in the rumen of grazing animals mediated by plant enzymes? *Br. J. Nut.* **75**: 507–510.

Vierstra, R.D. (1996) Proteolysis in plants: mechanisms and functions. *Plant Mol. Biol.* **32**: 275–302.

23

Postscript: some concluding observations and speculations

Steve G. Hughes and John M. Garland

While examining the concepts of apoptosis and programmed cell death as they relate to animals and plants, we have witnessed the significance of a pattern of differing activities of cells and of cellular behaviour in development and in the life-cycle of the organism. A re-examination of these differences provides a logical framework for the postscript to this volume.

Plants tend to have a more flexible body plan than animals, reflecting a sedentary lifestyle and a requirement for developmental responses to herbivory, environmental limitations and competition for resources. This requirement for architectural self-reinvention is manifest, in turn, in the loss and replacement of organs such as leaves, and in the development of structural elements, for example branches, to support the quest for individual space, and the development of conductive elements for water, solutes and air (xylem and aerenchema). In this context it is hard to make a sharp distinction between developmental (determined) and environmental (responsive) influences on differential growth. Indeed, both of these influences enlist programmed cell death as a contributory mechanism, often coupled to the reinitiation of cell division. The latter mechanism whereby differentiated or quiescent cells are recruited to become stem cells (meristems in the language of plant anatomy) is termed totipotency and is more common in plants than animals. This emphasizes the potential for reprogramming and the reversibility and diversity of plant cell fates.

In the processes leading to the loss of organs, as in deciduous leaf fall, cell death is seen as a mechanism for conservation of resources, often termed senescence. This may be viewed as a further consequence of a sedentary lifestyle. Unable to move away from environmental stress, plants have evolved and exhibit a variety of life cycle strategies to deal with seasonal change. An extreme form of this is termed monocarpic senescence which is found among annual, biennial and ephemeral plant species. In this form of senescence, the entire plant progressively dies following the production of a single round of seed. Much of the assimilates accumulated during plant growth is reallocated to seed reserves and the seed remains dormant until conditions are appropriate for

Programmed Cell Death in Animals and Plants, edited by J.A. Bryant, S.G. Hughes and J.M. Garland.
© 2000 BIOS Scientific Publishers Ltd, Oxford.

germination, outgrowth and another round of flowering. The processes of seed development and germination contain many instances of cell turnover and programmed cell death. Tissues and structures such as endosperm, aleurone, cotyledon and haustorium, appear and disappear as vital components in supporting the nutrition and outgrowth of the embryo. Nutritional deficiencies may accelerate the monocarpic cycle; alternatively, the process may be synchronized with seasonality by day length activated controls. Given the need for an orderly withdrawal and reassignment of assimilates, one might expect that under these circumstances cellular disassembly is tightly phased and controlled. Equally, however, given the plastic body plan and the need for environmental responsiveness, one might also expect the process to be to a great extent reversible if, for example, there is an amelioration of an environmental 'stress'. This is certainly borne out by the contributions to this volume. In this context, it is interesting to note the differences between the effector proteases that support programmed cell death in plants, and the proteases of the caspase cascade in animal cells which, while appearing to potentiate turnover, are really engaged at a higher regulatory level to secure irreversible commitment to death.

In contrast to plants, animals manifest a more invariable body plan, with a clearer and stricter pattern of determined cell fates and a sharper distinction between germline cells, stem cells, structural cells and specialized effector cells. Within this higher order organization and determination, there is a need for effective housekeeping to eliminate cells which have outlived their utility or cells that have started to behave aberrantly, for instance by escaping from cell cycle control. This is especially true of mobile cells where the dynamic equilibrium of the population of circulating cells needs to be held in check to maintain a correct balance of oxygen-carrying cells, scavenging cells and defence activators and mediators, for example, lymphocytes produced in a specific immune response. In this scenario, given the potential for damage once a cell is targeted for destruction one would expect programmed cell death, once initiated, to be nonreversible.

Plants and animals do share, however, a need to resist invading organisms, and both enlist localized cell death as part of their strategy of containment and damage limitation. Mobile scavenger cells and a responsive cellular immune system are absent from plants. The plant strategy is to fix and isolate the invader at the site of infection and to limit access of the invader to the host's resources. In the classical plant hypersensitive response this involves selective cell death in the region around the infection site, the withdrawal of assimilates and the production of antimicrobial toxins. In contrast, animals tend to proliferate their defensive cellular systems but need to detune these after they have dealt with an attack to avoid chronic inflammation and to maintain their options on the clonal elaboration of other immune cells in response to the next invader. We have seen in this volume the complexity of the interacting regulatory systems which allow for regulation of the immune response while maintaining immune memory. Aberrant behaviour of this process is a basis for inflammatory disease and indicates the potential of malfunction of the control of apoptosis in animal disease. As a further illustration of this potential, the nonreversibility of programmed cell death can give rise to problems for the whole organism if the programme is aberrantly triggered within an essential organ. Aberrations in initiation of programmed cell death have been described in the recovery from ischaemia, for example, where during reperfusion, dramatic changes in ion fluxes, membrane pores and cytochrome-*c* release promote cell death and tissue injury. This kind of malfunction is likely to be less significant in plants where the programme is more reversible and the integrity of the body is less dependent upon single organs.

Thinking mechanistically, although apoptosis as a distinct cellular process has only recently been established, the idea of programmed cell death for animal cells has been known for a long time; for example, the cell death that accompanies conversion of keratinocytes into simple keratin, or the conversion of cartilage into bone, have been known for about 100 years. The idea of tissue sculpturing during development and wound healing/tissue repair is even older. When we analyse these various modes of cell death that occur in the normal development and functioning of a multicellular animal, we can see that there are actually two levels of programmed cell death. The first 'level 1' programmed cell death, is the topographical programme which determines what shape the finished product will be; this is related to when genes will be turned off and on and when cells will divide, stay quiescent or die. Although there may be specific genes, such as homoeotic genes, controlling this process, it is the exquisite accuracy of the control mechanisms that is so remarkable; in rapidly developing embryos for example, the decisions for level 1 programmed cell death must be accurate to within a few hours if not minutes. The programme here for life and death is deterministic overall, but cannot be gleaned or predicted from studying the structure of the genes themselves. The second, 'level 2' programmed cell death is seen at the level of the molecular events and controls that operate in individual cells to decide whether a cell survives or not. This is the area currently being attacked with such vigour and creating various conundrums and apparent contradictions but also revealing amazing truths, many of which are apparent in earlier chapters of this book. Both these levels of cell death have their functional equivalents in plants, although, as pointed out earlier in this brief chapter, the two levels are less distinct from each other in plants than in animals. Whether or not we may equate level 2 programmed cell death in plants with that in animals may be just a matter for semantics. The important feature is that certain forms of programmed cell death in plants serve the same function for the organism as some occurrences of apoptosis in animals.

Intuitively, it is expected that level 2 apoptosis in animal cells, particularly in more complex animals is bound to be much more complicated (at least in its regulation) than programmed cell death in plant cells. In this context, one may consider cancer. Is it primarily a disease of apoptosis disregulation, of cell division disregulation, of pertubations in signalling or in environmental adaptation? One must bear in mind that apoptosis occurs extensively in many tumours, even if cell proliferation outstrips it. By the time one is looking at cancer cells, the end-product has been made and it is not surprising that all the cell regulators are changed, because the entire cell is not doing what a cell in that position usually does. And here is another thought. Is cancer also a 'disease'/malfunction of topographical, level 1 programmed cell death? In other words, is the cancer cell following the rules for its eventual destiny according to a different topographical blueprint (level 1) or has it lost the rules completely? If so, it will be programmed not to do things expected of normal cells, and we should expect to find a host of molecular regulators all intertwined in apparent confusion (at least to us as investigators) but functioning smoothly for the cancer cell in what it has been reprogrammed to achieve. That tumour cells invariably have significant alterations in their DNA organization suggests that level 1 programmed cell death may play an active role. The fact that tumour cells are extremely difficult to grow outside the body also demonstrates they are still responding to environmental level 1 cues.

Nevertheless, it has been recently demonstrated that in human cells, only three genes are required to generate a malignant phenotype (*telomerase, T* and *ras*). The telomerase protein ensures continuation of DNA replication integrity and fidelity of gene position; the *T* gene product regulates a large number of gene promoters and *ras* regulates many of the components of internal cell architecture (including cytoskeleton and nuclear envelope) which determine extracellular signalling responses. Put all three together, and the cell can construct its own new destiny even to the extent of doing without many environmental cues. Even here though, the number of malignant cells generated is relatively small, so selection at level 1 may still operate. One only has to dwell on the literally hundreds of thousands of papers on individual molecular events in cancer cells to realize that *everything* in a cancer cell is the 'same but different'. There is thus an argument that if we knew what kind of structure the cancer cell was trying to make (level 1), we would know how to activate level 2 apoptosis and thus delete the tumour.

Finally, returning to plants and looking for a moment to the future, it is worthy of note that the role of programmed cell death in selective cell loss during the reproductive development of plants is as yet an unexplored area. How in some species are three of the four meiotic products of early megasporogenesis programmed for death and the other programmed to continue development and cell division to provide the cell complement (egg cell, synergids, central cell and antipodals) of the embryo sac? What is the basis for the modulation of this pattern in other species? How is the sexual embryo sack programmed for premature death and engulfment by an asexual one in some apomictic species (asexually reproducing through seed)? Clearly, the study of single cells within the developing ovum is a difficult prospect which will require the novel application of methods in cell biology, molecular biology and genetics. The apparently very precise determination of cell movement and cell fates within the embryo sac may indicate a closer relationship between this form of plant cell death and animal cell apoptosis than seen elsewhere in plants. If so, the accounts of the genes controlling apoptosis in nematodes and other animals reported in this volume will offer a strong guide and offer a justifiable speculative entrée. The importance of plant reproductive development to the modification of plant breeding systems is such that this will be one of the major challenges of the next decade.

As Chapter 22 shows, programmed cell death is even seen to be playing a role not only in the subtleties of the bovine diet but also in the plant/animal relationship in terms of digestibility and the recycling of assimilates and ultimately in the global carbon and nitrogen cycles. This puts programmed cell death onto a larger stage than the survival of the organism itself and further confirms it as a valid quest of front-line biology across the whole spectrum of the biota.

> in the random cascade of an organism
> nothing is certain but death and duties
>   duties may be necessary
>   they are welcome to stay
> provided they accept a minor role,
> but death must live with dominion,
>   one of life's little essentials

Index